International Conference on Information and Communication Technology (IconICT 2017)

Journal of Physics: Conference Series Volume 930

Medan, Indonesia
25 – 26 August 2017

ISBN: 978-1-5108-5363-8
ISSN: 1742-6588

Printed from e-media with permission by:

Curran Associates, Inc.
57 Morehouse Lane
Red Hook, NY 12571

CURRAN ASSOCIATES INC.
proceedings
.com

Some format issues inherent in the e-media version may also appear in this print version.

Printed by Curran Associates, Inc. (2018)

For permission requests, please contact the Institute of Physics
at the address below.

Institute of Physics
Dirac House, Temple Back
Bristol BS1 6BE UK

Phone: 44 1 17 929 7481
Fax: 44 1 17 920 0979

techtracking@iop.org

Additional copies of this publication are available from:

Curran Associates, Inc.
57 Morehouse Lane
Red Hook, NY 12571 USA
Phone: 845-758-0400
Fax: 845-758-2633
Email: curran@proceedings.com
Web: www.proceedings.com

Table of contents

Volume 930

International Conference on Information and Communication Technology (IconICT)

25–26 August 2017, Medan, Sumatera Utara, Indonesia

Accepted papers received: 14 November 2017
Published online: 14 December 2017

Preface

Papers

Artificial Intelligence

Data Mining

Information Security

Numerical Analysis Science

INTRODUCTION

On behalf of the Organizing Committee, we delighted to welcome all participants to the International Conference on Information and Communication Technology (IConICT) 2017 which took place from July 15 to August 25. This conference is the first special conference that combines the field of information and communication technology which was held in Medan, Indonesia. With a very interesting program, this forum became the main medium of discussions, dissemination and research achievement in related fields.

Nearly 200 academics, researchers, practitioners and presenters from 6 countries have gathered and 106 papers abstract and full papers have been submitted in this year's conference. Each paper was reviewed with tight criteria from our invited and voluntary multi-national reviewers. Based on the review results, 52 papers have been accepted, which lead to an acceptance rate of 49.05%. All accepted papers were orally presented in 3 technical sessions and was published in the conference proceeding.

The conference is a prestigious event that would not been successful without extensive effort and great dedication from many volunteers. I would like to express my sincere appreciation and thanks to all participants and supporters, especially the members of Information Technology Study Programmed of STT Harapan Medan, APTIKOM Chapter North Sumatera, Indonesian Mathematics Society (Indo-MS) Chapter Aceh-North Sumatera, Indonesian Computer Society (IPKIN) Chapter North Sumatera. I would also like to thank the keynote speakers, Prof. Gerhard Wilhelm Weber, Prof.Abdel Salhi, Dr. Muhammad Fermi Pasha, who were willing to meet the IConICT 2017 invitation. My sincere gratitude to Rector of Sekolah Tinggi Teknik Harapan (STTH) Medan, and Ir M. Zulfin, MT.

I would also like to give my appreciation to many others who are involved in the success of this event, including the reviewers, student volunteers, and staffs at the Sekolah Tinggi Teknik Harapan Medan.

Lastly, on behalf of the organizing committee, I would like to welcome you to our first international conference on Information and Communication Technology, IConICT. Our hope is that you have an enjoyable and inspiring moment during our conference. Thank you for your participation on IConICT 2017.

Prof. Dr. Herman Mawengkang
Chair, IConICT 2017

IOP Conf. Series: Journal of Physics: Conf. Series **930** (2017) 011001 doi:10.1088/1742-6596/930/1/011001

IOP Conf. Series: Journal of Physics: Conf. Series **930** (2017) 011001 doi:10.1088/1742-6596/930/1/011001

IOP Conf. Series: Journal of Physics: Conf. Series **930** (2017) 011001 doi:10.1088/1742-6596/930/1/011001

IOP Conf. Series: Journal of Physics: Conf. Series **930** (2017) 011001 doi:10.1088/1742-6596/930/1/011001

IOP Conf. Series: Journal of Physics: Conf. Series **930** (2017) 011001 doi:10.1088/1742-6596/930/1/011001

IOP Conf. Series: Journal of Physics: Conf. Series **930** (2017) 011002 doi:10.1088/1742-6596/930/1/011002

Conference Information

Dates	: August 25, August 26, 2017
Organizer	: Program Study Teknik Informatika Sekolah Tinggi Teknik Harapan
Venue	: Hotel Garuda Plaza Medan Jl. Sisingamangaraja No.18, Mesjid, Medan Kota, Kota Medan, Sumatera Utara 20213
Phone	: +62 (61) 736 1111 / 7361234 .
Fax	: +62 (61) 736 4411
Email	: reserv@garudaplaza.com
Website	: //hotel.lewatmana.com /goto.html?url=/asia/indonesia/medan /garuda_plaza_hotel.html
Official Language	: English
Secretariat	: Program Study Teknik Informatika Sekolah Tinggi Teknik Harapan Jl. H.M Joni No.70 Medan, 20217, Sumatera Utara, Indonesia.
Email	: biro@stth-medan.ac.id.
Confrence Website	: http://iconict-stth.com

IOP Conf. Series: Journal of Physics: Conf. Series **930** (2017) 011002 doi:10.1088/1742-6596/930/1/011002

Committee

Steering Committee

Prof. Dr. Jasni Mohd.Zain (University Tun Hussein On Malaysia)

Prof. Dr. Gerhard Wilhelm Weber (Midle East Technical University Turkey)

Prof. Dr. Abdel Salhi (University Of Essex, United Kingdom)

Prof. Dr. Kok Lay Teo (Curtin University, Perth Australia)

Prof. Dr. Zainal A Hasibuan (University Of Indonesia, Indonesia)

Prof. Dr. Opim Salim Sitompul (University of Sumatera Utara, Indonesia)

Prof. Dr. Muhammad Zarlis (University of Sumatera Utara, Indonesia)

Prof. Dr. Saib Suwilo (University of Sumatera Utara, Indonesia)

Prof. Dr. Benny Mutiara (Gunadarma University, Indonesia)

Prof. Dr. Tulus (University of Sumatera Utara, Indonesia)

Dr. Syahril Efendi, M.IT (University of Sumatera Utara, Indonesia)

Dr. Rizal Munadi, MT (Syah Kuala University, Indonesia)

Dr. Benny B Nasution (Medan State Politechnic, Indonesia)

Dr. Rahmat W Sembiring (Medan State Politechnic, Indonesia)

Dr. Muhammad Fermi Pasha (Monash University, Malaysia)

Dr.Tapi Rondang Ni Bulan (Sekolah Tinggi Teknik Harapan, Indonesia)

Ir. Muhammad Zulfin, MT (Sekolah Tinggi Teknik Harapan, Indonesia)

Editorial Board

Prof. Dr. Herman Mawengkang (University of Sumatera Utara, Indonesia)

Prof. Dr. Tulus (University of Sumatera Utara, Indonesia)

Prof. Dr. Opim Salim Sitompul (University of Sumatera Utara, Indonesia)

Prof. Dr. Gerhard Wilhelm Weber (Midle East Technical University Turkey)

Prof. Dr. Abdel Salhi (University Of Essex, United Kingdom)

Prof. Dr. Kok Lay Teo (Curtin University, Perth Australia)

Dr. Muhammad Fermi Pasha (Monash University, Malaysia)

Sajadin Sembiring, S.Si, M.Comp,Sc (University of Sumatera Utara, Indonesia)

IOP Conf. Series: Journal of Physics: Conf. Series **930** (2017) 011002 doi:10.1088/1742-6596/930/1/011002

Chair

Prof. Dr. Herman Mawengkang (University of Sumatera Utara, Indonesia)

Co Chair

Sajadin Sembiring, S.Si, M.Comp,Sc (University of Sumatera Utara, Indonesia)

Ummul Khair, S.Kom, M.Kom (Sekolah Tinggi Teknik Harapan, Indonesia)

Executive Member

Abdul Jabbar Lubis, ST, M.Kom (Sekolah Tinggi Teknik Harapan, Indonesia)

Ilham Faisal, ST, M.Kom (Sekolah Tinggi Teknik Harapan, Indonesia)

Dra. Suriati, M.Kom (Sekolah Tinggi Teknik Harapan, Indonesia)

Eddy ST, M.T (Sekolah Tinggi Teknik Harapan, Indonesia)

Local Member

Mardiana, ST, M.Kom (Sekolah Tinggi Teknik Harapan, Indonesia)

Yuyun Dwi Lestari, ST, M.Kom (Sekolah Tinggi Teknik Harapan, Indonesia)

Haida Dafitri ST, M.Kom (Sekolah Tinggi Teknik Harapan, Indonesia)

Dodi Siregar M.Kom (Sekolah Tinggi Teknik Harapan, Indonesia)

Ari Usman, ST, M.Kom (Sekolah Tinggi Teknik Harapan, Indonesia)

Sayuti Rahman, ST, M.Kom (Sekolah Tinggi Teknik Harapan, Indonesia)

Ade Zulkarnain, ST, M.Kom (Sekolah Tinggi Teknik Harapan, Indonesia)

Ihsan Lubis, ST, M.Kom (Sekolah Tinggi Teknik Harapan, Indonesia)

Arnes Sembiring ST, M.Kom (Sekolah Tinggi Teknik Harapan, Indonesia)

T. M Diansyah ST, M.Kom (Sekolah Tinggi Teknik Harapan, Indonesia)

Rachmat Aulia, S.Kom, M.Sc,IT (Sekolah Tinggi Teknik Harapan, Indonesia)

Venue Map

Hotel Garuda Plaza Medan

Jl. Sisingamangaraja No.18, Mesjid, Medan Kota, Kota Medan, Sumatera Utara 20213

Phone : +62 (61) 736 1111 / 7361234 .

Fax : +62 (61) 736 4411

IConICT 2017 Venue

FLOOR PLAN

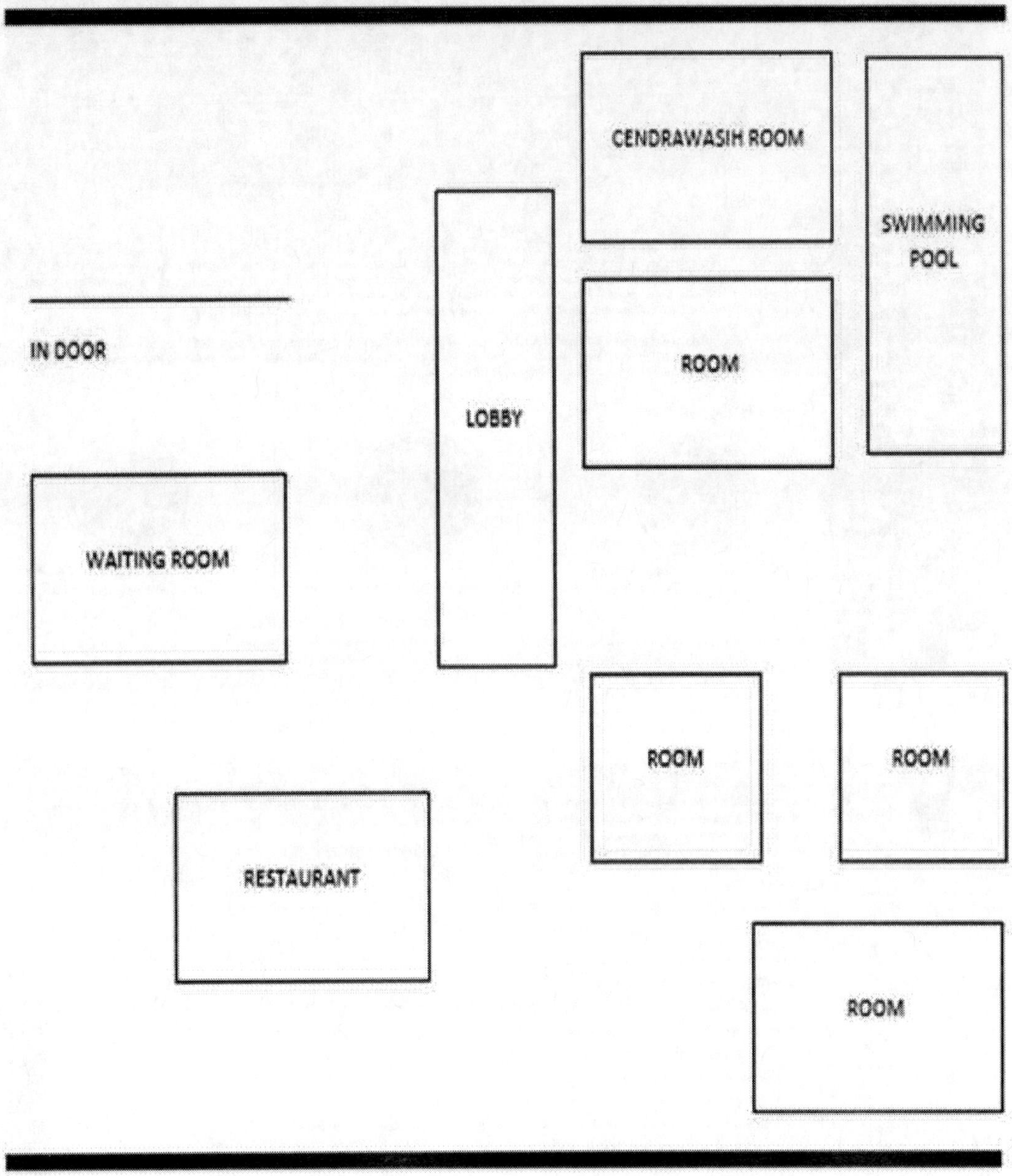
CENDRAWASIH ROOM
SWIMMING POOL
ROOM
IN DOOR
LOBBY
WAITING ROOM
ROOM
ROOM
RESTAURANT
ROOM

IOP Conf. Series: Journal of Physics: Conf. Series **930** (2017) 011003 doi:10.1088/1742-6596/930/1/011003

IOP Conf. Series: Journal of Physics: Conf. Series **930** (2017) 011004 doi:10.1088/1742-6596/930/1/011004

Peer review statement

All papers published in this volume of *Journal of Physics: Conference Series* have been peer reviewed through processes administered by the proceedings Editors. Reviews were conducted by expert referees to the professional and scientific standards expected of a proceedings journal published by IOP Publishing.

Visual Approach of Searching Process using Boyer-Moore Algorithm

Robbi Rahim[1*], Ansari Saleh Ahmar[2], Ayu Putri Ardyanti[3] and Dicky Nofriansyah[4]

[1]Department of Health Information, Akademi Perekam Medik dan Infokes Imelda, Jl. Bilal Ujung Medan 20116, Indonesia
[1]Postgraduate Ph.D. Student Universiti Malaysia Perlis, Universiti Malaysia Perlis, Malaysia
[2]Departement of Statistics, Universitas Negeri Makassar, Daeng Tata Street, UNM Campus, Makassar City, Indonesia
[3]Departement of Informatics Engineering, STMIK Primakara, Jalan Tukad Badung No.135 Denpasar Bali, Indonesia
[4]Department of Information System, STMIK Triguna Dharma, Jl. Jenderal Abdul Haris Nasution No.73, Medan 20219, Indonesia

*usurobbi85@zoho.com

Abstract. This research shows the process of string matching using a Boyer-Moore algorithm in the form of visual simulation by doing string searching of a pattern string, the result of visualization is recognized that Boyer-Moore algorithm will compare a n string from the right and will test to n-1 String to get the appropriate character of the desired input string.

1. Introduction

Today's digital era is very easy to get information, this is because the higher the advancement of technology such as the internet, smartphones that can be used by many users, to get information users can use search engines like Yahoo, Google or other search engines [1] [2], one of the techniques that can be used to search is the Boyer-Moore algorithm.

The Boyer-Moore algorithm has a string matching process from the right of the pattern [3] [4] [5], the initial concept of the string-handling process from the right to get more information [3] [5], this research tries to model in the simulation of string matching process using Boyer-Moore algorithm, it is expected for other researchers who wish to study or developed the Boyer-Moore algorithm to find out how the algorithm works.

2. Theory

Searching is a work that is usually performed in everyday life, the search process is done a job to find a word or replace a particular word [6] [7], and Boyer-Moore algorithm is one algorithm that could use for string matching.

Boyer Moore's algorithm was published by Robert S. Boyer and J. Strother Moore in 1977. The character matching made using the Boyer-Moore algorithm starts from the right-hand strings of the pattern. The idea behind this algorithm is to start matching characters from the right, and then more information will be obtained.

The Boyer-Moore algorithm includes the most efficient string matching algorithm compared to other string matching algorithms due to an efficient nature of the algorithm, many string matching algorithms developed based on the concept of Boyer-Moore algorithm, some of which are the Turbo BM algorithm and Quick Search algorithm, as for the string search steps on the Boyer-Moore algorithm as follows:

a. First, we need two tables with Match Heuristic (MH), and Occurrence Heuristic (OH) approaches to determine the number of shifts that will be performed on a pattern (P) if there are unsuitable characters in the matching process of characters in text (S).
b. If in the comparison process there is character mismatch between the characters on P and S, then the shift is done by looking at both tables with the largest shift value selected.
c. The possibility of completion in shifting to P is that if in previous matching there is no matching character then the shift is done by looking at the value of shift in occurrence heuristic table. If the character being compared does not exist in P, then the shift is done as much as the number of characters contained in P, but if the unsuitable character is contained in the P string, then the shift is done based on the table.
d. If the characters in the matching text match the characters on the P string, then the character checking position on P and S shifts each left position 1 from the previous position, then proceed with matching at that position and so on, then if there is a character mismatch In P and S, the shift is done by looking at the heuristic match table and the occurrence heuristic where the largest shift value to be selected is reduced by the number of matching characters.
e. If all the characters have a match, that means P has been found in S, then move the pattern by one character, continue until the end of the S pattern.

3. Result and Discussion

Here is an example of Boyer-Moore algorithm analysis for word search process, it is known that the input string is DAELYMMAKRINAAMRSQ and the target string is RINA, and here is the process of applying the Boyer-Moore algorithm to find the target string from the string of entry.

a. Declare input strings and target strings

D	A	E	L	Y	M	M	A	K	R	I	N	A	A	M	R	S	W

R	I	N	A

b. First Step

The last character of the target string 'A' does not match the 'L' character of the word 'DAELYMMAKRINAAMRSQ', the first checking process is done by counting the length of the target string n and then the last letter of the target string is matched to the nth character of the input string.

D	A	E	L	Y	M	M	A	K	R	I	N	A	A	M	R	S	W

R	I	N	A

Seen in the process above the letter A != L then the next process is to test L with all the letters that exist in the target string as below

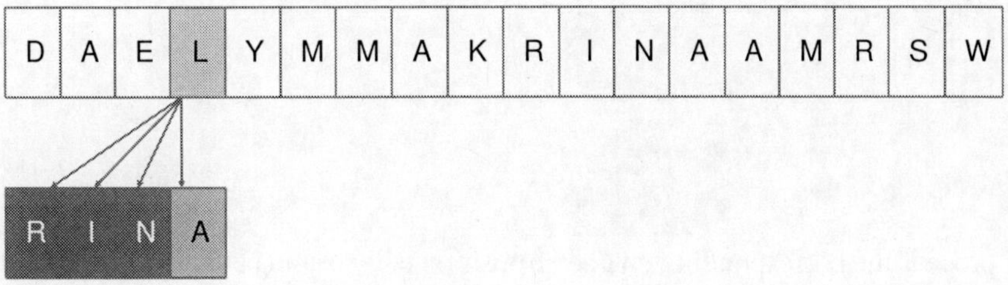

Since nothing matches then the target string shifts to the right as much as the length of the input string

c. Second Step

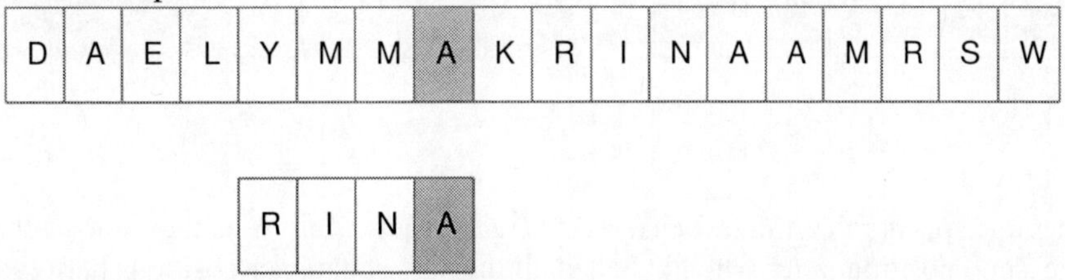

It appears in the process above the letter A == A then the next process is to test all the characters contained in the target string as much as n the target string of the input string, the following results

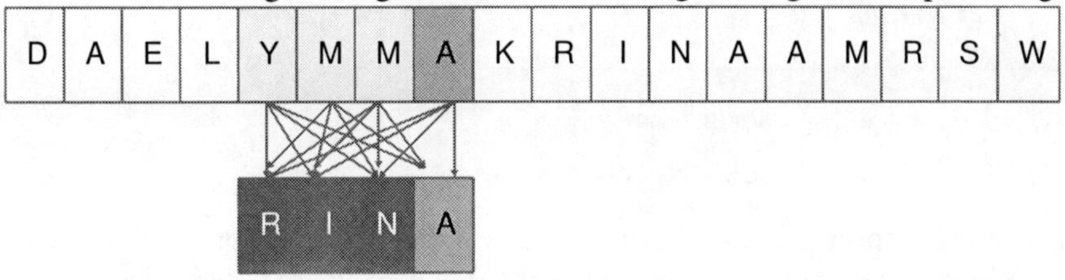

It appears in the above process that the corresponding characters are only characters A and not for other characters then the next process of the target string goes forward as many as the target string n characters

d. Third Step

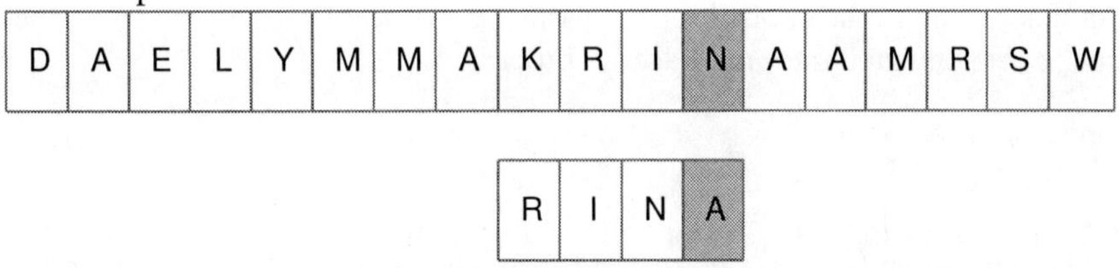

Seen in the process above the letter A != N then the next process is to test all the characters contained in the target string as much as n the target string of the input string, the following results.

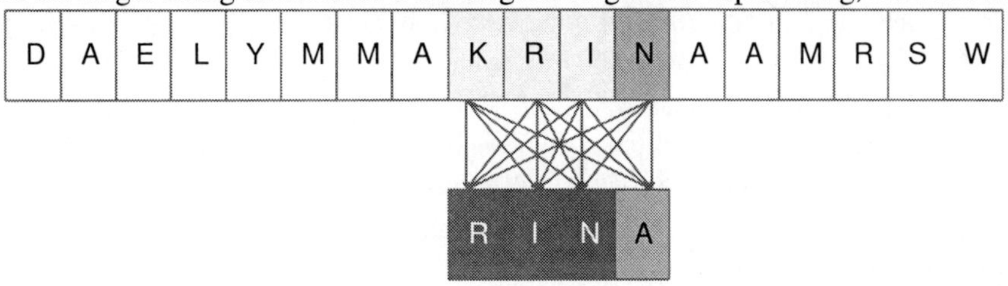

in the above process there are some similar characters between the target strings with the string of entry, based on the function of the Boyer-Moore algorithm if there is the same character then the character position between the target string and the goal string must be equated as below

D	A	E	L	Y	M	M	A	K	R	I	N	A	A	M	R	S	W

R	I	N	A

Seen in the above process the corresponding character position is the same between the target string position and the input string, then the above process is checked again to test whether the target string character is the same as the character of the input string

D	A	E	L	Y	M	M	A	K	R	I	N	A	A	M	R	S	W

R	I	N	A

In the above process, it appears that the last character of the target string 'A' is the same as the 'A' input string for the same position, after which it tested all the same character positions between the target string and the input string.

D	A	E	L	Y	M	M	A	K	R	I	N	A	A	M	R	S	W

R	I	N	A

Appear in the table above is appropriate then the process is stopped, and the word "RINA" is located in characters 10 to 13, then the remaining characters of the input string are not processed.

4. Conclusion

The process visualization of the Boyer-Moore algorithm allows researchers or scholars to learn how the Boyer-Moore algorithm works, and in the application development it will be easier to create a function for word search and can be implemented into many search processes.

References

[1] N. R. Dalal and P. Jadhav, "A Composite Algorithm for String Matching," *International Journal of Modern Trends in Engineering and Research (IJMTER),* vol. 2, no. 7, pp. 68-73, 2015.

[2] R. Janani and S. Vijayarani, "An Efficient Text Pattern Matching Algorithm for Retrieving Information from Desktop," *Indian Journal of Science and Technology,* vol. 9, no. 43, pp. 1-11, 2016.

[3] A. Z. M. Saleh, N. A. Rozali, A. G. Buja, K. A. Jalil, F. H. Mohd Ali and T. F. A. Rahman, "A Method for Web Application Vulnerabilities Detection by Using Boyer-Moore String Matching Algorithm," in *The Third Information Systems International Conference*, 2015.

[4] A. A. Alsaheel, A. H. Alqahtani and A. M. Alabdulatif, "Analysis of Parallel Boyer-Moore String Search Algorithm," *Global Journal of Computer Science and Technology Hardware & Computation,* vol. 13, no. 1, pp. 43-47, 2013.

[5] R. Y. Tsarev, A. S. Chernigovskiy, E. A. Tsareva, V. V. Brezitskay, A. Y. Nikiforov and N. A. Smirnov, "Combined string searching algorithm based on Knuth-Morris-pratt and Boyer-Moore algorithms," in *XIX International Scientific Conference Reshetnev Readings*, Rusia, 2015.

[6] G. Pandey and G. L. Prajapati, "Applying Bi-Directional Search Strategy in Selected String Matching Algorithms," *International Journal of Computer Applications,* vol. 143, no. 11, pp. 40-43, 2016.

[7] C. S. Rao, K. B. Raju, and S. V. Raju, "Parallel String Matching with Multi-Core Processors - A Comparative Study for Gene Sequences," *Global Journal of Computer Science and Technology Hardware & Computation,* vol. 13, no. 1, pp. 26-41, 2013.

IOP Conf. Series: Journal of Physics: Conf. Series **930** (2017) 012002 doi:10.1088/1742-6596/930/1/012002

Forecasting Error Calculation with Mean Absolute Deviation and Mean Absolute Percentage Error

Ummul Khair[1*], Hasanul Fahmi[2], Sarudin Al Hakim[3] and Robbi Rahim[4]

[123]Department of Informatics Engineering, Sekolah Tinggi Teknik Harapan, Jl. H.M. Joni No. 70 C, Medan 20216, Indonesia
[4]School of Computer Engineering and Communication, Universiti Malaysia Perlis, Malaysia

*ummul.kh@gmail.com

Abstract. Prediction using a forecasting method is one of the most important things for an organization, the selection of appropriate forecasting methods is also important but the percentage error of a method is more important in order for decision makers to adopt the right culture, the use of the Mean Absolute Deviation and Mean Absolute Percentage Error to calculate the percentage of mistakes in the least square method resulted in a percentage of 9.77% and it was decided that the least square method be worked for time series and trend data.

1. Introduction
The use of forecasting or prediction has been widely employed in organizational activities to prepare conditions that may occur in the future[1][2]. Predictions are also the basis for all business decisions even if they are not precise, but the organization can get a picture for future decisions. In made the prediction does not close the possibility of the organization less attention to the pattern of data owned, so finally using predictive methods that are less of the data pattern[2][3]. The predicted results obtained are not maximal in helping the organization determine the steps in the future, so the understanding of the pattern of data owned is essential[2].

Prediction errors are common and almost all forecasting methods have errors in predicted results[1], one of the forecasting methods that can be used is the least square method that performs calculations with time series data and has a seasonal trend[4], with time series calculated data error possibility Prediction will also occur frequently[4], these prediction errors can be computed using the Mean Absolute Deviation method and Mean Absolute Percentage Error.

Mean Absolute Deviation and Mean Absolute Percentage Error is a method that can be used to calculate margin error from predicted least square method of data [5], both approaches have different concepts in performing calculations with different results, the use of both of these ways in calculating the predicted error gives The organization's choice to consider the utilization of a method of prediction..

2. Theory
Forecasting models are then validated using some indicators, indicator used is Mean Absolute Deviation and Mean Absolute Percentage Error, and here is the explanation:

a. Mean Absolute Deviation

The method for evaluating forecasting methods uses the sum of simple mistakes. Mean Absolute Deviation (MAD) measures the accuracy of the prediction by averaging the alleged error (the absolute value of each error). MAD is useful when measuring prediction errors in the same unit as the original series[5][6]. The value of MAD can be calculated using the following formula.

$$MAD= \frac{\Sigma|y1- yt'|}{n}$$

b. Mean Absolute Percentage Error

Mean Absolute Percentage Error (MAPE) is calculated using the absolute error in each period divided by the observed values that are evident for that period. Then, averaging those fixed percentages. This approach is useful when the size or size of a prediction variable is significant in evaluating the accuracy of a prediction[7][8]. MAPE indicates how much error in predicting compared with the real value.

$$MAPE= \frac{\Sigma\frac{|y1- yt'|}{y1}}{n} \text{ x } 100\ \%$$

3. Result and Discussion

Testing validation error performed actual data below before the prediction process, the data displayed has time series and trends, and data below is the request data items in Kg:

Table 1. Actual Data before Prediction

No	Month	Year		
		2011	**2012**	**2013**
1	January	41,2	39,5	42,81
2	February	43,7	45,9	52,3
3	March	38,8	47,9	55,4
4	April	42,6	35,76	54,69
5	May	36,7	37,8	45,9
6	June	39,8	41,34	52,79
7	July	45,9	45	57,6
8	August	42,89	42	49,87
9	September	41,89	37,5	42,9
10	October	39,8	31,4	32
11	November	43,05	38,2	39,87
12	December	45,87	40,3	45,8
Σ		502.2	482.6	571.93

The actual data above is then calculated by the least square method to get prediction every month for the year 2014-2015 starting from September and got a result as follows:

Table 2. Data Prediction

No	Year	Month	Number of Requests
1	2014	September	41,77
2	2014	October	26,6
3	2014	November	37,19
4	2014	December	43,92
5	2015	January	43,585
6	2015	February	60,2
7	2015	March	72,26
8	2015	April	62,485
9	2015	May	53,93
10	2015	June	64,125

11	2015	July	67,05
12	2015	August	55,39
Total Request			628.505

Based on original data contained in Table 1 and predicted data using the least square method in Table 2, error calculation is done by using Mean Absolute Deviation and Mean Absolute Percentage Error. The details of forecasting error calculation using Mean Absolute Deviation (MAD) and Mean Absolute Percentage Error (MAPE) is as follows:

a. Period of September

Table 3. Forecasting Error Analysis September

Year	Data Actual (y1)	Forecasting (yt')	(|y1- yt'|)
2011	41,89	42,275	0,385
2012	37,5		4,775
2013	42,9		0,625
∑	122,29		5,785

$$\text{MAD} = \frac{\sum|y1 - yt'|}{n} = \frac{5,785}{3} = 1,92$$

$$\text{MAPE} = \frac{\sum\frac{|y1 - yt'|}{y1}}{n} \text{ x } 100\ \% = \frac{0,04}{3} \text{ x } 100\ \% = 4,7\ \%$$

Then the calculation error is 4,7% by using MAPE method

b. Period of October

Table 4. Forecasting Error Analysis October

Year	Data Actual (y1)	Forecasting (yt')	(|y1- yt'|)
2011	39,8	42,275	17,1
2012	31,4		8,7
2013	32		9,3
∑	103,2		35,1

$$\text{MAD} = \frac{\sum|y1 - yt'|}{n} = \frac{35,1}{3} = 11,7$$

$$\text{MAPE} = \frac{\sum\frac{|y1 - yt'|}{y1}}{n} \text{ x } 100\ \% = \frac{0,34}{3} \text{ x } 100\ \% = 11,3\ \%$$

Then the calculation error is 11,3% by using MAPE method

c. Period of November

Table 5. Forecasting Error Analysis November

Year	Data Actual (y1)	Forecasting (yt')	(|y1- yt'|)
2011	43,05	35,6	7,45
2012	38,2		2,6
2013	39,87		4,27
∑	121,12		14,32

$$\text{MAD} = \frac{\sum|y1 - yt'|}{n} = \frac{14,32}{3} = 4,77$$

$$\text{MAPE}=\frac{\sum\frac{|y1-yt'|}{y1}}{n}\text{ x } 100\ \%=\frac{0{,}11}{3}\text{ x } 100\ \%= 11{,}8\ \%$$

Then the calculation error is 11,8% by using MAPE method

d. Period of December

Table 6. Forecasting Error Analysis December

Year	Data Actual (y1)	Forecasting (yt')	(|y1- yt'|)
2011	45,87		1,985
2012	40,3	43,885	3,585
2013	45,8		1,915
∑	131,97		7,485

$$\text{MAD}=\frac{\sum|y1-yt'|}{n}=\frac{7{,}485}{3}= 2{,}495$$

$$\text{MAPE}=\frac{\sum\frac{|y1-yt'|}{y1}}{n}\text{ x } 100\ \%=\frac{0{,}05}{3}\text{ x } 100\ \%= 1{,}8\ \%$$

Then the calculation error is 1,8% by using MAPE method

e. Period of January

Table 7. Forecasting Error Analysis January

Year	Data Actual (y1)	Forecasting (yt')	(|y1- yt'|)
2011	41,2		2,385
2012	39,5	43,585	4,085
2013	42,81		0,775
∑	123,51		7,245

$$\text{MAD}=\frac{\sum|y1-yt'|}{n}=\frac{7{,}245}{3}= 2{,}415$$

$$\text{MAPE}=\frac{\sum\frac{|y1-yt'|}{y1}}{n}\text{ x } 100\ \%=\frac{0{,}058}{3}\text{ x } 100\ \%= 1{,}93\ \%$$

Then the calculation error is 1,93% by using MAPE method

The above calculation process is some process of fault validation process that achieved by using MAD and MAPE method, here is the result of error checking until August 2015:

Table 8. MAD and MAPE Calculation results

Year	**Month**	**Method**	
		MAD	**MAPE**
2014	September	1,92	4,7 %
2014	October	11,7	11,3 %
2014	November	4,77	11,8 %
2014	December	2,495	1,8 %
2015	January	2,415	1,93 %
2015	February	12,9	9,09 %
2015	March	24,95	17,5 %
2015	April	18,135	13,6 %

2015	May	13,79	11,45 %
2015	June	19,48	14,54 %
2015	July	17,55	11,81 %
2015	August	10,47	7,7 %

The results above are the percentage of error that obtained by using Mean Absolute Deviation and Mean Absolute Percentage Error, as for the resulting graph as below

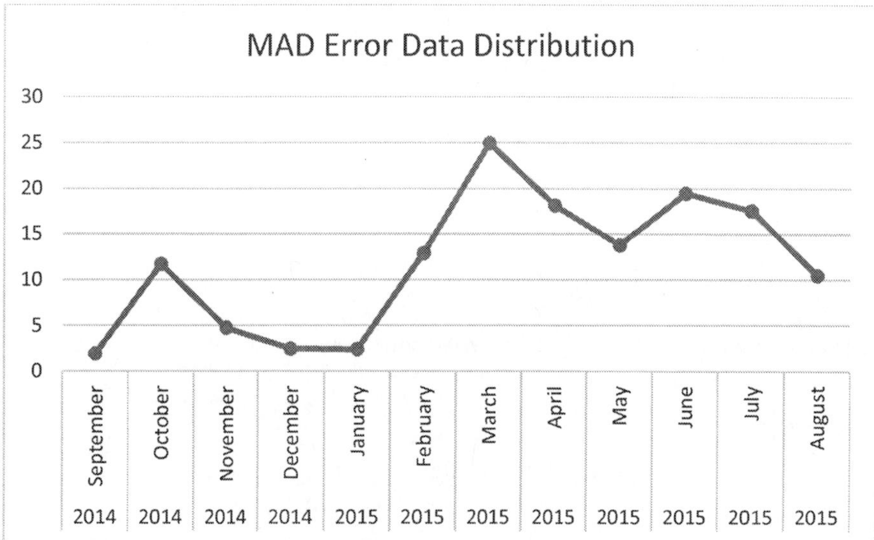

Figure 1. MAD Error Graph Distribution

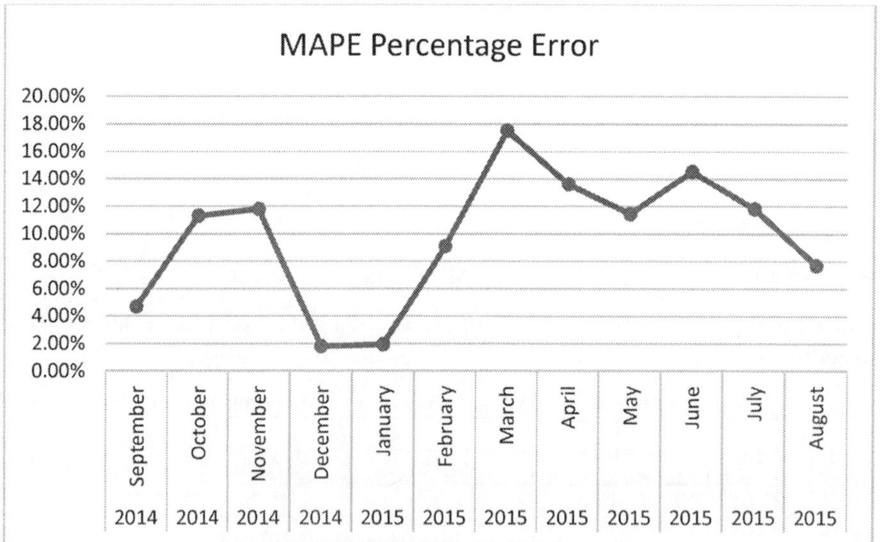

Figure 2. MAPE Percentage Error

4. Conclusion

Based on forecasting error testing conducted on predicted data with least square method found that the average error percentage 9.77%, this figure is quite rational because not exceeding 10% margin of error, the method of Mean Absolute Deviation and Mean Absolute Percentage Error can be applied well for

Checking the result of prediction result of least square method and also do not close possibility to another method.

References

[1] K. M. Kitani, B. D. Ziebart, J. A. Bagnell, and M. Hebert, "Activity forecasting," in *Lecture Notes in Computer Science (including subseries Lecture Notes in Artificial Intelligence and Lecture Notes in Bioinformatics)*, 2012, vol. 7575 LNCS, no. PART 4, pp. 201–214.

[2] G. S. Atsalakis and K. P. Valavanis, "Surveying stock market forecasting techniques - Part II: Soft computing methods," *Expert Syst. Appl.*, vol. 36, no. 3 PART 2, pp. 5932–5941, 2009.

[3] Y. K. Wu and J. S. Hong, "A literature review of wind forecasting technology in the world," in *2007 IEEE Lausanne POWERTECH, Proceedings*, 2007, pp. 504–509.

[4] S. Ismail and A. Shabri, "Time series forecasting using least square support vector machine for canadian lynx data," *J. Teknol.*, vol. 70, no. 5, pp. 11–15, 2014.

[5] Y. Moon and T. Yao, "A robust mean absolute deviation model for portfolio optimization," *Comput. Oper. Res.*, vol. 38, no. 9, pp. 1251–1258, 2011.

[6] C. Leys, C. Ley, O. Klein, P. Bernard, and L. Licata, "Detecting outliers: Do not use standard deviation around the mean, use absolute deviation around the median," *J. Exp. Soc. Psychol.*, vol. 49, no. 4, pp. 764–766, 2013.

[7] J. McKenzie, "Mean absolute percentage error and bias in economic forecasting," *Econ. Lett.*, vol. 113, no. 3, pp. 259–262, 2011.

[8] A. de Myttenaere, B. Golden, B. Le Grand, and F. Rossi, "Mean Absolute Percentage Error for regression models," *Neurocomputing*, vol. 192, pp. 38–48, 2016.

Analysis of Multiple Data Hiding Combined Coloured Visual Cryptography and LSB

[1]Halim Maulana, [2]Edy Rahman Syahputra

[12]Department of Information System Sekolah Tinggi Teknik Harapan, Medan-20217, Indonesia

[1]thekazagami@gmail.com, [2]ydeaja@yahoo.com

Abstract. Currently the level of data security becoming a major factor in data transfer. As we know every process of sending data through any medium the risk of that data gets hacked will still be there. Some techniques for securing data such as steganography and cryptography also often used as a solution for securing data. But it does not last long because it has been fou out the weaknesses of the algorithm so that the security be assured. So, in need of variety of new algorithms to be able to protect the data so that data security can be guaranteed. In this study tries to combine two visual algorithms that steganography and cryptography. Where in these experiments will try to secure two pieces of data type that is the type of image data and text data where both the data is regarded as a message so to obtain the correct information receiver should get that two types of data.

1. Introduction

In the communication between the two sides, there is no guarantee that states that the communication occurs has been safe from the threat of a third party. The presence of a third party in a communication can interfere with the comfort of both parties. The third party can retrieve important information from the communication that is going on. This course will be very detrimental to the first and second parties. Based on this was the need for a technique to secure information that is considered important to avoid the threat of a third party. Steganography is an art as well as techniques to hide information such as messages and data on digital media [1]. Securing steganography techniques to hide the message by means of a medium, so that third parties do not suspect messages in the media. Digital media are typically used to hide the message can be text files, digital images, audio, and video. One method of steganography is the Least Significant Bit (LSB) is done by modifying bits including LSB bits in each byte of colour in a pixel. [2][3][4][5]. However, the data security techniques by using steganography methods have emerged many weaknesses and shortcomings in the use of which may cause data to be very unsafe. It can be seen from several previous studies [6][7][8]. In cryptography is also known in the visual cryptography which in visual basic concepts of cryptography works by breaking down an image into sections picture that looks like white noise, but when the parts are united then the hidden image will be displayed [9][10][11]. But based on research-research that past steganography and cryptography techniques are becoming less secure and thus require a breakthrough or other variations in a secure data. In another study [12] conducted a combination of visual methods of steganography and cryptography. In this process, the original image file will be processed by steganography and then encrypted using visual cryptography produce multiple images. In the result, even if the image file managed in the reconstruction process but when done visual cryptography hidden data being lost when detected with a hex editor. The data is lost due to the imperfection of the reconstruction process of visual cryptography. So, it takes a more optimal reconstruction process so that confidential data is not lost either in the form of image or message text. Based on that can be concluded that both the technique of steganography and cryptography have lack in securing the data so that the data becomes less secure.

2. Proposed Methodology

2.1. *LSB Process*

Step 1: Read the values of the primary colour components are red, green and blue.
Step 2: Read the text of the message that will be hidden
Step 3: Compares the message size and image size. the process will only run if the size of the image larger than the size of the message.
Step 4: For a character message, all i up to n, perform the checks smallest pixel value of the components of R, G and B by using Operations AND / OR
Step 5: Message Successfully embedded and stego image was generated

2.2 *Visual Cryptography Process*

Step 1 : Read the value of pixels for each - each component of R, G, B
Step 2 : Make a histogram for each of the intensity of the main components of the R, G, B of the original image. Wherein for each - each histogram R, G, B has 256 levels of intensity.
Step 3 : As we saw in the previous step, each colour component has 256 levels of intensity. From the histogram that have been formed in the previous step will be determined Colour Level of each colour component R, G, B. Where the process of determining the Colour, Level is determined by finding the standard deviation value of the components of R, G, B and Colour Levels are divided into 8 categories as follows.

Category 2 = Value Standard Deviation range 20001-30000
Category 3 = Value Standard Deviation range 15001-20000
Category 4 = Value Standard Deviation range 10001-15000
Category 6 = Value Standard Deviation range 8001-10000
Category 8 = Value Standard Deviation range 5001-8000
Category 16 = Value Standard Deviation range 4001-5000
Category 24 = Value Standard Deviation range 2001-4000
Category 32 = Value Standard Deviation <2000

Step 4 : Group the histogram into equal parts. For each colour component X ∈ {R, G, B}, will be

made on a histogram of the Nx groups X. To do so, will be determined in advance the color intensity of the boundary between each pair of adjacent groups are defined as K0, ..., KNX -2, In other words, the histogram X will be divided into regions Nx, for example [0, K0), [K0, K1), ..., [KNX -2, 255]. The division is made to make Nx each region has the same size in the area. So, for each group, there will be the same number of pixels in the image are included in each group.
Step 5: Next is the stage of making share. To do so, the authors apply the following method for each primitive colour independently. First, take the k-of-n VCS black-and-white. The basis for Boolean Matrix Bb (for b = 0.1), in Denote Bb= [Bb0, Bb1, …, Bbm-1]

Bbi shows column i (0=i=m-1) of the Bb. Second, for each color component X ∈ {R, G, B} done in

stages as follows:

1. Assuming if the intensity of the color pixel corresponding to the color components of X falls in a group of k-th (where 0=k = NX -1) then it will be calculated the probability P = k / (NX-1).

2. With a probability P, the process will be conducted as follows:
   a. Considering the B0 and randomly select a column, for example, where 0 =j =m B0J-1.
   b. Assume B0J as an n-bit vector. For the first bit will be set in black if the bit value is 1, otherwise it will oversee red / red. The process is repeated until all the pixels have been set with a certain colour to n share.

3. For the probability 1-P, will do a stage like the previous step but by replacing B0 to B1. Based on the value of P, in Table 1 shows a summary of the probability distribution B0 and B1 for groups of individuals. Because columns B0 and B1 is selected at random, then the chances of getting any column in B0 and B1 for the group k becomes k / ((NX -1) m) to B0 and (1-k / (NX -1)) / m) to B1. Finally, we have put Share R to the i-th with Share G to the i-th and Share B to the i-th, for i = 1, ..., n, to form the final value of i-th share ¬yang comprising components of R, G and B.

3. Encryption and Decryption Process

3.1 *Encryption*
A. *Visual Cryptography*

Image file is assumed to be in the process by using visual cryptography steganography has experienced before. We assume each pixel is divided into multiple subpixels, but the human eye still sees the pixel as one pixel. A little review about VCS-based Black & White (k, n). Visual Cryptography process used is to use VCS for coloured image. In assuming that the image n is the image that will be used to process the VCS. An image of a coloured image will be used to generate n share the image. Image that will be processed using the VCS will be in previously resize with a size of 800 x 600. The stages than encryption algorithms VCS that his adaptation of the results of previous studies [13][14][15][16] which consists of several steps, namely Histogram Generation Determination Quality Colour, Grouping, and Share Creation.

B. *LSB bit Encryption*

Share 1 and share 2 on visual cryptography get from the process will be used as a medium to hide text messages using LSB techniques. The insertion process is done by first converting messages into binary form and then in binary form will be divided into two parts where the half of the first bit will be inserted in the share1 and the half of the other bits will be inserted in share2 using the AND operation and OR [17] [18].

3.2 *Decryption Process*
A. *LSB Extraction*

The LSB bits of the steganography image is extracted at first. Then the last pixel value LSB number is identified to the share positions. The LSB bits of other pixels are arranged according the number identified. The share 1 and share 2 are generated using this LSB extraction.

B. *Visual Cryptography Decryption*
The obtained shares from the LSB extraction is fed as input to the visual decryption process. The shares are combined by alternating the pixels from each share to generate the secret message image.

4. Experimental Results
At this stage it will be a test to hide a secret message and the secret image by using a combination of algorithms Visual Cryptography and LSB. Suppose there is an image that will fund a message is secured using the algorithm. Then the message will be processed in advance using lsb then continued using visual cryptography techniques.

(a) (b) (c)
Figure 1. (a) Secret Image, (b) Share1 (c) Share2

The output of the algorithm is a combination of two share the image that share 1 and share 2 in which the secret message has been inserted into the two shares to regain a secret message and the secret image in need both share the image. After both share the image collected, to be able to get back data and images is carried out secret decryption process which will generate a secret message that is stored in the secret.txt and will also generate a secret image derived from the share 1 and share 2 which has been in combine.

(a) (b)
Figure 2. (a) Overlapped Image, (b) Regenerated Secret Message

4.1 *Performance Analysis*
Various parameters recommended by the researchers to evaluate the performance of visual cryptography Naor and Shamir (1995) suggest two main parameters: pixel expansion and contrast a m. Expansion pixel m refers to the number of subpixels in the resulting shares representing a pixel of the original input image. It presented losses resolution than the original image with the results of the share. Instead a is the relative difference between the combination of the share coming from the white pixels and black pixels in the original image. Pixel expansion that occurs between the original image and share the image it will affect the image quality. In table 1 shows that there is no change of the dimensions of the image, either Height or Weight image of Secret Image, share1, Share2 and Overlapped image, but the image size after the process has been reduced which is significant, but the image restoration still can recognize well as the secret image. From the results obtained in Table 1 it can be concluded that the results are not experiencing visual cryptography pixel expansion [19][20] [21]. On the table shows that every visual cryptography scheme has a pixel expansion in addition to algorithms used in this study.

Tabel 2. Other Visual Cryptography Algorithm

ImageProperty	Secret Image	Share 1	Share 2	Retr. Image
Dimension	450 x 600	450 x 600	450 x 600	450 x 600
Height (pixels)	450	450	450	450
Width (pixels)	600	600	600	600
Size	345KB	172 KB	173 KB	107 KB
Bit Depth	24	24	24	24

No.	Author	Year	Sec.Image	Exp.	Format
1	Naor & Shamir	1995	1	4	Biner
2	Nakajima	2002	1	4	Biner
3	Young-Chang Hou	2003	1	4	Color
4	Z.Zhou	2006	1	4	Biner
5	Inkoo Kang,	2009	1	4	Color
6	Jagdeep Verma & Dr.Vineeta	2012	1	4	Color
7	Dr.D.Vas umathi & M.Surya Prakash	2012	1	4	Color
8	**Proposed Method**	2016	1	1	Color

In tables 3 show that the PSNR for the CVC is the largest compared with VC and EVC, where it is means that the value of the reconstruction will be of high quality. However, MSE that produced very high caused by the insertion of a message contained therein. It also means that the degree of similarity between the secret image and the image in the lower encryption because in the encrypted image does not have enough information about the secret images.

Table 3. Performance Analysis

Picture	PSNR			NCC			Analisis		
	VC	EVC	CVC	VC	EVC	CVC	VC	EVC	CVC
Lena	13.321	18.215	25.256	1	1	1	17.821	12.453	195.436
Pepper	16.274	21.296	25.949	1	1	1	17.154	12.134	165.230
SailBoat	11.263	16.319	25.362	1	1	1	16.731	11.318	189.143
Barbara	14.126	19.615	25.495	1	1	1	17.153	12.126	183.473

4.2 *Histogram Analysis*
In the field of image processing histogram shows the distribution of pixel values in an image. Histogram used attacker (attacker) to perform cryptanalysis by utilizing the frequency of appearance of pixels in the histogram. Attackers hope pixel values that often appear in the plain-image is correlated with pixel values that often appear in the cipher-image. By analysing the frequency of occurrence of pixel values, the attacker deduces the key or the pixels in the plain-image.

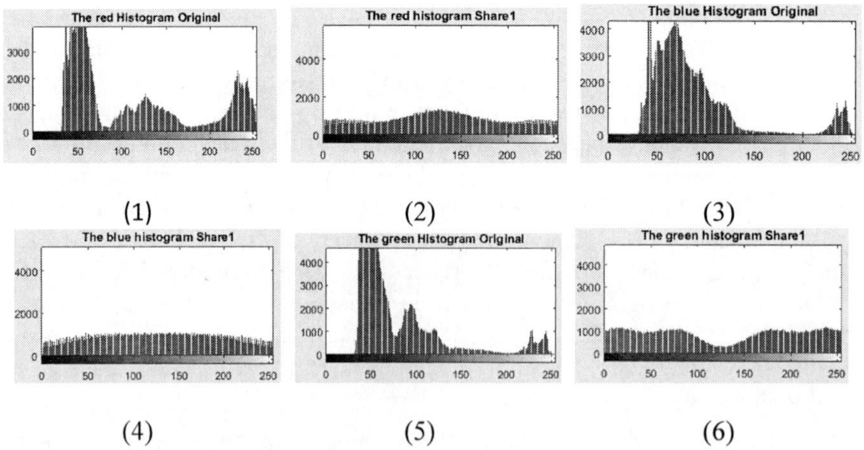

Figure 3. Original and Share Image Histogram Compare

5. Conclusion

The results from the combination of UCS and LSB algorithm has managed well with can hide the image data and messages and can restore the data with good results. Secret image that retrieved might not like the image to the original before being processed, but they can recognize as the same image. High PSNR value means high quality image generated great surplus of other methods. The value of the Mean Squared Error to be very high due to the message that imbed in the image. However, higher values also mean MSE between the original image and the image is not there a similarity that can be considered as a different picture.

References

[1] Channalli, S & Jadhav, A. Steganography An Art of Hiding Data. International Journal on Computer Science and Engineering. 1(3) (2009) 137-141
[2] Laskar, S.A & Hemachandran, K. High Capacity data hiding using LSB Steganography and Encryption. International Journal of Database Management Systems (IJDMS). 4(6) (2012) 57-68
[3] Champakamala. B.S, Padmini, Radhika.D. K. Least Significant Bit algorithm for image steganography. International Journal of Advanced Computer Technology. 3(4) (2012) 34-38
[4] Gupta, S., Gujral, G & Aggarwal, N. Enhanced Least Significant Bit Algorithm for Image Steganography. IJCEM International Journal of Computational Engineering & Management. 15(4) (2012) 40-42.
[5] Reddy, L.V., Subramaniam & Reddy, C.P. Implementation of LSB Steganography and its Evaluation for Various File Formats. Int. J. Advanced Networking and Applications. 2(5) (2011) 868-872
[6] J. Fridrich &M. Goljan. Practical Steganalysis—State of the Art. Proc. SPIE Photonics Imaging. Security and Watermarking of Multimedia Contents.4675 (2002) 1–13.
[7] Provos, N & Honeyman, P. Detecting Steganographic Content on the Internet. Network and Distributed System Security Symp., Internet Soc. (2002)
[8] Westfeld, A. F5—A Steganographic Algorithm: High Capacity Despite Better Steganalysis, Proc. 4th Int'l Workshop Information Hiding. Springer-Verlag. (2001) 289–302.
[9] Dhole, A.B & Janwe, N.J. An Implementation of Algorithms in Visual Cryptography in Images. 3(3) (2013) 1-5
[10] Noar, M & Shamir, A. Visual cryptography. Advances in Cryptology - EUROCRYPT'94. (1995) 1-12.
[11] I. Kang, G. R. Arce and H. K. Lee. Color Extended Visual Cryptography Using Error Diffusion," in IEEE Transactions on Image Processing, 20(1) (2011) 132-145
[12] G. Abboud, J. Marean and R. V. Yampolskiy. Steganography and Visual Cryptography in Computer Forensics. Systematic Approaches to Digital Forensic Engineering (SADFE), 2010 Fifth IEEE International Workshop on, Oakland, CA, (2010) 25-32
[13] X. Wu, D. S. Wong and Q. Li. Extended Visual Cryptography Scheme for color images with no pixel expansion. Security and Cryptography (SECRYPT), Proceedings of the 2010 International Conference on, Athens. (2010) 1-4
[14] F. Liu, C. K. Wu and X. J. Lin. Colour visual cryptography schemes. IET Information Security, 2(4) (2008) 151-165

[15] X. Wu, D. S. Wong and Q. Li. Threshold Visual Cryptography Scheme for Color Images with No Pixel Expansion. Proceedings of the Second Symposium International Computer Science and Computational Technology (ISCSCT '09). (2009) 310-315
[16] S. J. Shyu and H. W. Jiang. General Constructions for Threshold Multiple-Secret Visual Cryptographic Schemes. IEEE Transactions on Information Forensics and Security. 8(5) (2013) 733-743
[17] K. Joshi, P. Dhankhar and R. Yadav. A new image steganography method in spatial domain using XOR," 2015 Annual IEEE India Conference (INDICON), New Delhi. (2015) 1-6
[18] X. Gui, X. Li and B. Yang. Steganalysis of LSB Matching Based on Local Binary Patterns. Intelligent Information Hiding and Multimedia Signal Processing (IIH-MSP), 2014 Tenth International Conference on, Kitakyushu. (2014) 475-480.
[19] Chettri, L. Visual Cryptography Scheme Based on Pixel Expansion for Black & White Image. International Journal of Computer Science and Information Technologies. 5(3) (2014) 4190-4193
[20] Kumar, P., Agarwal, S & Shivani, S. Halftone Visual Cryptography with Pixel Expansion through Error Diffusion. International Journal of Information & Computation Technology. 4(14 (2014) 1419-1427
[21] Blundo, C., Cimato, S & De Santis, A. Visual Cryptography with Optimal Pixel Expansion. Theoretical Computer Science 369 (2006) 169-182

IOP Conf. Series: Journal of Physics: Conf. Series **930** (2017) 012004 doi:10.1088/1742-6596/930/1/012004

Measuring the Accuracy of Simple Evolving Connectionist System with Varying Distance Formulas

Al-Khowarizmi[1], O S Sitompul[2*], Suherman[3], E B Nababan[4]

[1,2] Department of Computer Science
[4] Department of Information Technology
Faculty of Computer Science and Information Technology
Universitas Sumatera Utara, Medan – 20155, Indonesia
[3] Department of Electrical Engineering
Faculty of Engineering, Universitas Sumatera Utara, Medan – 20155, Indonesia

[*]Corresponding Author. E-mail address: opim@usu.ac.id

Abstract. Simple Evolving Connectionist System (SECoS) is a minimal implementation of Evolving Connectionist Systems (ECoS) in artificial neural networks. The three-layer network architecture of the SECoS could be built based on the given input. In this study, the activation value for the SECoS learning process, which is commonly calculated using normalized Hamming distance, is also calculated using normalized Manhattan distance and normalized Euclidean distance in order to compare the smallest error value and best learning rate obtained. The accuracy of measurement resulted by the three distance formulas are calculated using mean absolute percentage error. In the training phase with several parameters, such as sensitivity threshold, error threshold, first learning rate, and second learning rate, it was found that normalized Euclidean distance is more accurate than both normalized Hamming distance and normalized Manhattan distance. In the case of beta fibrinogen gene -455 G/A polymorphism patients used as training data, the highest mean absolute percentage error value is obtained with normalized Manhattan distance compared to normalized Euclidean distance and normalized Hamming distance. However, the differences are very small that it can be concluded that the three distance formulas used in SECoS do not have a significant effect on the accuracy of the training results.

1. Introduction

Artificial neural networks (ANN) as one of the information processing systems are inspired from biological network of neurons and have been proven to have high performance [1]. As a modern applied computational mathematics, ANN is used to solve problem by learning process from data patterns [2]. One known method in ANN is Simple Evolving Connectionist System (SECoS) [3] which is also known as simple evolving multi-layer perceptron as a minimal implementation of Evolving Connectionist Systems (ECoS) [4]. ECoS is a computational intelligent method based on neural network, but is using a technique that differ from those of computational intelligence techniques. The ECoS operated continuously, adapting to some structure and functionality on every iterative interaction with the environment and the rest of the systems. By this manner, ECoS is able to overcome the commonly encountered ANN weaknesses in terms of longer training time and difficulty in determining the system architecture [5].

By using normalized Hamming distance in the activation function, [6] confirmed that SECoS is very sensitive to parameters such as first and second learning rate and could recognize data better and adapted faster to new data while in the learning process. Furthermore, in [4] SECoS was used to classify voice recognition by optimizing the parameter value. In the learning process normalized Hamming distance was used in the activation function to obtain smallest error. In addition to classification in medical case, [7] also used SECoS to classify gene data from cancer tissues. In [8] ECoS was implemented to develop a system which can help radiologists to identify the symptoms of breast cancer by using Breast Imaging Reporting and Database System as a standard system.

Adaptive capability of SECoS in the learning process is also studied in [9] in which the activation function is also calculated using normalized Hamming distance. Evolving Multi-layer Perceptron [10] for learning process in big data using normalized Hamming distance in the activation function obtained smallest error rate. Furthermore, using a different method, [11] compared Manhattan distance and Euclidean distance as two most popular distance formulas and came to a conclusion that Euclidean distance had more iteration compared to Manhattan distance. To optimize discrete wavelet frame (DWF) configuration, [12] used texture feature extraction method on images which used Manhattan distance, Euclidean distance, normalized Manhattan distance and normalized Euclidean distance. They concluded that normalized Euclidean distance gave an optimal value.

Various applications of SECoS had been conducted in the field of medical, environment and health studies [7]. In this research, a calculation of the activation value in SECoS is performed not only using normalized Hamming distance formula but also using normalized Manhattan distance and normalized Euclidean distance in order to compare the smallest error value and the best learning rate obtained on the case of beta fibrinogen gene -455 G/A polimorphism patients. The accuracy of the SECoS training results from the three distance formula is then caculated using mean absolute average error (MAPE) as one of the accuracy measure commonly used to calculate percentage error in statistical forcasting [13].

2. Methodology

2.1. Dataset Used

Data used in this research is data from beta fibrinogen gene -455 G/A polimorphism patients obtained from the examination of neurological and fibrinogen concentration rate [14]. The data belong to 136 patients which have 9 characteristics such as age, Hb, Ht, Leucosite, KGD, cholesterol, HDL, LDL and fibrinogen concentration rate.

2.2. General Architecture

The general architecture of this research is depicted in Figure 1. Data from the database is firstly normalized in the range of 0-1. These data are then given to the SECoS training network, whereby each of the distance formula is used in the calculation. Lastly, The accuracies of the output of the SECoS network are calculated using the mean absolute percentage error (MAPE).

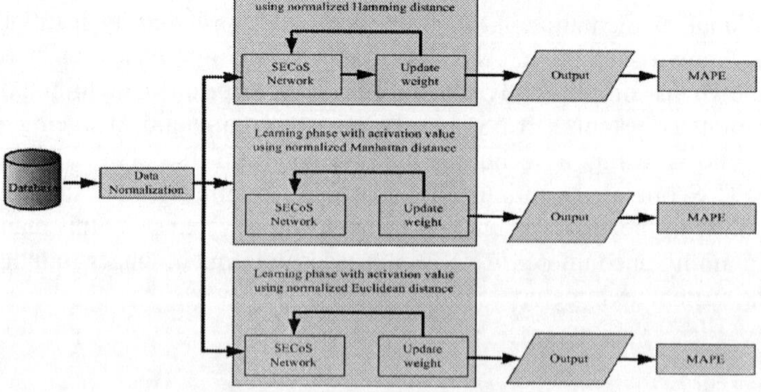

Figure 1. The general architecture

2.3. Simple Evolving Connectionist Systems

SECoS consists of three neuron layers [6], in which the first layer is an input layer with linear or other activation functions, the second layer is an evolving hidden layer, and the third layer is the output layer. The linear activation function A_n on the evolving node n is defined by [4] as in (1).

$$A_n = 1 - D_n \tag{1}$$

where A_n is the activation value on node n and D_n is normalized distance value between input vector and incoming weight vector on corresponding node. In general, distance value D_n is calculated using the following normalized Hamming distance as suggested in [4]:

$$D_n = \frac{\sum_1^K |I_i - W_i|}{\sum_1^K |I_i + W_i|} \tag{2}$$

where K is the number of input nodes on SECoS, I is the input vector, and W is the input weight matrix on evolving layer.

In this paper, an analysis on the activation function is performed with distance values are calculated not only using normalized Hamming distance but also using normalized Manhattan distance and normalized Euclidean distance. Formula to calculate the distance values as adapted from [12] are shown in (3) and (4), respectively.

$$D_n = \frac{1}{k}\left(\sum_{i=1}^{K} |I_i - W_i|\right)^2 \tag{3}$$

$$D_n = \sqrt{\frac{1}{k}\left(\sum_{i=1}^{K} |I_i - W_i|^2\right)} \tag{4}$$

A learning process is then conducted using the SECoS method on patients data of beta fibrinogen gene -455 G/A polymorphism. The SECoS algorithm is adapted from [15] as in the following:

Step 1. Propagate input vector I into the network.

Step 2. If the maximum activation (A_{max}) from one node is less than the sensitivity threshold (S_{thr}) coefficient, then add a new node. Otherwise,

a. Calculate the error value between the learning output of output vector O_c and the actual value of output vector O_d.
b. If the error value is greater than the error threshold (E_{trh}) coefficient or output node desired is not active, then:
    i Add a new node, otherwise
    ii Change the connection weight on the winning hidden node.

Step 3. Repeat Step 1 and Step 2 for each input vector.

Whenever a node is added, the input weight is initialized correspond to input vector I and the output weight is initialized corresponds to output vector O_d. Propagation process from hidden layer to output layer can be performed in two ways. First is using One-of-N propagation method in which propagation is performed by the hidden node with highest activation value. Second is using Many-of-N propagation method where propagation is peformed by hidden node with activation value above the activation threshold.

2.4. Mean Absolute Percentage Error (MAPE)

The amount of result errors are calculated using MAPE as shown in (5). Among various accuracy criteria, MAPE is a measuring tool which incorporating the best characteristics [13]. Apart from some its disadvantages, MAPE is a popular and widely used accuracy measure used in businesses and organizations [16].

$$MAPE = \frac{1}{n}\left(\sum_{t=1}^{n} \frac{|A_t - F_t|}{|A_t|} \right) \times 100 \qquad (5)$$

where A_t is the actual data, F_t is calculated data, and n is the amount of data.

3. Result and Discussion

In this section, result of comparing normalized Hamming distance, normalized Manhattan distance and normalized Euclidean distance on the SECoS activation function will be examine to obtain the smallest error value and the best learning rate in the training stage.

3.1. Test results with three distance formulas

In performing a training stage using SECoS method, the first thing to do is determining the parameter values, namely learning rate 1, learning rate 2, sensitivity threshold, error threshold and the number of training data. The parameter values are then normalized in the range of 0-1, continued by propagation from input layer to hidden layer by calculating activation values using normalized Hamming distance and finally performing propagation from hidden layer to output layer.

The activation functions of normalized Hamming distance, normalized Manhattan distance and normalized Euclidean distance in the SECoS training are tested using beta fibrinogen gene -455 G/A polymorphism patient data. Testing for those distance formulas will be compared using various combination of parameter values. The first input parameter values are calculated with sensitivity threshold = 0.6, error threshold = 0.1, learning rate 1 = 0.3, and learning rate 2 = 0.3, resulted in a MAPE value of 0.56618 and 24 hidden nodes, 0.57179 MAPE value and 24 hidden nodes, and 0.60317 MAPE value and 28 hidden nodes, for normalized Hamming distance, normalized Manhattan distance, and normalized Euclidean distance, respectively (*see* the first row of Table 1). As the parameter values used on the training stage are affecting the resulted error values, therefore some combination of the parameter values are performed in order to obtain the smallest error. The complete combination of parameter values can be seen in Table 1.

Table 1. Combination of parameter values using three distance formulas

Test No	**Sensitivity Threshold**	**Error Threshold**	**Learning Rate 1**	**Learning Rate 2**	**Normalized Hamming Distance**		**Normalized Manhattan Distance**		**Normalized Euclidean Distance**	
					MAPE	**Hidden Node**	**MAPE**	**Hidden Node**	**MAPE**	**Hidden Node**
1	0.6	0.1	0.3	0.3	0.56618	24	0.57179	24	0.60317	28
2	0.6	0.1	0.3	0.6	0.57644	27	0.59153	33	0.66081	33
3	0.6	0.1	0.3	0.9	0.58201	34	0.60809	34	0.63439	37
...	...	...	...	...	...	...	...	...	...	...
19	**0.3**	**0.1**	**0.3**	**0.3**	**0.52377**	**22**	-	-	-	-
...	...	...	...	...	...	...	...	...	...	...
22	**0.3**	**0.1**	**0.6**	**0.3**	-	-	**0.52657**	**25**	**0.51691**	**24**
...	...	...	...	...	...	...	...	...	...	...
34	0.3	0.05	0.9	0.3	0.55399	27	0.55280	29	0.54024	28
35	0.3	0.05	0.9	0.6	0.56651	33	0.58345	36	0.59051	35
36	0.3	0.05	0.9	0.9	0.55603	36	0.57230	37	0.54533	37

It is shown in Table 1 that the smallest error obtained in the SECoS training with activation function using normalized Hamming distance obtained at test number 19 where sensitivity threshold = 0.3, error threshold = 0.1, learning rate 1 = 0.3, and learning rate 2 = 0.3, with a MAPE value of 0.52377 and 22 hidden nodes. Training stage with the activation function using normalized Manhattan distance resulting in the smallest error at test number 22 where sensitivity threshold = 0.3, error threshold = 0.1, learning rate 1 = 0.6, and learning rate 2 = 0.3, with a MAPE value of 0.52657 and 25 hidden nodes. Using normalized Euclidean distance the smallest error obtained at test number 22 where sensitivity threshold = 0.3, error threshold = 0.1, learning rate 1 = 0.6, and learning rate 2 = 0.3, with a MAPE value of 0.51691 and 24 hidden nodes.

3.2. Discussion

On the test with normalized Hammning distance, it can be observed that the smaller sensitivity threshold value used, the smaller the error value obtained. In term of error threshold, the larger the value is given, the smaller the error value obtained. The same pattern is also hold for the learning rate 1 and learning rate 2, that is the smaller the values given, the smaller the error value. On the sensitivity threshold and error threshold the same patterns are also obtained from the test on normalized Manhattan distance and normalized Euclidean distance. However, in normalized Manhattan distance the smaller the value of learning rate 1 and learning rate 2 used, the smaller the error value obtained, whereas with normalized Euclidean distance the smaller learning rate 1 and learning rate 2 used, the smaller the error value obtained.

Result of parameter combination and test with normalized Hamming distance as shown in Table 1, give the smallest error value of 0.52377 obtained with a combination of parameter learning rate 1 = 0.3, learning rate 2 = 0.3, sensitivity threshold = 0.3 and error threshold = 0.1. As for the largest error value obtained is 0.63086 with combination of parameter learning rate 1 = 0.9, learning rate 2 = 0.6, sensitivity threshold = 0.6 and error threshold = 0.05. Meanwhile, result of combination and test with normalized Manhattan distance shows that the smallest error obtain is 0.52657 using parameter combination of learning rate 1 = 0.6, learning rate 2 = 0.3, sensitivity threshold = 0.3 and error threshold = 0.1. As for the largest error value obtain is 0.68776 with combination parameters learning rate 1 = 0.9, learning rate 2 = 0.9, sensitivity threshold = 0.6 and error threshold = 0.1. Finally, parameter combination and test using normalized Euclidean distance shows the smallest error value of 0.51691 with combination of parameters learning rate 1 = 0.6, learning rate 2 = 0.3, sensitivity threshold = 0.3 and error threshold = 0.1. The largest error obtained is 0.67534 with combination parameters learning rate 1 = 0.3, learning rate 2 = 0.6, sensitivity threshold = 0.6 and error threshold = 0.05. Summary of the test result on the beta fibrinogen gene -455 G/A polymorphism patient data is given in Table 2.

Table 2. Comparison of Distance Formulas

No	Distance Formula	MAPE	Parameter Value			
			Sensitivity Threshold	Error Threshold	Learning Rate 1	Learning Rate 2
1	Normalized Hamming distance	0.52377	0.3	0.1	0.3	0.3
2	Normalized Manhattan distance	0.52657	0.3	0.1	0.6	0.3
3	Normalized Euclidean distance	0.51691	0.3	0.1	0.6	0.3

4. Conclusion

The smallest MAPE value in the test using beta fibrinogen gene -455 G/A polimorphism patient data is obtained using normalized Euclidean distance. However, the differences is very small that it can be concluded that the three distance formulas used in SECoS do not have a significant effect on the accuracy of the training results. The smallest *error* on the SECoS training is also affected by *sensitivity threshold* and *error threshold.* In term of *hidden nodes* generated, the number of nodes are always different since SECoS algorithm perform the training continuously based on the input data given.

Acknowledgement

The author would like to thanks Lembaga Penelitian Universitas Sumatera Utara (LP USU) for partly supporting the dissemination of this research.

References

[1] Aram Z, Jafari S, Ma J, Sprott JC, Zendehrouh S and Pham V-T 2017 Using chaotic artificial neural network to model memory in the brain *Commun Nonlinear Sci Numer Simulat* **44** 449–459

[2] Khaze SR, Masdari M and Hojjatkhah S 2013 Application of artificial neural networks in estimating participation in election. *Int. J. of Technology Information, Modelling and Computing* **1** 23–32

[3] Watts MJ 2006 Nominal-scale evolving connectionist systems *Proc. of the 2006 IEEE Int. Joint Conf. on Neural Network* 2055–59

[4] Watts MJ and Kasabov NK 2002 Evolutionary optimisation of evolving connectionist systems. *Proc. of the Congress on Evolutionary* 606–9

[5] Kasabov NK 2007 *Evolving Connectionist System The Knowledge Engineering Approach.* 2nd Ed. Springer: London

[6] Watts MJ and Kasabov NK 2000 Simple evolving connectionist systems and experiment on isolated phoneme recognation *Proc. of the 2000 IEEE Symposium on Combinations of Evulotionay Cumputation and Neural Network* 232–9

[7] Futschik ME, Reeve A and Kasabov N K 2003 Evolving connectionist systems for knowledge discovery from gene expression data of cancer tissue. *Artificial Intelligence in Medicine* **28** 165–189

[8] Nababan EB, Iqbal M and Rahmat RF 2016 Breast cancer identification on digital mammogram using evolving connectionist systems 2016 IEEE International Conference on Informatics and Computing (ICIC) 132–6

[9] Ghobakhlou A, Watts MJ and Kasabov NK 2003 Adaptive speech recognition with evolving connectionist systems *Information Science* **156** 71–83

[10] Rahmat RF, Rizki A, Alharthi AF and Budiarto R 2016 Big data forecasting using evolving multi-layer perceptron *2016 4th Saudi Int.Conf. on Information Technology* 1–5

[11] Sinwar D and Kaushik R 2014 Study of Euclidean and Manhattan distance metrics using simple k-means clustering *Int. J. for Res.in Applied Science and Engineering Technology* **2** 270–4

[12] Fauzi MFA 2009 Optimal Discrete Wavelet Frames Features for texture-based image retrieval applications *Proc. of the First International Visual Informatics Conference* 66–77

[13] Makridakis S 1993 Accuracy measures: theoretical and practical concerns *Int J.of Forecasting* 9 527–9

[14] Ritarwan K 2014 Effect of beta fibrinogen gene -455 G/A polymorphism on aspirin treatment and its effect to the barthel index score and modified ranking scale on ischemic stroke based on the age groups. PhD Dissertation. Universitas Sumatera Utara

[15] Watts MJ 2016 Sleep learning and max-min aggregation of evolving connectionist systems *International joint conference on Neural Network (IJCNN)* 4325–30

[16] Tofallis C 2014 A better measure of relative prediction accuracy for model selection and model estimation *Journal of Operational Research Society* 1–11

Symmetric Stream Cipher using Triple Transposition Key Method and Base64 Algorithm for Security Improvement

Heri Nurdiyanto[1], Robbi Rahim*[2] and Nur Wulan[3]

[1]Department of Informatics Engineering, STMIK Dharma Wacana, Jl.Kenanga No.03 Mulyojati 16C Metro Barat, Metro Lampung, Indonesia
[2]School of Computer Engineering and Communication, Universiti Malaysia Perlis, Malaysia
[3]Department of Informatics Engineering, Sekolah Tinggi Teknik Harapan, Jln. H.M. Joni No. 70 C Medan, Medan, Indonesia

*usurobbi85@zoho.com

Abstract. Symmetric type cryptography algorithm is known many weaknesses in encryption process compared with asymmetric type algorithm, symmetric stream cipher are algorithm that works on XOR process between plaintext and key, to improve the security of symmetric stream cipher algorithm done improvisation by using Triple Transposition Key which developed from Transposition Cipher and also use Base64 algorithm for encryption ending process, and from experiment the ciphertext that produced good enough and very random.

1. Introduction
Data security is very important in maintaining the confidentiality of information, especially containing sensitive information that should only be known by a particular party, especially if the delivery is arranged through the public network, if the data is not secured it will be easy to tap and known contents Information by unauthorized parties[1][2]. One of the ways used to secure data is to use the cryptographic system by encoding the substance of the information (plaintext) into the contents that are not understood through the encryption process and to recover the original information using the decryption process by using the correct key[1].

The symmetric stream cipher method in this research uses triple transposition key and Base64 algorithm to increase the security of symmetric stream cipher algorithm. Improvisation performs on symmetric stream cipher method, in this study received two key types namely the main key and Private Crypto Code. With the result forming three transposition keys containing 463, 251, 181 loop which is interconnected due to feedback output from key processing. With the method of the symmetric stream, cipher will produce output with high speed and have Private Crypto Code.

2. Theory
A. Symmetric Stream Cipher

A stream cipher is a type of symmetry encryption algorithm that is much faster than any block cipher algorithm, and stream ciphers are used for smaller data blocks in bit sizes[3]. A stream cipher produces what is called a key stream. The encryption process is achieved by combining key stream with plaintext usually with bitwise XOR operations and can be made independent of plaintext and ciphertext which results in what is called a synchronous stream cipher[3][4][5].

B. Triple Transposition Key Algorithm

In transposition cipher, plaintext remains the same, but the order is changed. In other words, this algorithm transfers the character set inside the text. Another name for this method is the permutation[6] because transposed each character in the text is the same as doing the character of the characters, in the results of this study the authors to improvise transposition algorithm that is named Triple Transposition Key specifically used to generate keys[6].

Triple Transposition Key is a technique used to form keys of 463, 251, and 181 which is connected because of the feedback output of key processing, the Triple Transposition Key process that the author uses is as follows:

Encodebyte= a byte XOR FnULTRA(FEEDBACK)
FEEDBACK=EncodeByte
Keys that have been formed then combined using the following functions:
Output = (OUT1 + OUT2 + OUT3) Mod 256

C. Base64 Algorithm

The Base64 transformation is one of the algorithms for encoding and decoding data in ASCII format, which are based on the number 64. The character generated from Base64 consists of A-Z, a-z and 0...9, plus the last two characters of / and + [7].

3. Proposed Method

In this experiment data security is perform using stream cipher algorithm, and the key utilized in the encryption process consists of the primary key and Private Crypto Code and then process with Triple Transposition Key algorithm, and next step is encryption string with Symmetric Stream Cipher, and Base64 algorithm, the use of 2 keys in the encryption process is expected to be able gives better ciphertext results.

The process of completion of cryptography method Symmetric Stream Cipher can be divide into three parts, such as:

a. Key Formation Process
b. Dummy String Formation Process
c. Encryption and Decryption Process.

For example the process of applying the Triple Transposition Key, first determines the primary key and private crypto key as follows:

Primary Key = IconICT
Private Crypto Key = Harapan

a. Primary Key = 'IconICT'
b. Initialization KEY1 (KEY1= Primary Key)
   KEY1 = 'IconICT'
   KEY1 =
   (73, 99, 111, 110, 73, 67, 84)
   KEY1Len = 7
c. Value Initialization K1(),P1,S1
   K1 (0 .. 462)
   P1 = 0
   S1 = 0
   Change K1 (j) with K1 (i) total of 463 times
   j = (j + K1 (i) + KEY1 (i Mod KEY1Len)) Mod 463
   Loop of – 0
   j = (0 + K1 (0) + KEY1 (0 Mod 7)) Mod 463
   j = (0 + 1 + KEY1 (0)) Mod 463

j = (0 + 1 + 73) Mod 463
j = (74) Mod 463
j = 73
Exchange K1 (73) WITH K1 (0) = 0 WITH 73

The above process is done up to 462 times round so as to produce the following value, thus obtained:
K1 =
(73,173,286,25,104,409,175,15,347,19,423,40,459,451,240,244,7,121,339,86,214,462,405,160,242,410,401,290,150,42,71,97,380,38,420,65,34,80,275,35,411,117,278,232,254,139,108,251,383,375,169,221,301,92,434,416,82,297,57,113,381,374,93,193,61,145,215,305,197,408,186,218,98,70,255,64,256,406,144,310,348,176,349,20,318,112,3,170,2,450,430,325,156,421,105,166,329,87,129,427,152,309,130,81,379,461,299,277,272,454,168,363,37,341,85,124,397,140,126,265,110,382,328,445,247,435,172,162,74,60,271,184,400,444,330,398,359,201,220,56,377,300,32,324,120,39,414,43,418,136,59,252,141,66,206,460,360,428,196,8,203,316,344,26,283,354,281,132,233,269,385,127,241,327,279,181,182,369,295,446,51,22,191,345,29,440,146,68,178,100,147,431,386,148,285,13,306,49,41,228,332,353,211,101,44,273,336,442,367,315,179,253,14,235,320,207,453,9,288,267,188,346,342,417,289,5,438,387,11,202,250,31,287,248,155,280,0,391,407,84,21,199,204,225,291,333,189,456,53,415,231,227,223,119,99,192,24,426,125,268,102,389,388,50,378,392,308,303,17,439,133,449,322,312,230,12,198,294,447,296,356,304,190,212,10,366,452,292,167,91,319,237,88,128,264,437,135,79,208,419,75,138,154,364,205,355,185,368,399,76,261,351,213,217,164,258,96,63,47,276,27,90,337,436,361,298,36,357,249,384,83,52,4,78,69,372,448,137,229,394,103,122,274,58,313,163,331,362,425,314,326,293,311,352,111,209,403,371,365,95,370,6,422,159,343,393,429,157,259,177,321,317,149,48,158,234,457,424,165,195,338,395,266,404,263,16,270,143,33,358,226,46,115,340,402,246,118,30,243,45,142,458,134,219,262,245,183,373,412,114,23,151,94,335,55,376,236,441,77,334,200,257,89,433,187,174,238,123,161,432,443,62,109,18,107,307,323,194,210,216,72,455,302,224,116,350,67,131,413,284,171,390,54,396,106,222,1,28,239,153,282,180,260)

The result of K1 above is the outcome of the process of Triple Transposition Key 462, for keys with 251 and 181 also in the same way to obtain a different result, and each result will combine with function (OUT1 + OUT2 + OUT3) Mod 256 to get a key.

The second process generates the dummy string to add ciphertext security, for the dummy string used is the word HERI, the formulas and processes used for the dummy string are as follows:

a. SeedString = 'HERI' = (72, 69, 82, 73)
   Length of SeedString = 4
b. Initialize Random SizeDummy
   SizeDummy = Int(224*rnd)+32
   SizeDummy = 124
c. Change the value of Variable Size Dummy with
   SizeDummy = SizeDummy Xor Asc(Mid(SeedString, k, 1)
   As much as Len (seedstring) times
   01. SizeDummy = 124 Xor 72 = 52
   02. SizeDummy = 52 Xor 69 = 113
   03. SizeDummy = 113 Xor 82 = 35
   04. SizeDummy = 35 Xor 73 = 106
   SizeDummy = 106
   If SizeDummy> 255 then SizeDummy = SizeDummy - X * 224
   If SizeDummy <32 then SizeDummy = SizeDummy + 224
   SizeDummy = 106
d. Forming Random Dummy with SizeDummy size – 1
   Random Dummy = (220, 143, 128, 6, 28, 205, 12, 204, 197, 35, 93, 30, 61, 169, 59, 115, 180, 37, 103, 87, 19, 43, 46, 178, 47, 137, 61, 77, 67, 225, 183, 146, 157, 152, 133, 147, 151, 239, 237, 79, 201, 112, 103, 48, 125, 140, 172, 215, 99, 130, 15, 28, 134, 130, 16, 81, 127, 19, 47, 199, 233, 178,

254, 148, 21, 188, 122, 147, 254, 142, 155, 171, 127, 234, 42, 85, 195, 156, 211, 88, 253, 94, 123, 215, 90, 250, 211, 142, 52, 174, 46, 89, 137, 212, 143, 246, 171, 194, 7, 11, 57, 240, 62, 107, 77)

e. Randomize Dummy with Triple Transposition Key 16 times
1st loop, RndKey Formation Process throughout 16 Bytes
Q = Random 0 s/d 255 = 88
Q = Q Xor SeedString(1) = 88 Xor 72 = 16
Q = Random 0 s/d 255 = 7
Q = Q Xor SeedString(2) = 79 Xor 69 = 10
Q = Random 0 s/d 255 = 149
Q = Q Xor SeedString(3) = 149 Xor 82 = 199
Q = Random 0 s/d 255 = 247
Q = Q Xor SeedString(4) = 247 Xor 73 = 190
Q = Random 0 s/d 255 = 35
Q = Q Xor SeedString(1) = 35 Xor 72 = 107
Q = Random 0 s/d 255 = 217
Q = Q Xor SeedString(2) = 217 Xor 69 = 156
Q = Random 0 s/d 255 = 125
Q = Q Xor SeedString(3) = 125 Xor 82 = 47
Q = Random 0 s/d 255 = 235
Q = Q Xor SeedString(4) = 235 Xor 73 = 162
Q = Random 0 s/d 255 = 129
Q = Q Xor SeedString(1) = 129 Xor 72 = 201
Q = Random 0 s/d 255 = 82
Q = Q Xor SeedString(2) = 82 Xor 69 = 23
Q = Random 0 s/d 255 = 144
Q = Q Xor SeedString(3) = 144 Xor 82 = 194
Q = Random 0 s/d 255 = 118
Q = Q Xor SeedString(4) = 118 Xor 73 = 63
Q = Random 0 s/d 255 = 217
Q = Q Xor SeedString(1) = 217 Xor 72 = 145
Q = Random 0 s/d 255 = 159
Q = Q Xor SeedString(2) = 159 Xor 69 = 218
Q = Random 0 s/d 255 = 242
Q = Q Xor SeedString(3) = 242 Xor 82 = 160
Q = Random 0 s/d 255 = 183
Q = Q Xor SeedString(4) = 183 Xor 73 = 254
rndKey = (16, 10, 199, 190, 107, 156, 47, 162, 201, 23, 194, 63, 145, 218, 160, 254)

The randomization process above is performe as much as 16 times and generate random dummy string as follows:
Random Dummy = (106, 20, 46, 19, 40, 23, 159, 135, 228, 243, 64, 37, 237, 84, 168, 167, 98, 67, 134, 30, 124, 163, 91, 147, 183, 7, 101, 50, 68, 75, 248, 1, 190, 227, 82, 153, 165, 168, 46, 195, 118, 85, 228, 5, 212, 36, 77, 20, 113, 237, 93, 225, 91, 159, 216, 157, 109, 199, 26, 195, 243, 55, 192, 24, 167, 196, 89, 130, 85, 152, 156, 71, 29, 162, 73, 236, 160, 188, 130, 234, 95, 237, 250, 57, 213, 126, 154, 251, 168, 196, 79, 52, 48, 127, 115, 40, 39, 33, 134, 213, 27, 203, 145, 77, 228, 110)

After determining the key with Triple Transposition Key and dummy string algorithms, next is to encrypt the process for the word "Conference", here is the process:

a. PlainText = 'Conference'
Primary Key= 'IconICT'
PCC Key = 'Harapan

IOP Conf. Series: Journal of Physics: Conf. Series **930** (2017) 012005 doi:10.1088/1742-6596/930/1/012005

This information is then converted into ASCII form as follows, let call this is PPC Text: (72, 69, 51, 13, 62, 10, 0, 0, 0, 7, 0, 67, 3, 99, 3, 101, 2, 102, 3, 110, 3, 111, 3, 114, 3, 114, 60, 6, 194, 217, 186, 3)

b. Forming DummyString
SeedString = HERI
DummyString in ASCII: (121, 10, 27, 70, 38, 249, 7, 74, 181, 143, 122, 53, 250, 247, 67, 124, 125, 47, 18, 191, 169, 149, 82, 219, 190, 67, 26, 250, 158, 145, 230, 121, 151, 93, 222, 201, 73, 58, 186, 181, 162, 245, 186, 59, 97, 86, 241, 181, 88, 49, 123, 44, 171, 216, 87, 235, 103, 153, 216, 201, 168, 193, 141, 237, 213, 254, 97, 117, 171, 90, 12, 216, 186, 6, 103, 117, 191, 110, 223, 87, 246, 132, 127, 137, 1, 19, 221, 84, 133, 133, 135, 75, 81, 82, 132, 232, 112, 123, 33, 11, 205, 250, 250, 49, 250, 100, 5, 179, 142, 168, 220, 90, 157, 211, 112, 135, 232, 77, 46, 128, 58)

c. Combine DummyString with PPCText
NewPlainText = DummyString & PPCText & (The last 2 bytes of DummyString)
NewPlainText in ASCII: (121, 10, 27, 70, 38, 249, 7, 74, 181, 143, 122, 53, 250, 247, 67, 124, 125, 47, 18, 191, 169, 149, 82, 219, 190, 67, 26, 250, 158, 145, 230, 121, 151, 93, 222, 201, 73, 58, 186, 181, 162, 245, 186, 59, 97, 86, 241, 181, 88, 49, 123, 44, 171, 216, 87, 235, 103, 153, 216, 201, 168, 193, 141, 237, 213, 254, 97, 117, 171, 90, 12, 216, 186, 6, 103, 117, 191, 110, 223, 87, 246, 132, 127, 137, 1, 19, 221, 84, 133, 133, 135, 75, 81, 82, 132, 232, 112, 123, 33, 11, 205, 250, 250, 49, 250, 100, 5, 179, 142, 168, 220, 90, 157, 211, 112, 135, 232, 77, 46, 128, 58, 72, 69, 51, 13, 62, 10, 0, 0, 0, 7, 0, 67, 3, 99, 3, 101, 2, 102, 3, 110, 3, 111, 3, 114, 3, 114, 60, 6, 194, 217, 186, 3, 128, 58)

d. Forming key encryption with Triple Transposition Key
Encryption Process per Byte
Number of Bytes = 154
The 0th iteration
Process FnUltra(FEEDBACK)
Input = Feedback = 0
P1 = (P1+1) Mod 463
P1 = (0+1) Mod 463
P1 = 1
S1 = (S1 + K1(P1) + FB) Mod 463
S1 = (0 + K1(0) + 0) Mod 463
S1 = (0 + 73 + 0) Mod 463
S1 = 73
OUT1 = K1((K1(P1) + K1(S1)) Mod PR1) Mod 256
OUT1 = K1((K1(1) + K1(173)) Mod 463) Mod 256
OUT1 = K1((327 + 173) Mod 463) Mod 256
OUT1 = K1((500) Mod 463) Mod 256
OUT1 = K1(37) Mod 256
OUT1 = 80 Mod 256
OUT1 = 80

P2 = (P2 + 1) Mod 251
P2 = (0 + 1) Mod 251
P2 = 1
S2 = (S2 + K2(P2) + OUT1) Mod 251
S2 = (0 + K2(0) + 80) Mod 251
S2 = (0 + 152 + 80) Mod 251
S2 = 232
OUT2 = K2((K2(P2) + K2(S2)) Mod PR2) Mod 256
OUT2 = K2((K2(1) + K2(4)) Mod 251) Mod 256

OUT2 = K2((137 + 175) Mod 251) Mod 256
OUT2 = K2((312) Mod 251) Mod 256
OUT2 = K2(61) Mod 256
OUT2 = 231 Mod 256
OUT2 = 231

The above calculation process is done by testing the New Plaintext with the key result of Triple Transposition Key, the above process is continued until all New Plaintext encryption using lock result of Triple Transposition Key, the end result of the process are:
Encrypted String in ASCII: (56, 252, 110, 156, 57, 138, 131, 13, 88, 14, 97, 227, 64, 202, 168, 159, 16, 109, 216, 74, 171, 244, 175, 67, 255, 213, 220, 218, 250, 188, 12, 146, 78, 94, 49, 59, 158, 225, 195, 97, 196, 168, 95, 188, 251, 144, 50, 146, 39, 101, 138, 214, 197, 128, 64, 228, 134, 121, 167, 210, 86, 106, 48, 73, 176, 118, 123, 39, 137, 248, 242, 223, 29, 134, 115, 39, 215, 230, 130, 242, 138, 35, 202, 239, 77, 254, 46, 149, 73, 6, 240, 115, 197, 26, 69, 56, 46, 53, 64, 209, 50, 47, 164, 7, 12, 79, 73, 209, 51, 180, 201, 32, 27, 117, 67, 251, 50, 0, 248, 137, 178, 93, 125, 107, 198, 105, 20, 123, 55, 243, 206, 75, 156, 26, 45, 236, 140, 28, 59, 51, 91, 162, 132, 41, 209, 251, 126, 188, 6, 180, 51, 231, 27, 46, 137)
Ciphertext in ASCII is then encoded with Base64 with the following results=
OPxunDmKgw1YDmHjQMqonxBt2Eqr9K9D/9Xc2vq8DJJOXjE7nuHDYcSoX7z7kDKSJ2WK1sW
AQOSGeafSVmowSbB2eyeJ+PLfHYZzJ9fmgvKKI8rvTf4ulUkG8HPFGkU4LjVA0TIvpAcMT0nR
M7TJIBt1Q/syAPiJsl19a8ZpFHs3885LnBot7IwcOzNbooQp0ft+vAa0M+cbLok

4. Conclusion

Conclusion on the results of this study, the addition of security to Symmetric Stream Cipher with Triple Transposition Key and Base64 produce a ciphertext that is very random and not easily penetrated even with some techniques commonly used by cryptanalyst because the encryption process has some fairly complicated process let alone the key used Raised with the Triple Transposition Key used by the author.

References

[1] R. von Solms and J. van Niekerk, "From information security to cyber security," *Comput. Secur.*, vol. 38, no. October 2013, pp. 97–102, 2013.
[2] W. Stallings, *Network security essentials : applications and standards*. 2011.
[3] C. De Canniere, "Trivium: A stream cipher construction inspired by block cipher design principles," *Inf. Secur.*, vol. 507932, pp. 171–186, 2006.
[4] S. Manna and S. Dutta, "A Stream Cipher based Bit-Level Symmetric Key Cryptographic Technique using Chen Prime Number," *Int. J. Comput. Appl.*, vol. 107, no. 12, pp. 975–8887, 2014.
[5] M. I. Alam and M. R. Khan, "Performance and Efficiency Analysis of Different Block Cipher Algorithms of Symmetric Key Cryptography," *Int. J. Adv. Res. Comput. Sci. Softw. Eng.*, vol. 3, no. 10, pp. 2277–128, 2013.
[6] Q.-A. Kester, "A HYBRID CRYPTOSYSTEM BASED ON VIGENÈRE CIPHER AND COLUMNAR TRANSPOSITION CIPHER," *Int. J. Adv. Technol. Eng. Res.*, vol. 3, no. 1, pp. 141–147, 2013.
[7] A. T. SHOLEH, E. GUNADHI, and A. D. SUPRIATNA, "Mengamankan Skrip Pada Bahasa Pemograman Php Dengan Menggunakan Kriptografi Base64," *J. Algorithm. Sekol. Tinggi Teknol. Garut*, vol. 10, no. 1, pp. 1–9, 2013.

Analysis of Accuracy and Epoch on Back-propagation BFGS Quasi-Newton

Herlan Silaban[1], Muhammad Zarlis[2], Sawaluddin[3]

[1,2,3]Department of Computer Sciences, Faculty of Computer Sciences and Information Technology, Universitas Sumatera Utara, Medan, Indonesia. 21055

E-mail : herlan76ban@gmail.com, m.zarlis@yahoo.com, sawal@usu.ac.id

Abstract. Back-propagation is one of the learning algorithms on artificial neural networks that have been widely used to solve various problems, such as pattern recognition, prediction and classification. The Back-propagation architecture will affect the outcome of learning processed. BFGS Quasi-Newton is one of the functions that can be used to change the weight of back-propagation. This research tested some back-propagation architectures using classical back-propagation and back-propagation with BFGS. There are 7 architectures that have been tested on glass dataset with various numbers of neurons, 6 architectures with 1 hidden layer and 1 architecture with 2 hidden layers. BP with BFGS improves the convergence of the learning process. The average improvement convergence is 98.34%. BP with BFGS is more optimal on architectures with smaller number of neurons with decreased epoch number is 94.37% with the increase of accuracy about 0.5%.

1. Introduction

Back-propagation is one of the training algorithms on artificial neural networks. The network architecture on back-propagation is multilayer, it makes the output of the model more accurate. Architecture will greatly affect the time of training and the accuracy of network training results. If the artificial neural network more complex, it will take a long time of training time [1]. Suciati [2] classify batik by using back-propagation, and the number of epoch is highest for about 10000. Back- propagation optimization has been done to improve the performance of artificial neural network back- propagation either to decrease the convergent level of training or even improve the accuracy of back- propagation. Quasi-Newton BFGS is one of the optimization methods that can be used in the optimization of back-propagation training by making changes in the weight contained in the network [3]. This research uses Glass Identification dataset from UCI Machine Learning repository [4]. Glass dataset is a classification glass based on the oxide content of the glass to identify the process of making the glass whether using the float method or not. The glass type that exist on the dataset are : Float Processed (Building_windows_float_processed; 70 rows & Vehicle_windows_float_processed;17 rows) and Non Float Processed (Building_windows_non_float_processed; 76 rows, Vehicle_windows_non_float_processed, Containers; 13 rows, Tableware; 9 rows & Headlamps; 29 rows). The purpose of this research is to analyze the number of epoch and accuracy on backpropagation network training with BFGS Quasi- Newton on some network architectures so that will be find the suitable architecture for classification problem.

2. Backpropagation

Back-propagation is one of the most popular training on artificial neural networks for predictive and pattern recognition cases. Back-propagation is widely used because of its multilayer network so that the resulting output is more accurate. Back-propagation was first introduced by Rumelhart and McCelland in 1986. BP was formed by generalizing the training rules in the network model by adding a hidden layer. One of the BP networks that will be used in this research is a network with architecture

9-7-2. With 1 input layer (X) consisting of 9 input neurons, 1 hidden layer (Z) with 7 neurons and 1 output layer (Y) with 2 output neurons float processed (Y1) or non float processed (Y2).
Base on the study, the number of input neurons is 9, the output neuron is 2 and the number of neurons in the hidden layer are using the formula [5].

$$\#1^{st}\ hidden\ layer = \sqrt{(\mathrm{m}+2)\mathrm{N}} + 2\frac{\sqrt{\mathrm{N}}}{\mathrm{m}+2}$$

$$\#2^{nd}\ hidden\ layer = \mathrm{m}\frac{\sqrt{\mathrm{N}}}{\mathrm{m}+2}$$

N is the number of datasets whereas m is the number of outputs. Number of datasets used N = 214 and output m = 2.
In this study, the determination of the number of neurons also concerns the rule-of-thumb number of neurons [6].

a. The number of hidden neurons should be between the size of the input layer and the size of the output layer.
b. The number of hidden neurons should be 2/3 the size of the input layer, plus the size of the output layer.
c. The number of hidden neurons should be less than twice the size of the input layer.

The BP architectures that will be used in this research can be seen on Table 1.

Table 1. Architectures of Back-propagation

Architecture	Description
9-59-2	X = 9, Z = 59 & Y=2
9-43-2	X = 9, Z = 43 & Y=2
9-44-2	X = 9, Z = 44 & Y=2
9-8-2	X = 9, Z = 8 & Y=2
9-7-2	X = 9, Z = 7 & Y=2
9-4-2	X = 9, Z = 4 & Y=2
9-44-15-2	X = 9, Z^1 = 44, Z^2 = 15 & Y=2

3. BFGS Quasi-Newton

The BFGS algorithm (Broyden-Fletcher-Goldfarb-Shanno) is an algorithm that developed from the Quasi-Newton algorithm. BFGS algorithm is one of the algorithms that is used to solve nonlinear optimization problems without any constraints. Quasi-Newton equation [7] [8].
The BFGS Algorithm :

Step 0. Given x1 ∈Rn; B1 ∈Rn×n positive definite;
Compute g1 = ∇f(x1). If g1 = 0, stop; otherwise, set k := 1.
Step 1. Set dk = − gk.
Step 2. Carry out a line search along dk, getting αk > 0,
xk+1 = xk + αkdk, and gk+1 = ∇f(xk+1);
If gk+1= 0, stop.
Step 3. Set

$$\mathrm{B}_{\mathrm{k+1}} = B - \frac{BkSkSkBk}{(Sk)tBkSk} + \frac{YkYkt}{(Sk)TyK}$$

where
$s_k = \alpha_k d_k$,
$y_k = g_{k+1} - g_k$.
Step 4. $k := k+1$; go to Step 1.

4. Experiments & Results

The test will be conducted first on the backpropagation standard with learning rate 0.5. Base on the test result, the smallest number of epoch that is obtained from 9-4-2 architecture is 231 with MSE 0.0619 and 91.6% accuracy. While the largest number of epoch is 10000 on architecture 9-59-2, 9-43- 2, 9-8-2 and 9-7-2. The highest accuracy in standard BP testing is 98.6% with MSE 0.0203 while lowest accuracy is 91.6% with MSE 0.0619.

Table 2. Standard Backpropagation Testing

BP's Architecture	Epoch	MSE	Accuracy (%)
9-59-2	10000	0.0203	98.6
9-43-2	10000	0.0252	98.1
9-44-2	800	0.0413	94.9
9-8-2	10000	0.0434	93.5
9-7-2	10000	0.0349	96.3
9-4-2	231	0.0619	91.6
9-44-15-2	1474	0.0323	95.8

In BP testing with Quasi-Newton BFGS the number of epochs on the BP network with Quasi-Newton BFGS decreased significantly. The highest epoch value is 31 and the lowest epoch value is 13. While the accuracy on BP with BFGS decreases, the accuracy decrease in BFGS is not significant from the accuracy that had been produced by BP Standard. The lowest accuracy on BFGS is 92.1% with MSE 0.0483 and the highest is 95.8% with MSE 0.0335.

Table 3. Backpropagation Testing with BFGS Quasi-Newton

BP's Architecture	Epoch	MSE	Accuracy (%)
9-59-2	16	0.0431	95.3
9-43-2	21	0.0409	95.8
9-44-2	24	0.0447	94.9
9-8-2	33	0.0483	92.1
9-7-2	22	0.0457	93.9
9-4-2	13	0.0586	92.1
9-44-15-2	31	0.0335	95.8

Based on the test, there was a significant decrease in epoch value. The greatest decrease in epoch value occurred on architecture 9-59-2 from 10000 (BP Standard) to 16 (BP BFGS) with 98.1% accuracy (BP Standard) and 95.8% (BP BFGS), 99.84% decrease epoch. The average decrease in the epoch number is 98.34%. The presentation of the decrease in epoch number from standard BP and BP with BFGS can be seen in Table 4.

Table 4. Improvement epoch BP with BFGS

BP's Architecture	Epoch BP Standard	Epoch BP BFGS	Decrease of epoch (%)
9-59-2	10000	16	98.6
9-43-2	10000	21	98.1
9-44-2	800	24	94.9
9-8-2	10000	33	93.5
9-7-2	10000	22	96.3
9-4-2	231	13	91.6
9-44-15-2	1474	31	95.8

Based on the research that has been done by testing 7 network architecture using BP standard and BP with BFGS Quasi-Newton weight changes, there is a decrease in the number of epoch in all network architecture accompanied by the increase of accuracy value on 1 architecture while on 6 others decreased accuracy. From the 7 networks tested there was a decrease in the epoch value in all the architectures. The comparison graph of the epoch value in standard BP with BP BFGS can be seen in Figure 2.

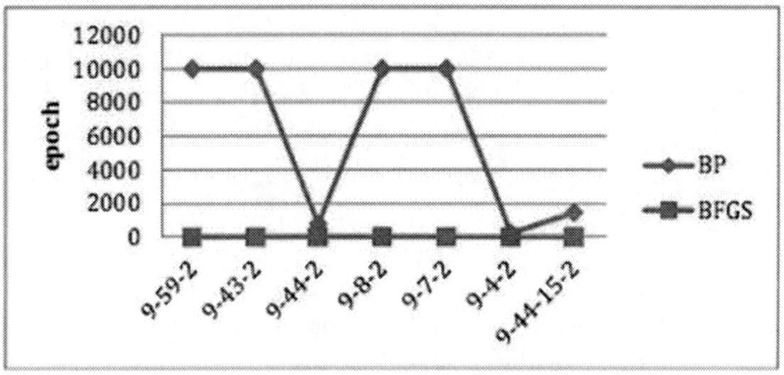

Figure 1. Comparison Epoch BP VS BP BFGS

5. Conclusion

BP using BFGS method will be optimal with small number of neurons with decreased epoch 94.37% and accuracy improved by 0.5%. BP artificial neural networks can be used to classify glass dataset with proper architectural election. In a network architecture that has a smaller number of neurons, there is an increase in accuracy when using BP with BFGS. Base on the test results, it can be concluded that there is a significant decrease in the epoch number when using BFGS on back- propagation. Future task, the accuracy on BP with BFGS is still in reduction, so further research is needed to avoid the decreasing of accuracy.

References

[1] Panchal, G., Ganatra, A., Shah, P., & Panchal, D. 2011. Determination of over-learning and over-fitting problem in backpropagation nueral network. International Journal on Soft Computing (IJSC) 2, pp. 40-51.

[2] Suciati, N. 2014. Batik Motif Classification using Color-Texture-Based Feature Extraction and Backpropagation Neural Network. Proceeding of IIAI 3rd International Conference on Advanced Applied Informatics, pp 517-512.

[3] Nazri, M.N., Meghana, R., & Rajesh, R. 2006. An Improved Learning Algorithm Based on The Broyden-Fletcher-Goldfarb- Shanno (BFGS) Method For Back Propagation Neural Networks. Proceedings of the Sixth International Conference on Intelligent Systems Design and Applications (ISDA'06).

[4] UCI Machine Learning Repository. 2017. Glass Identification Dataset. (Online) https://archive.ics.uci.edu/ml/datasets/glass+identification (5 Juni 2017).

[5] Stathakis, D. 2009. "How many hidden layers and nodes?", International Journal of Remote

Sensing, vol. 30, pp. 2133–2147.

[6] Heaton, J. Introduction to Neural Networks for Java, Second Edition. Heaton Research: St. Louis.

[7] Dai, Y. H. 2002. Convergence Properties of the BFGS Algoritm. Proceeding of Society for Industrial and Applied Mathematics (SIAM), pp. 693–701.

[8] Dai, Y. H. 2012. A perfect example for the BFGS method. Proceeding of Mathematical Programming, pp. 501–530.

[9] Caruana, R. 2001. Overfitting in Neural Nets : Backpropagation, Conjugate Gradient, and Early Stopping. (Online) https://papers.nips.cc/paper/1895-overfitting-in-neural-nets- backpropagation-conjugate-gradient-and-early-stopping.pdf (18 Mei 2017).

[10] Cui, Z., Yuan, G., Sheng. Z., Liu, W., Wang. X., & Duan, X. 2015. A Modified BFGS Formula Using a Trust Region Model for Nonsmooth Convex Minimizations. (Online) http://journals.plos.org/plosone/article/authors?id=10.1371/journal.pone.0140606 (15 Juni 2017).

[11] Lawrence, S. & Lee, C. 2000. Overfitting and Neural Networks: Conjugate Gradient and Backpropagation. Proceeding of International Conference on Neural Networks, pp. 114-119.

[12] Heaton, J. 2015. Artificial Intelligence for Humans, Volume 3: Deep Learning and Neural Networks. Heaton Research: St. Louis.

[13] Koscak, J. & Jaksa, R. 2010. Stochastic Weight Update in The Backpropagation Algorithm on Feed-Forward Neural Networks. IEEE.

[14] Liu, S., L. Xu & D. Li. 2016. Multi-scale prediction of water temperature using empirical mode decomposition with back-propagation neural networks. Journal Computer and Electrical Engineering 49: 1-8.

[15] Rich, L., Y. Zeng & T. Chen. 2015. Backpropagation neural network with adaptive differential evolution algorithm for time series forecasting. Journal Expert Systems with Applications, pp. 855-863.

Comparison Searching Process of Linear, Binary and Interpolation Algorithm

Robbi Rahim*[1], Saiful Nurarif[2], Mukhlis Ramadhan[2], Siti Aisyah[3] and Windania Purba[3]

[1]Department of Health Information, Akademi Perekam Medik dan Infokes Imelda, Jl. Bilal Ujung Medan 20116, Indonesia
[1]Student Universiti Malaysia Perlis, Universiti Malaysia Perlis, Malaysia
[2]Department of Computer System, STMIK Triguna Dharma, Jl. Jenderal Abdul Haris Nasution No.73, Medan 20219, Indonesia
[3]Department of Information System, Prima Indonesia University, Jl. Sekip Simpang Sikambing, Medan 20111, Indonesia

*usurobbi85@zoho.com

Abstract. Searching is a process that cannot be issued for a transaction and communication process, many search algorithms that can be used to facilitate the search, linear, binary, and interpolation algorithms are some searching algorithms that can be utilized, the comparison of the three algorithms is performed by testing to search data with different length with pseudo process approach, and the result achieved that the interpolation algorithm is slightly faster than the other two algorithms.

1. Introduction

Searching is a process that is usually performed in everyday activities, the search process is usually done to look for something with the purpose only to know whether the data is in a set of data or not [1] [2] [3], while at other times we may want the position of the data sought [3] [4] [5].

In computer science, there are various algorithms for search methods, such as the linear search method (Linear/Sequential Search), binary search and interpolation search [6]. Each algorithm has different prerequisites and different ways and execution time [6] [7]. The choice of search method meant on the user's circumstances, and the desire of the method which usually depends on the amount of data, data type and data structure used [8] [9].

This research tried to compare linear, binary and interpolation search algorithms with pseudo process approach for a gradual testing process.

2. Theory

Searching is a process with the aim to know whether the data is in a set of data or not, or maybe at other times the position of the data sought is needed for a particular purpose [1] [2] [3], or if the appearance of data more than once then all positions and frequency of appearance to be displayed [1] [2] [10], The simplest search can be described as follows [1]:

Suppose a sequence of data X [1] ... X [n], then the problem is whether Y (any data with the same data type as the existing data type in the X sequence and usually in input or already known) X [1] ... X [n] or whether Y is inside or not with Boolean result true / false or success / fail.

In computer science there are many algorithms for search method; generally, the data search methods split into several sections [4] such as:

1. Linear search method.
2. Binary search method.
3. Interpolation search method.

Each search algorithm has different prerequisites and ways and time of implementation, the selection of search methods is based on the circumstances, and desires of the user of the method which usually depends on the amount of data, data type and data structure used [8].

A. Linear Search Method

Linear Search performed on sequences of numbers that are ascending or descending or unordered. The linear search was given by comparing the data sought (X) with data in row A [1] ... A [n] starting with the first element data in sequence A. If the comparison is equal, then the search is stopped and declared successful [4] [6] [7]. Whereas if the comparison is not the same value then,

1. If the data not sorted (random data), then the search will proceed to the next data.
2. If the data sequentially sorted, then the search process will proceed to the next data which is to the right of the data, given that the data sought (X) is greater than the data being compared.

If data is descending, then the search will only continue to the next data which is to the right of the data being compared if the data sought (X) is smaller than the data being compared.

B. Binary Search Method

Binary Search can only perform on sequences of numbers that have been sorted either ascending or descending. Binary Search performs a search of X data in row A [1] ... A [n] with starting from intermediate data in sequence A. If X data value is equal to the middle data value of row A [4] [5], then the search is stopped and declared success. While if not the same then,

1. For ascending data, the search will be continued to the left ½ if the data value of X is smaller than the middle data value in sequence A. whereas if the data value of X is greater than the middle data value in sequence A, Will continue to ½ right.
2. For descending data, the search will be continued to the left ½ if the X data value is greater than the middle data value in sequence A. whereas if the X data value is less than the middle data value in sequence A, Will continue ½ left.

The search will be stopped and declared to fail if ½ left or right half is a single data and the data is not the same as the X data being searched [5].

C. Interpolation Search Method

Interpolation search looks for data by guessing (guessing) the position of the data searched by using a certain formula. Interpolation search can only perform on rows of numbers that have been sorted either ascending or descending [9] [10]. The alleged position as the data position sought (X) can be calculated using the following formula:

$$\text{Pos} = \text{ceiling}\,[BB + ((X - A\,[BB]) / (A\,[BA] - A\,[BB])) * (BA - BB)]$$

If the data in that position is equal to the value of data sought (X), then the search is stopped and declared successful [10]. Whereas if the value of data in the position of suspected is not equal to the value of data sought (X) [2] [10], then:

1. For ascending data, if the value of the data sought (X) is smaller than the data value in the suspected position, then the search is continued by changing the upper boundary of the search area. Whereas

IOP Conf. Series: Journal of Physics: Conf. Series **930** (2017) 012007 doi:10.1088/1742-6596/930/1/012007

if the data value sought (X) is greater than the value of data in the suspected position, then the search is continued by changing the lower boundary of the search area.

For descending data, if the value of the data sought (X) is greater than the value of the data in the suspected position, then the search is continued by changing the upper boundary of the search area. Whereas if the data value sought (X) is smaller than the data value in the suspected position, then the search is continued by changing the lower boundary of the search area.

3. Result and Discussion

Experiment process comparision of linear, binary and interpolation algorithm are using random number data, and the experiment perform to each algorithm, for the value used as follows

Data = 10, 8,5,15,19,35,4,2, 65, 13, 17, 32

For the first test searching process are using linear search algorithm, the result as below:
Data: 10, 8, 5, 15, 19, 35, 4, 2, 65, 13, 17, 32
X = Data searched
I = the position of the data compared against X
N = Amount of data
F = variable Boolean

A (I) = Data with the 1st position

X = 4, I = 1, N = 12, F = False, A (1) = 10
While (1 <= 12) And Not (False) -> True, Search resumed.
If (4 = 10) -> False
I = I + 1 = 1 + 1 = 2

X = 4, I = 2, N = 12, F = False, A (2) = 8
While (2 <= 12) And Not (False) -> True, Search resumed
If (4 = 8) -> False
I = I + 1 = 2 + 1 = 3

X = 4, I = 3, N = 12, F = False, A (3) = 5
While (3 <= 12) And Not (False) -> True, Search resumed
If (4 = 5) -> False
I = I + 1 = 3 + 1 = 4

X = 4, I = 7, N = 12, F = False, A (7) = 4
While (7 <= 12) And Not (False) -> True, Search resumed.
If (4 = 4) -> True
Found = True

X = 4, I = 7, N = 12, F = True, A (7) = 4
While (7 <= 12) And Not (True) -> False, Search not resumed.

F = True
Data '4' Found!

The process of searching data with the pseudo-linear process has been performed, following test search data with the binary algorithm, the result as below:
Data: 2, 4, 5, 8, 10, 13, 15, 17, 19, 32, 35, 65
X = Data searched
N = Amount of data

Left = the left variable value of the binary algorithm
Right = the right variable value of the binary algorithm
Middle = the middle variable value of the binary algorithm
Found = variable Boolean
A (Mid) = Data with position to - < middle value >

X = 4, Left = 1, Right = 12, Middle = 0, Found = False
While (1 <= 12) and Not (False) -> True,
Middle = (Left + Right) \ 2 = (1 + 12) \ 2 = 6
If (4 = 13) -> False
(4 < 13) -> True
Right = Middle - 1 = 6 - 1 = 5

X = 4, Left = 1, Right = 5, Middle = 6, Found = False, A (6) = 13
While (1 <= 5) And Not (False) -> True, Search resumed.
Middle = (Left + Right) \ 2 = (1 + 5) \ 2 = 3
If (4 = 5) -> False
(4 < 5) -> True
Right = Middle - 1 = 3 - 1 = 2

X = 4, Left = 1, Right = 2, Middle = 3, Found = False, A (3) = 5
While (1 <= 2) And Not (False) -> True, Search resumed.
Middle = (Left + Right) \ 2 = (1 + 2) \ 2 = 1
If (4 = 2) -> False
(4 < 2) -> False
Left = Middle + 1 = 1 + 1 = 2

X = 4, Left = 2, Right = 2, Middle = 1, Found = False, A (1) = 2
While (2 <= 2) And Not (False) -> True, Search resumed.
Middle = (Left + Right) \ 2 = (2 + 2) \ 2 = 2
If (4 = 4) -> True
Found = True

X = 4, Left = 2, Right = 2, Middle = 2, Found = True, A (2) = 4
While (2 <= 2) And Not (True) -> False, Search not resumed.

Found = True
Data '4' Found!

The last using with interpolation search algorithm as follows:
Data: 2, 4, 5, 8, 10, 13, 15, 17, 19, 32, 35, 65
X = Data searched
N = Amount of Data
BB = Lower Boundary Value
BA = Upper Limit Value
Pos = the position of the data being compared
CekPos = CekPos Variable Value
Found = Variable Boolean
A (Pos) = Data with position to - <pos value>

X = 4, BB = 1, BA = 12, Pos = 0, CekPos = 0, Found = False

IOP Conf. Series: Journal of Physics: Conf. Series **930** (2017) 012007 doi:10.1088/1742-6596/930/1/012007

While (1 <= 12) And Not (False) -> True, Search resumed.
Temp = BB + ((X - A (BB)) / (A (BA) - A (BB))) * (BA - BB)
Temp = 1 + ((4 - 2) / (65 - 2)) * (12 - 1) = 1.349206
Pos = Int (Temp) = Int (1.349206) = 1
If (1.349206 - 1 > 0) -> True
Pos = Pos + 1 = 1 + 1 = 2
If (0 = 2) -> False
(4 = 4) -> True
Found = True

X = 4, BB = 1, BA = 12, Pos = 2, CekPos = 0, Found = True, A (2) = 4
While (1 <= 12) And Not (True) -> False, Search not resumed

Found = True
Data '4' found!

The last process of the interpolation search has been completed, from the above three search algorithm process using same length of data and same value, based on the pseudo process of each algorithm and by re-testing for longer lengths get the result as follows:

Table 1. Result Searching

No	Length Number	Algorithm		
		Linear (Millisecond)	Binary (Millisecond)	Interpolation (Millisecond)
1	500	3.35	3.03	2.54
2	1000	5.12	6.19	4.51
3	2000	7.54	7.41	6.42

Based on the above process with a search of different lengths results obtained under the interpolation search algorithm more quickly than linear and binary algorithms

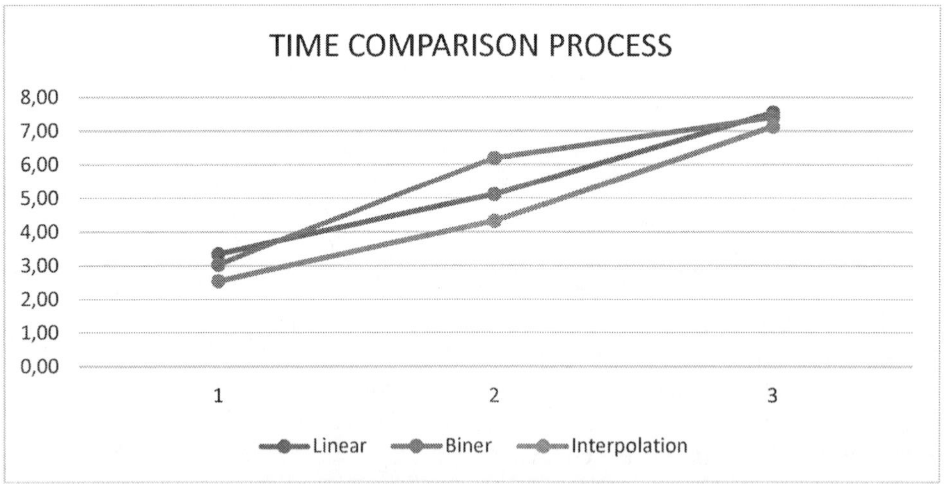

Figure 1. Time Comparison

Experiment was perform using Python programming because have a stable memory and large memory cache so that the experiment process can be done as objective as possible.

4. Conclusion

The comparison of linear, binary and interpolation search algorithms is made by testing with data lengths of 500, 1000 and 2000, based on the authors experiment it appears that the interpolation search algorithm is faster than the other two algorithms, the results of this test are not absolute for all cases that can be implemented, but this test using pseudo code approach process with random data that the author did and re-examined until a couple of times by using a program created by the Python language, the results that interpolation search algorithm faster

References

[1] S. Pushpa and P. Vinod, "Binary Search Tree Balancing Methods: A Critical Study," *International Journal of Computer Science and Network Security,* vol. 7, no. 8, pp. 237-243, 2007.

[2] J. z. Li and H. K. Wong, "Batched Interpolation Searching on databases," in *IEEE Third International Conference on Data Engineering*, Los Angeles, CA, USA, USA, 1987.

[3] S. Kareem B, "Matrix Search Algorithm Using Binary Search Trees," *Research & Reviews: Journal of Engineering and Technology,* vol. 5, no. 2, pp. 1-6, 2016.

[4] N. Arora, G. Bhasin and N. Sharma, "Two way Linear Search Algorithm," *International Journal of Computer Applications,* vol. 107, no. 21, pp. 6-8, 2014.

[5] B. G. Balogun and J. S. Sadiku, "Simulating Binary Search Technique Using Different Sorting Algorithms," *International Journal of Applied Science and Technology* , vol. 3, no. 6, pp. 67-75, 2013.

[6] K. K. Pandey and N. Pradhan, "A Comparison and Selection on Basic Type of Searching Algorithm in Data Structure," *International Journal of Computer Science and Mobile Computing,* vol. 3, no. 7, pp. 751-758, 2014.

[7] V. P.Parmar and CK Kumbharana, "Comparing Linear Search and Binary Search Algorithms to Search an Element from a Linear List Implemented through Static Array, Dynamic Array and Linked List," *International Journal of Computer Applications,* vol. 121, no. 3, pp. 13-17, 2015.

[8] A. Mehta, A. Saxena, J. Patel and A. Thanna, "A REVIEW ON COMPARISION OF BINARY SEARCH AND LINEAR SEARCH," *International Journal of Engineering Sciences & Management Research,* vol. 2, no. 10, pp. 85-89, 2015.

[9] Y. Perl and L. Gabriel, "Arithmetic interpolation search for alphabet tables," *IEEE Transactions on Computers,* vol. 41, no. 4, pp. 493 - 499, 1992.

[10] G. MARSAGLIA and B. NARASIMHAN , "Simulating Interpolation Search," *Computers Mathematics Application,* vol. 26, no. 8, pp. 31-42, 1993.

PERFORMANCE ANALYSIS OF COMBINED METHODS OF GENETIC ALGORITHM AND K-MEANS CLUSTERING IN DETERMINING THE VALUE OF CENTROID

[1]Adya Zizwan Putra, [2]Muhammad Zarlis, [3]Erna Budhiarti Nababan

[123]Master of Informatics Engineering - Fasilkom TI - University of North Sumatera
[1]E-mail : putrazizwanadya@gmail.com

Abstract. The determination of Centroid on K-Means Algorithm directly affects the quality of the clustering results. Determination of centroid by using random numbers has many weaknesses. The GenClust algorithm that combines the use of Genetic Algorithms and K-Means uses a genetic algorithm to determine the centroid of each cluster. The use of the GenClust algorithm uses 50% chromosomes obtained through deterministic calculations and 50% is obtained from the generation of random numbers. This study will modify the use of the GenClust algorithm in which the chromosomes used are 100% obtained through deterministic calculations. The results of this study resulted in performance comparisons expressed in Mean Square Error influenced by centroid determination on K-Means method by using GenClust method, modified GenClust method and also classic K-Means.

Keywords: *Centroid, K-Means, GenClust, Chromosome*

1. Introduction

Clustering is a process in searching and classifying data that has similarity characteristics between one data with other data. Clustering is widely used in areas such as social network analysis, software engineering, and crime detection. There are several clustering algorithms that can be used, but the K- Means and Fuzzy C-Means algorithms are commonly used algorithms because they are quite simple (Bai et al., 2011). Clustering is one of a group of data mining (Larose, 2006). In the K-Means algorithm, determining the number of clusters and centroid determination is quite difficult to do. Determination of the number of clusters and centroid determination directly affects the quality of the clustering process (Maitra, et al., 2010).

The GenClust method is considered good enough to determine the number of clusters and also the centroid of each cluster as it allows for increased clustering performance for each generation. However, what needs to be taken into consideration is the experiments performed by Rahman and Islam (2014) using 50% of the chromosomes obtained through deterministic calculations and 50% of chromosomes obtained through random numbers. Rahman and Islam (2014) at the end of the study suggest that the GenClust method will experience computational constraints when applied to large datasets. Therefore, Rahman and Islam (2014) at the end of his research suggested taking samples of data on the dataset and implementing GenClust to obtain the best chromosome and then implement it as an initial centroid on K-Means. However, there is no further research that discusses the performance comparison of the GenClust method and the modified GenClust method in accordance with the advice of Rahman and Islam (2014).

This study will discuss the comparison between met This study will discuss the comparison between the GenClust method, the modified GenClust method and the classical K-Means in centroid determination especially in the comparison to measure the measured performance value of Mean Square Error occurring for a dataset.

1.1. Formulation of The Problem
The placement of a data within a dataset in the K-Means algorithm is based on the proximity of the data with the centroid of each cluster, so that the centroid value is very influential on the clustering results by using the K-Means algorithm. The problem is in the K-Means algorithm centroid determination is done by generating random numbers, so the clustering results can show unfavorable results. This study will use the modified GenClust method, in which the chromosomes used are all obtained through deterministic calculations. Through this research we will get the performance comparison expressed in Mean Square Error (MSE) value between classic K-Means algorithm, GenClust method, and also modified GenClust method. The smaller Mean Square Error (MSE) means the better the performance of the obtained K-Means algorithm.

1.2. Scope of Problem
Due to the extent of the problem and the limited time and knowledge of the researchers, the researchers limit the problems to be discussed in this study as follows.
1. This research will discuss the effect of determining the number of clusters and centroid on a large dataset by comparing GenClust method with classic K-Means algorithm.
2. Comparison of performance in this study is based on the Mean Square Error value obtained in each generation (iteration).

1.3. Benefits of Research
The benefits of research are as follows.
1. Through this research the researcher will get the result of analysis about the relation between cluster number and centroid determination on clustering result.
2. Knowing the achievement of performance obtained by using GenClust method when compared with using classic K-Means algorithm.

2. Literature Review

2.1. Genetic Algorithm
The genetic algorithm is a stochastic algorithm that modeled the evolutionary process of biological species through natural selection (Konar, 2005). In general, this population is generated randomly and the solution that is generated after the consecutive stage of the crossover and mutation process. Each individual of the population has an associated value into a fitness value, in relation to solving a problem (Rabunal, 2006). The block diagram of the classical genetic algorithm can be seen in Figure 1.

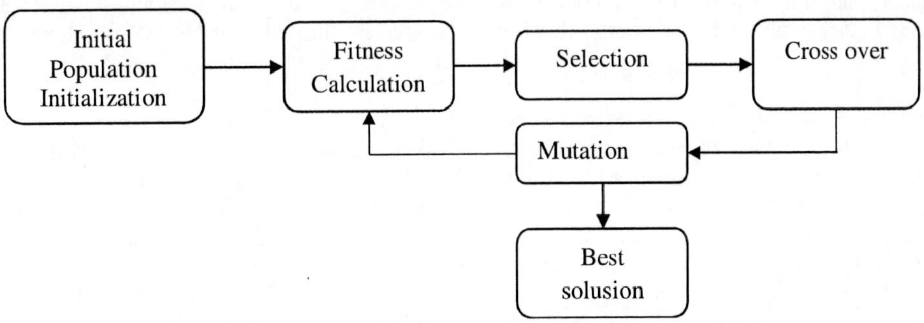

Figure 1. Block diagram of the Genetic Algorithm (Negnevitsky, 2005)

2.2. GenClust

The GenClust algorithm is a type of Hybrid Clustering that combines genetic algorithms with K-Means. The study of GenClust was first performed by Rahman and Islam (2014). Collaboration between Genetic Algorithm and K-Means can prevent K-Means from being trapped under Local Minima and can achieve Global Optima conditions. The GenClust method is considered good enough to determine the number of clusters and also the centroid of each cluster as it allows for increased clustering performance for each generation. The process stages of the GenClust Algorithm can be seen in Figure 2.

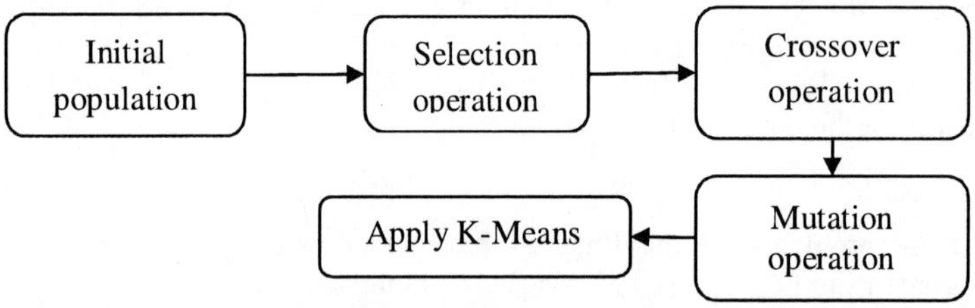

Figure 2. Process Stages of the GenClust Algorithm

The basic idea of determining the number of clusters is to use random numbers generated with a range of cluster numbers of 2 to, where n is the sum of the objects in the dataset. The position of centroid generated in chromosome form is 50% of chromosomes obtained through deterministic calculations and 50% of chromosomes are obtained through random numbers, eg the number of chromosomes is 60, then 30 chromosomes are obtained from deterministic calculations and 30 chromosomes are obtained randomly. The steps in obtaining chromosomes through deterministic calculations are as follows (Rahmad and Islam, 2014).

1. Find the random number coordinates of rx which express the coordinates of the centroid used as references.
2. Take the coordinates of the position of each object in the dataset.
3. Calculate the distance from each object to random number rx using euclidean distance.
4. Take 30 objects with the smallest distance to the random number rx (suppose it takes 30 chromosomes).
5. Coordinate positions of 30 selected objects fill coromosomes obtained through deterministic calculations.

As for 50% of other chromosomes to be obtained at random, the coordinates that will fill the gene on the chromosome are obtained by generating random numbers that are between the smallest position coordinates and the largest position coordinates.The genetic algorithm is a stochastic algorithm thatmodeled the evolutionary process of biological species through natural selection (Konar, 2005).

2.3. Performance Measurement on K-Means Algorithm

This study will examine the performance of the K-Means algorithm in relation to centroid determination. Centroid is the center of each cluster. This Centroid will greatly affect the performance of the K-Means algorithm because the placement of a data in a dataset in the K-Means algorithm is based on the proximity of the data with the centroid of each cluster, so that the value of centroid greatly influences the clustering results by using K-Means algorithm.

The researchers in this study will use the modified GenClust method where the overall generated chromosome (100%) is obtained through a deterministic calculation, then the centroid for each cluster

IOP Conf. Series: Journal of Physics: Conf. Series **930** (2017) 012008 doi:10.1088/1742-6596/930/1/012008

will be obtained based on the best chromosome for use in the clustering process using the K-Means algorithm.

3. Research Methodology

In the K-Means algorithm, determining the number of clusters and centroid determination is quite difficult to do. Determining the number of clusters and centroid determination directly affects the quality of the clustering process. The Hybrid Clustering method known as GenClust combines the use of the K-Means algorithm with the Genetic Algorithm. Genetic Algorithm is used to determine the number of clusters and also the centroid of each cluster. The use of the GenClust method may prevent the K-Means algorithm from being trapped under local optima.

It is necessary to test the determination of the number of clusters and also the centroid determination using GenClust and compare it with the classical K-Means on a large dataset. The existing dataset uses UCI Machine Learning Repository. UCI Machine Learning Repository is a collection of database, domain theory, and data generator used by community learning machine learning, for the purpose of empirical analysis of machine learning algorithm. The dataset available at UCI Machine Learning Repository is used by students, educators, and researchers around the world as the primary source of data sets in machine learning. The amount of data sets available at UCI Machine Learning Repository at this time has amounted to 320 data sets that can be used in accordance with the needs of machine learning learning.This study will discuss the comparison between the GenClust method with the classic K-Means algorithm especially in the comparison to measure the measured performance value of Mean Square Error occurring for a large dataset.

3.1. Data Used

The data used are Iris Data Set benchmark data taken from UCI Machine Learning Repository. Iris Data set is a data set that is widely used in pattern recognition problems. The attributes of information contained in the Iris Data Set are composed of: Sepal Length, Sepal Width, Petal Length, and Petal Width. Iris Data Set has 3 classes: Iris Setosa, Iris Versicolour, and Iris Virginica. Measurement of performance in this research using Mean Square Error (MSE) method.

4. Analysis Results

In this research will be shown the results of performance assessment in connection with centroid determination on K-Means algorithm. This study will compare the performance between centroid determination using the K-Means algorithm using random centroid determination, centroid determination using GenClust algorithm, and centroid determination using the modified K-Means algorithm. Measurement accuracy based on Mean Square Error. The accuracy value will be expressed in terms of Mean Square Error value which is the average value for error based on the test by using the number of iterations which are varied, that is: 50, 75, and 100. Small MSE value indicates that theclustering result using K-Means Clustering has been able to recognize existing patterns, whereas large MSE values indicate that clustering results using K-Means Clustering still have not achieved the desired results. The test will be performed using Iris dataset sourced from UCI Machine Learning Repository. The results of tests conducted by researchers will be presented in tabular form.

Table 1. Testing Results Using Classical K-Means Algorithms, GenClust Algorithms, and Modified GenClust Algorithms

	50 iterations		75 iterations		100 iterations	
	Average MSE	*Best* MSE	*Average* MSE	*Best* MSE	*Average* MSE	*Best* MSE
K-Means Algorithm	1.041	0.6	0.893	0.43	0.503	0.24
*GenClust*Algorithm	0.9637	0.36	0.818	0.4	0.506	0.36
Modified *GenClust*Algorithm	0.673	0.34	0.606	0.35	0.468	0.11

Based on Table 4.13 it can be seen that in general centroid determination influences the performance of the K-Means algorithm shown in the Mean Square Error (MSE) value obtained based on the classification result. Determination of centroid effect on the performance of genetic algorithm because the placement of a data into a dataset based on the proximity between the coordinates of the object with the coordinates of the centroid where the distance calculation used is Euclidean Distance. In the classic K-Means algorithm centroid determination is done by using random numbers. In the GenClust algorithm the centroid determination is based on the genetic algorithm, where 6 chromosomes are used, ie 3 chromosomes are obtained using random numbers and 3 chromosomes are obtained by using deterministic calculations, then the chromosomes will undergo selection, crossover, and mutation The best chromosome that will be used as a centroid on K-Means Algorithm.

Better performance of the modified GenClust algorithm is shown in terms of Average MSE values as well as better Best MSE values, and obtained both on tests using 50, 75, and 100 iterations.The results in this study may vary due to the random numbers generated, especially within the classic K-Means algorithm and the GenClust algorithm. In addition, the generation of random numbers can also define a number of processes within genetic algorithms such as selection, crossover, and mutation.

5. Conclusions

The conclusions that can be drawn from this research are as follows.

1. The performance shown in terms of the average MSE value and also the Best MSE score indicates that centroid determination determines the performance of the K-Means algorithm. Determination of centroid using K-Means algorithm has been modified better than centroid determination using GenClust algorithm and also K-Means Classic algorithm.
2. The results showed that the larger number of iterations will also provide better performance results on the K-Means algorithm. This happens because at each iteration stage of the K-Means algorithm will be adjusted to the value of centroid based on data items placed in a cluster.

References

[1] Ahmad, A. and L. Dey. 2007. A K-Mean Clustering Algorithm for Mixed Numeric and Categorical Data. Data & Knowledge Engineering63: pp. 503-527

[2] Bai, L., J. Liang, & C. Dang. 2011. An Inititalization Method to Simultaneously Find Initial Cluster Centers and The Number of Clusters for Clustering Categorical Data. Knowledge-Based Systems24: pp. 785–795.

[3] Cao, F., J. Liang, and L. Bai. 2009. A New Initialization Method for Categorical Data Clustering. Expert Systems with Applications36: pp. 10223-10228

[4] Fayyad, U., G.P. Shapiro, and P. Smyth. 1996. From Data Mining to Knowledge Discovery in Database. AI Magazine: pp. 37-53

[5] Konar, Amit. 2005. Computational Intelligence Principles, Techniques, and Applications. Springer: Calcutta, India
[6] Larose, D.T. 2006. Data Mining Methods and Models. John Wiley and Sons, Inc: New Jersey.
[7] Li, R., X. Q. Li and G. Wang. 2015. Global Optimal and Minimal Solutions to K-Means Cluster Analysis, Proceedings of the International Conference IP, Computer Vision, and Pattern Recoginition, pp. 274-283
[8] Maitra, R., A.D. Peterson, A.P. Ghosh. 2010. A Systematic Evaluation of Different Methods for Initializing the K-Means Clustering Algorithm. IEEE Trans. Knowl. Data Eng.
[9] Muzid, Syafiul. 2014. Dinamisasi Parameter AlgoritmaGenetikaMenggunakan Population Resizing on Fitness Improvement Fuzzy Evalutionary Algorithm (PROFIFEA). Prosiding SNATIF 2014, pp. 471-478
[10] Negnevitsky, Michael. 2005. Artificial Intteligence-A Guide to Intelligent Systems. Addison Wesley: Edinburg
[11] Rabunal, J.R. & Dorado, J. 2006. Artificial Neural Networks in Real-Life Applications, Ideal Group Publishing: Hershey, United States of America.
[12] Rahman, M.A. and M.Z. Islam. 2012. CRUDAW: A Novel Fuzzy Technique for Clustering Records Following User Defined Attribute Weights, Proceedings of the Tenth Australasian Data Mining Conference (AusDM), pp. 27-41
[13] Rahman, M.A and M.Z. Islam. 2014. A Hybrid Clustering Technique Combining a Novel Genetic Algorithm with K-Means. Knowledge-Based Systems71: pp. 345-365
[14] Rokach, L. And O. Maimon. 2005. Data Mining and Knowledge Discovery Handbook. Springer: Tel Aviv.
[15] Salman, M.S., O. Kukrer, and A. Hocanin. 2017. Recursive Inverse Algorithm: Mean-Square-Error Analysis. Digital Signal Processing 66: pp. 10-17
[16] Serapiao, A.B.S., G.S. Correa, F.B. Goncalves, and V.O. Carvalho. 2016. Combining K-Means and K-Harmonic With Fish School Search Algorithm for Data Clustering Task on Graphics Processing Units. Applied Soft Computing41: pp. 290-304
[17] Shapiro, G.P. 1991. Knowledge Discovery in Real Databaes: A Report on The International Joint Conference in Artificial Intelligence (IJCAI). AI Magazine: pp. 68-70
[18] Steinbach, Michael, Ertoz, Leventdan Kumar, Vipin. 2014. The Challenges of Clustering High Dimensional Data. Minnesota Supercomputing Institute

The Combination of RSA And Block Chiper Algorithms To Maintain Message Authentication

Sepri Yanti Tarigan[1], Dewi Sartika Ginting[2], Melva Lumban Gaol[3], Kristin Lorensi Sitompul[4]
[1,2,3,4]Magister of Information Technology, Universitas Sumatera Utara

Email:saveyanti@gmail.com

Abstract. RSA algorithm is public key algorithm using prime number and even still used today. The strength of this algorithm lies in the exponential process, and the factorial number into 2 prime numbers which until now difficult to do factoring. The RSA scheme itself adopts the block cipher scheme, where prior to encryption, the existing plaintext is divide in several block of the same length, where the plaintext and ciphertext are integers between 1 to n, where n is typically 1024 bit, and the block length itself is smaller or equal to log(n)+1 with base 2. With the combination of RSA algorithm and block chiper it is expected that the authentication of plaintext is secure. The secured message will be encrypted with RSA algorithm first and will be encrypted again using block chiper. And conversely, the chipertext will be decrypted with the block chiper first and decrypted again with the RSA algorithm. This paper suggests a combination of RSA algorithms and block chiper to secure data.

Keywords: RSA algorithm, Block chiper, Message authentication

1. Introduction

The security of information in this global age has become a vital necessity in some aspects of life. However, in the rapid development of information technology recently have a positive and negative impact where it becomes a convenience to everyone to get an information or a data without having to require a long process. Various ways do a lot of people to get it. Therefore, there should be a method used to secure information from unauthorized parties. One of the methods used is cryptography.

According to Request for Comments (RFC), cryptography is a mathematical science that made data incomprehensible (to hide its meaning), to prevent it from unauthorized change, or to prevent it from unauthorized use. If the transformation can be restored, cryptography means the process of converting back encrypted data into understandable form. Cryptography itself has the goal of confidentiality, data integrity, message authentication and non-repudiation. In this paper will discuss how to maintain message authentication.

One of the algorithms in cryptography is RSA. RSA algorithm is public key algorithm using prime number and even still used today. The power of this algorithm lies in the exponential process, and the number of factorials into 2 prime numbers which until now difficult to perform is known. The

RSA scheme itself adopts a block cipher scheme, where before encryption, the existing plaintext is divide in several block of the same length, where the plaintext and ciphertext are integers between 1 and n, where n is usually 1024 bits, and the block length itself more Small or equal to log (n) +1 with base 2. With combination of RSA algorithm and block chiper expected message authentication can be maintained.

2. Review Of Literature

2.1 Cryptography

Cryptography (cryptography) is derived from the Greek, kryptos means hidden and graphia means something written so that cryptography can also be called as something written in secret or hidden.

According to its terminology,cryptography is art and science of keeping messages safe when messages are sent from one place to another.

In cryptography, messages or information referred to plaintext or clear text. The process of transforming the original text (plaintext) into secret text (chipertext) is called encryption. Unreadable messages are called secret text (chipertext). The reverse process of encryption is called decryption. Decryption will return the secret text (chipertext) to the original text (plaintext). Both encryption and decryption processes require the use of some secret information, often called keys.

There are four fundamental objectives of this cryptographic science which is also an aspect of information security that is, confidentiality, data integrity, authentication and non-repudiation or non-denial.

Cryptography based on the key used in the process of encryption and decryption is divided into two parts:

a. Cryptography symmetry, this algorithm is also called classical algorithm because it use the same key for the process of encryption and decryption. Other terms of this cryptography are private key cryptography, secret-key cryptography, or conventional cryptography. In symmetry key cryptography it can be assumed that the recipient and the sender has first shared the key before the message is sent. The security of this system lies in the secrecy of the key. Scheme of symmetric cryptographic process is in the following figure.

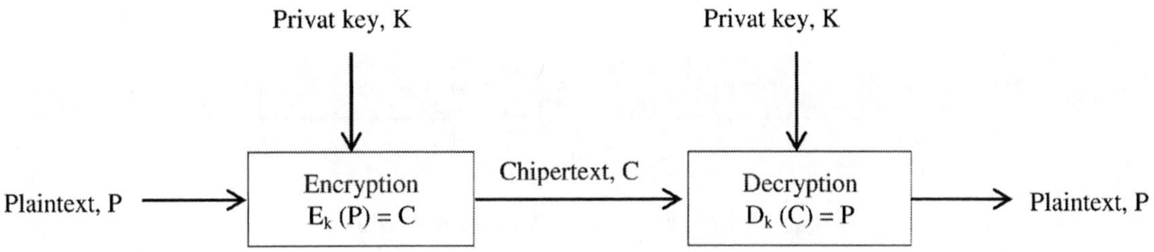

Figure 1. Symmetry Cryptography Scheme

There are two basic techniques commonly used in cryptography symmetry is a substitution technique in which the replacement of each plaintext character with other characters and transposition techniques (permutations) where this technique using character permutation. Some algorithms that use the same key (symmetry) include Data Encryption Standard (DES), International Data Encryption Algorithm (IDEA), Advanced Encryption Standard (AES).

b. Asymmetry Cryptography, Unlike the symmetry key cryptography, public key cryptography has two distinct keys to the encryption process and its description. Another name for this asymmetric key is key cryptography Public (public-key cryptography). The key to encryption in cryptographic asymmetry is not secret (publicly known), while the key to decryption is confidential (private key). The sender's entity will encrypt it by using the public key, while the receiving entity decrypts using the private key. Scheme of the cryptographic process this asymmetry is in the following figure.

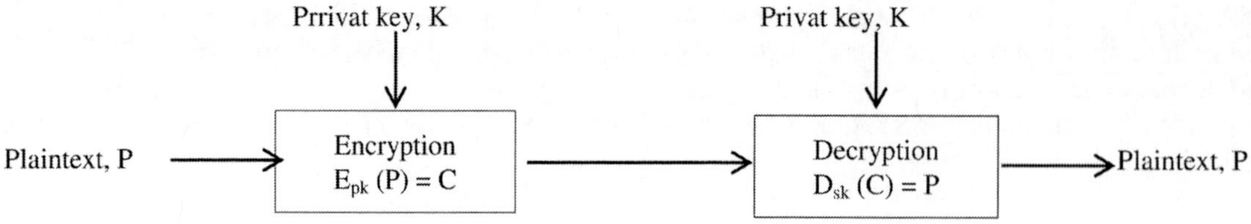

Figure 2. Asymmetry Cryptography Scheme

This asymmetry cryptography can be analogized like a locked mailbox and has a hole to insert mail. Everyone can put a letter into the mailbox, but only the owner of the letter has a key and can open the mailbox. The public key that have sent to the recipient through the same channel as the channel used to send the message, no need to fear because the unauthorized parties will not be able to decrypt the message, because it does not have a private key, or the recipient can provide the sender with the public key. Some algorithms that use different keys (asymmetry) include Digital Signature Algorithm (DSA), Rivest Shamir Adleman (RSA), Diffie-Helman (DH), Quantum Cryptography.

2.2 RSA Algorithm

The RSA algorithm comes from a combination of three names. The researchers are Ron RIvest, Adi Shamir and Leonard Adleman. This cryptographic method is classified into modern cryptography which consists of two keys, namely public key (public key) to perform encryption and privacy key (private key) to do the decryption.

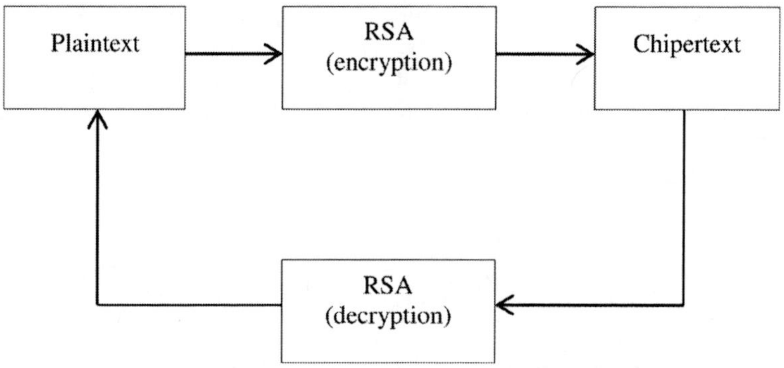

Figure 3. Process of RSA Algorithm Method

Encryption converts the original data (plaintext) to the encoded data (ciphertext), while the decryption is returning the ciphertext to the plaintext.

RSA algorithm is divided into 3 steps:

1. Key Generators
   a. Select 2 large primes for p and q
   b. Calculate the modulus value
   $$n = p x q \quad (1)$$
   c. Calculate using the Euler function
   $$\varphi (n) = (p 1) x (q-1) \quad (2)$$
   d. Select the random integer e as a public key, provided that it meets the Greater Common Divisor
   $$(e, \varphi (n)) = 1, 1 < e < \varphi (n) \quad (3)$$
   e. Calculate private key d so
   $$d x e = 1 (mod \varphi (n)) \quad (4)$$

2. Encryption

$$C = M^{e} \bmod n \quad (5)$$

3. Decryption

$$M = C^{d} \bmod n \quad (6)$$

2.3 Block Chiper

In the Cipher block, bits of plaintext is divided into blocks of the same length as 64-bits. The block cipher encryption process scheme can generally be described as follows:

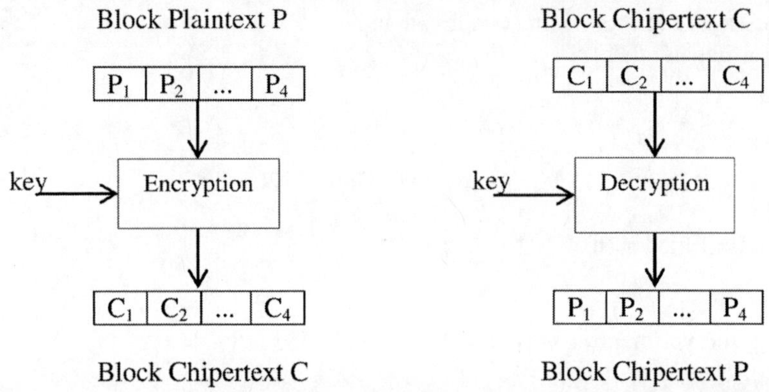

Figure 4. Encryption- Decryption of Block Chiper

Suppose the plaintext block (P) of n bits

$$P = (p_{1,} p_{2} \dots ..p_{n)} \quad (1)$$

The ciphertext block (C) then block C is

$$C = (c_{1,} c_{2,} \dots c_{n)} \quad (2)$$

Key (K) then key is

$$K = (k_{1,} k_{2,} \dots .k_{n}) \quad (3)$$

So the Encryption process is

$$E_{k} (P) = C \quad (4)$$

The decryption process is

$$D_{k} (C) = P \quad (5)$$

2.4 Message Authentication

Authentication is related to the identification/recognition, both unity of the system and the information itself. Two parties who make communication with each other should introduce each other. Information transmitted over channels must be authenticated, data, the delivery time, and so on. Data authentication problems in this cryptographic process are of primary concern. The success of the cryptographic process is greatly influenced by prime numbers on the generation of RSA keys. If there is a discrepancy of the required parameters then the process of returning the code file to the original file will not work.

3. Research Methode

The research method to obtain information and data required in this paper are as follows:

a. Library study
   Conducted by reading books related to the problem of book lending.
b. Observation
   Conducted to collect data by way of direct observation on matters relating to the library.
c. Analysis some cases, with take issues related to the discussion.

4. Analysis And Discussion

Message encryption with RSA algorithm is done using mathematical calculations, either in the process of generate keys, encrypting messages, or text decryption. RSA algorithm is based on the concept of prime numbers, modulo arithmetic, and Euler's theorem. In summary, RSA algorithm consists of three parts, namely generate keys, encryption, and decryption.

An example of encryption-decrypting a message with an RSA algorithm is: if an A wants to send a message to user B, then the first time A must send his public key to user B. After that user B will encrypt messages sent using the public key B. Then the user B perform encryption by means of $C = M^{e}$ (mood n). After that user A will receive the message C and decryption by means of $M = C^{d}$ (n mood). Noted that the message length of M must be less than n.

1. Choose 2 different prime numbers
   P = 61 and q = 53
2. Calculate n = pq
   N = 61 * 53 = 3233
3. Totient Calculate φ (n) = (p-1) (q-1) Φ (n) = (61-1) (53-1) = 3120
   E = 17
4. Select d to satisfy ed = 1 mod φ (n)
   D = 2753
   17 * 2753 = 46801 = 1 + 15 * 3120
5. If m = 123, then the encryption process is: $C = 123^{17} \bmod 3233 = 855$
6. For the decryption process are:
   $M = 855^{2753} \bmod 3233 = 123$

Suppose the message (plaintext) to be sent is "HARI INI" or in the decimal system (ASCII encoding) is 7265827332737873.

How to encrypt it is:

1. Break the m into smaller blocks, for example m split into six blocs that are 3 digits:
   $M_1 = 726$ $M_4 = 273$
   $M_2 = 582$ $M_5 = 787$
   $M_3 = 733$ $M_6 = 003$
   These m values are still located within the interval [0, 3337-1] for the transformation to be one.
2. If the public key is e = 79 and n = 3337, then the plaintext blocks can be encrypted to be:
   $C_1 = 726^{79} \bmod 3337 = 215$; $C_2 = 582^{79} \bmod 3337 = 776$; $C_3 = 733^{79} \bmod 3337 = 1743$; $C_4 = 273^{79} \bmod 3337 = 933$; $C_5 = 787^{79} \bmod 3337 = 1731$; $C_6 = 003^{79} \bmod 3337 = 158$. So the resulting chipertext is c = 215 776 1743 933 1731 158
3. Decryption is done using private key d = 1019. The chipertext blocks are decrypted to be:
   $M_1 = 215^{1019} \bmod 3337 = 726$
   $M_2 = 776^{1019} \bmod 3337 = 582$
   $M_3 = 1743^{1019} \bmod 3337 = 733$
   $M_4 = 933^{1019} \bmod 3337 = 273$
   $M_5 = 1731^{1019} \bmod 3337 = 787$
   $M_6 = 158^{1019} \bmod 3337 = 003$
4. Finally recovered the original plaintext
   M = 7265827332737873, which in the ASCII encoding system is M = "HARI INI"

From the example above it can be seen that the whole process of encryption and decryption with RSA algorithm using mathematical calculation, either from key generation process, until encryption and decryption of message. In order to be calculated, the text you wish to keep a secret must be converted into numeric form first. The results of the encryption is a cipher text which is also a series of numbers.

5. Conclusions

RSA algorithm is public key algorithm using prime number and even still used today. The strength of this algorithm lies in the exponential process, and the factorial number into 2 prime numbers which until now difficult to do factoring.

The RSA scheme itself adopts the block cipher scheme, where prior to encryption, the existing plaintext is divide in several block of the same length, where the plaintext and ciphertext are integers between 1 to n, where n is typically 1024 bit, and the block length itself is smaller or equal to log(n)+1 with base 2. With the combination of RSA algorithm and block chiper it is expected that the authentication of plaintext is secure.

References

[1] Akiwate,Bahubali& Desai,Vennai. 2013. Artificial Neural Networks for Cryptanalysis of DES. International Journal of Innovations in Engineering and Technology (IJIET). Vol. 2

[2] Ashioba, N. C. & Yoro, R. E., 2014. RSA Cryptosystem using Object-Oriented Modeling Technique. *International Journal of Information and Communication Technology Research*. Vol. 4(2): 57-61.

[3] Goyal, Kashish & Kinger Supriya. 2013. Modified Caesar Cipher for better Security. *Enhancement. International Journal of Computer Applications* (0975 – 8887). Vol.73(3).

[4] Shika Mathur, Deepika Gupta, Vishal Goar, Manoj Kuri, 2017. Analysis And Design of Enhanced RSA Algorithm to Improve The Security. International Conference of Computational Intelligence & Communication Technology (CICT). Vol. 3: 1-11.

[5] Upadhyay, Sachin., Singh, Yasphal., and Jain, Kumar, Amit. 2012. "An Analysis of the Attack on RSA Sryptosystem Trhourgh Formal Methods". *International Journal of Soft Computing and Engineering (IJSCE)*. Vol(2): 2231-2307.

[6] Yan, Y Song. 2007. *Cryptanalystic Attack on RSA. Springer*: University of Bedfordshire, UK.

ANALYSIS OF BEARING CAPACITY PILE FOUNDATION WITH USING CAPWAP SOFTWARE FOR TESTING PILE DRIVING ANALYZER (PDA) AT FASFEL DEVELOPMENT PROJECT PARLIMBUNGAN KETEK SIKARA-KARA MANDAILING NATAL DISTRICT (NORTH SUMATERA)

Johan Oberlyn Simanjuntak[1], Diana Suita[2]

Lecturer S2, Universitas HKBP Nommensen Medan, Jl. Perintis Kemerdekaan No. 4 A Medan, INDONESIA
Lecturer S2, Sekolah Tinggi Teknik Harapan Medan (STTH Medan), Jl. H.M. Joni 7 C Medan, INDONESIA

E-mail: oberlyn.simanjuntak@yahoo.co.id, dns1301@ gmail.com

Abstract. Pile foundation is one type deep foundation that serves to distribute the load of hard soil structure loading which has a high bearing capacity that is located deep enough inside the soil. To determine the bearing capacity of the pile and at the same time control the Calendring results, the Pile Driving Analyzer (PDA) test at 8 pile sections from the 84 point piling section (10% of the number sections), the results were analyzed by CAPWAP SOFTWARE, and the highest bearing capacity of Ru 177 ton and the lowest bearing capacity of 111 tons, is bigger than the plan load which load plans that is 60,9 tons. Finally the PDA safe is bearing bearing capacity of the load planning .

Keywords: CAPWAP Sotware, PDA Test, Pile bearing capacity

1. Introduction

Referring to RI Law no. 17 of 2008 on shipping indicates the need for the provision of wharf/jetty infrastructure as a place for intra and inter-modal transport. Before construction wharf/jetty were built, Pile Driving Analyzer (PDA) testing was conducted to analyze the data pile bearing capacity. This function of PDA test to re- control the results of the calendering test. The PDA test was performed on eight (8) sections. Analysis obtained from PDA data using CAPWAP software. CAPWAP software processed the PDA data further by carrying out signal matching process. The program provides the distribution of soil resistance along pile. CAPWAP produces load vs setlement curve as well as capacity of the tested pile.

2. Literatur Review

2.1 Pile Driving Analyzer (PDA)

Pile Driving Analyzer (PDA) is a pile dynamic load test that aims to determine the pile maximum capacity, emergy transfer from hammer to pile etc To carry out PDA test driven pile hammer can be used. When the piling rig is no longer on site, a drop hammer with a minimum weight of 1.5 percent (%) of estimated ultimate pile capacity can be used. PDA program calculates the dynamic force and strain obatained from accelerometers and strain transducers mounted on the tested pile. Data from these sensors is transmited to PDA procesing unit through cable or wirelessly (Fellenius, 2017).

2.2 The CAPWAP

The CAPWAP program performs a signal matching or reverse analysis base on measurements taken ona deep foundation under an impact load. The program is based on a one dimensional mathematical model. Under certain conditions, the model only crudely approximates the often complex dynamic situations. The CAPWAP analysis relies on the input of accurately measured dynamic data plus additional parameters describing pile and soil behavior. If the field measurements of force and velocity are incorrect or were taken under inappropriate conditions or if the input pile model is incorrect, then the solutions cannot represent the actual soil behavior. The CAPWAP analysis software is one of many means by which the capacity of a deep foundation can be assessed. The engineer performing the analysis is responsible for proper software application and the analysis results. Pile dynamics accepts no liability whatever of any kind for analysis solution or application of the analysis result (Fellenius, 2017).

3. Research Methodology

3.1 Project Name: Faspel Parlimbungan Ketek Sikara-kara , Mandailing Natal Distric Location: North Sumatera

Tabel 1. Project Information and Pile Properties

Data 1		**Data 1**	
Pile Name	TP-A 20	LE	43.7 m
DESC	Steel 50- Diesel 6.5 T	AR	217. 27 cm^2
File	TP-A 30 PDA	EM	2109 T/cm^2
Blow Number	22	SP	7.88 T/m^3
Serial Number	4690 LE	WS	5123.0 m/s
		WC	5.11 m/s
Data 2		EA/C	89.5 Tn- s/m
File Name	TP - B 12	2L/C	17.10 ms
DESC	Steel 50 – Diesel 6.5 T	JC	0.50
File	TP – B 12 PDA	LP	34.9 meter
Blow Number	17		
Serial Number	4690 LE	**Data 2**	
		LE	47.8 meter
Data 3		AR	213.75 T/m^2
File Name	TP A3-3	EM	2109 cm^2
DESC	Steel 50- Diesel 6.5 T	SP	7.88 T/m^3
File	TP A3-3 PDA	WS	5123.0 m/s
Blow Number	5	WC	5112.3 m/s
Serial Number	4523 LE	EA/C	88.0 Tn- s/m
		2L/C	18.70 ms
		JC	0.50
		LP	39.5 m
		Data 3	
		LE	47.5 m
		AR	213.75 cm^2
		EM	2109 T/cm^2
		SP	7.88 T/m^3
		WS	5123.0 m/s

WC	5107.5 m/s
EA/C	88.0 Tn-s/m
2L/C	18.50 ms
JC	0.50
LP	39.0 m

4. Analyis and Discuss

The results obtained from the analysis are:

Tabel 2. Bearing Capacity

Pile Name	PDA	CAPWAP				
	Bearing Capacity (RMX/RSU) Ton	Bearing Capacity (RU) Ton	Friction (SF) Ton	End Bearring (EB) Ton	Elastic Displacement (Dy) mm	Displacement Maximum (Dx) mm
TP-A20	163	158	157	2	46.39	71.36
TP-A21	125	143	140	3	61.22	97.65
TP-B12	173	177	159	18	16.50	40.37
TP-C12	140	133	125	8.0	37.33	55.19
TP- A3-3	199	174	172	2.0	18.90	28.29
TP- B2-2	111	111	110	1.0	35.22	72.21
TP- C4	177	153	145	8.0	16.31	33.18
TP-C8-1	161	148	148	0.0	11./53	43.49

5. Conclusions and recommendations

5.1 Conclusion

1. CAPWAP software calculation results for PDA test at 8 pile sections ie highest bearing capacity on TP - B12 Ru 177 ton and lowest bearing capacity TP - B22 Ru 111 ton.
2. loading plan of pile bearing capacity Qu 60.9 tons.
3. The CAPWAP software analysis results for the PDA test state that the bearing capacity of the pile is safe to ultimate the load of a planning bearing capacity , Ru> Qu.

5.2 Suggestion

In analyzing the bearing capacity of pile had better by compiling the result of bearing capacity PDA with manual /calendring.

References

[1] Maizir H, Gofar N, Kassim K A, (2015) "Artificial Neural Network Model for Prediction of Bearing Capacity of Driven Pile, Juurnal Teknik Sipil, Vol. 22 No.1 April

[2] Prabowo, D. S. P., 2012, Bearing Capacity:Meyerhof, Jurnal Teknik Sipil, Universitas Gunadarma, Depok.

[3] Fellenius B. H, (2017), "Basics of Foundation Design", Electronic Edition, January 2017

[4] Yusti. A, Fahriani. F, (2014), "ANALISIS DAYA DUKUNG PONDASI TIANG PANCANG DIVERIFIKASI DENGAN HASIL UJI PILE DRIVING ANALYZER TEST DAN CAPWAP" Jurnal Fropil, Vol. 2 Nomor 1 Januari - Juni 2014

[5] U.S. Department of Transportation Federal Highway Administration, "Design and Construction of Driven Pile Foundations – Volume I, Publication No. FHWA - NHI - 16 - 009 FHWA GEC 012 – Volume I September 2016

[6] Likins, G. E., Rausche, F., September, (2008), "What Constitutes a Good PDA Test?", Proceedings of the Eighth International Conference on the Application of Stress Wave Theory to Piles 2008: Lisbon, Portugal; 403-407.

IOP Conf. Series: Journal of Physics: Conf. Series **930** (2017) 012011 doi:10.1088/1742-6596/930/1/012011

Automatic Locker Key With Barcode Based Microcontroller Atmega 8535

M. Irfan Fahmi[1] and Jhonson Efendi Hutagalung[2]

[1,2]AMIK Royal Kisaran jalan Imam Bonjol No.179, 0623-42366 majoring of computer technik , AMIK Royal, Kisaran

e-mail: jhonefendi12@yahoo.co.id,

Abstract. MCB (miniature circuit breaker) is an electromagnetic device that embodies complete enclosure in a molded insulating material. The main function of an MCB is to switch the circuit, i.e., to open the circuit (which has been connected to it) automatically when the current passing through it (MCB) exceeds the value for which it is set. Unlike fuse, an MCB can be easily reset and thus offers improved operational safety and greater convenience without incurring large operating cost.The principal of operation is simple. In simple terms MCB is a switch which automatically turns off when the current flowing through it passes the maximum allowable limit. Generally MCB are designed to protect against over current and over temperature faults (over heating).sometimes the overload the current through the bimetal causes to raise the temperature of it. The heat generated within the bimetal itself enough to cause deflection due to thermal expansion of metals. This solution is used by LDR, and LM 35 as the sencor to control center. Therefore it is very important because it is related about local control switches, isolating switches against faults and overload protection devices for installations or specific equipments or appliances

1 INTRODUCTION

Electricity is one of the most important blessings that science has given to mankind. It has also become a part of modern life and one cannot think of a world without it. Electricity has many uses in our day to day life. But using the electricity we found that the fire as the reaction of a substance with oxygen. An electric spark is an area where the air conducts electricity by ionization. is the transfer of electrons along a wire, fire is the reaction of a substance with oxygen.

Nowadays we use more commonly miniature circuit breaker or MCB in low voltage electrical network instead of fuse. The MCB has some advantages compared to fuse. It automatically switches off the electrical circuit during abnormal condition of the network means in over load condition as well as faulty condition.

The operation of the MCB includes two stages, i.e., thermal operation and short circuit operation. The former operation based on the thermal effect of over current while the later operation is based on the electromagnetic effect of over current.All MCBs operate on the air-break principle where the arc between the contacts is forced into the splitter plates through arc runners. This causes to spilt single arc into a series of arcs and then extinguishes the arc by extracting the energy from the arc and by cooling it.

With the use of bimetallic strip, thermal operation is achieved in case overload conditions. When overload current flows through the MCB, the bimetallic strip gets heated and causes to deflect. In doing so, it moves the trip lever and releases the latch mechanism and hence the contacts open under spring mechanism. The electricity spark is to detected by LDR and LM 35 as the sensor to detected about temperature to the control center.

The AVR 8535 microcontroller and its new version ATmega8535 are versatile, high-performance but low-cost chips. This article is Integrated other electronics or called minimum system. The result is to detected the sensor and to input the Atmega 8535, when the electricity spark and over load in this control panel upper 35 so this indicator will be sounded and sparks will be displayed on the LCD.

2. Literature Review

The first popular microcontroller made by Intel in 1976, the 8 bit microcontroller Intel 8748. Microcontroller is part of the microcontroller family MCS 48. Previously, Texas instruments has marketed the first 4 bits microcontroller TMS 1000 in 1974. TMS 1000 that began to be made Since 1971 is a microcomputer on a chip, complete with RAM and ROM.

Microcontoler is an intelligent chip that becomes a trend in automatic control. With many family types, memory capacity, and various features, the microcontroller becomes an option in mini-processor applications for small-scale control. We can Seen the picture in the below it:

Figure 1. AVR ATMega 8535 Microcontroller

Configuration Pin Atmega 8535 Microcontroller with 40 pin DIP (dual in line package) package can be seen in the figure below To maximize performance and parallelism, AVR uses Harvard architecture (with separate memory and bus for program and data). The CPU architecture of the AVR is shown in Figure 6 The instructions on program memory are executed with 10 pipelining single levels. While an instruction is being done, the next instruction is taken from the program memory.

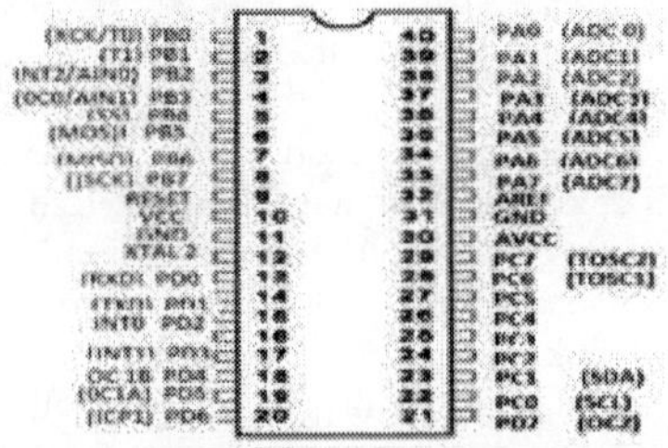

Figure 2. ATMega 8535 Microcontroller Configuration

LM35 temperature sensor is a temperature sensor in the form of integrated circuit and has a voltage output of a voltage change linear and proportional to temperature (scale celscius) that is 10 mV / oC. LM 35 does not require external calibration to produce ¼oC accuracy at temperatures to be measured from 2 to 40 o C (Andrianto Heri, 2015: 145).

Electronic components have a function to convert the amount of temperature into electrical quantities in the form of voltage. The LM35 temperature sensor used in this research is the electronic component produced by National Semiconductor. LM35 has high accuracy and ease of design when compared with other temperature sensors, LM35 also has low impedance accuracy and high linearity so it can easily be connected with a special control circuit and does not require advanced settings.

Although this voltage can reach 30 volts but given to the sensor is 5 volts, so it can be used with a single supply provided that LM35 only requires a current of 60 A this means that LM35 has the ability to generate heat (self heating) from the sensor Can cause a low reading error of less than 0.5 °C at 25 °C

In principle the sensors will perform sensing at the time of temperature change every 1°C temperature will show a voltage of 10 mV. In the placement LM35 can be affixed with adhesive or can also be cemented on the surface but the temperature will slightly decrease about 0.01 °C as it is absorbed at the surface temperature. In this way it is expected that the difference between air temperature and surface temperature can be detected by the LM35 sensor equal to the ambient temperature, if the ambient temperature is higher or lower than the surface temperature, then LM35 is at surface temperature and ambient temperature.

IOP Conf. Series: Journal of Physics: Conf. Series **930** (2017) 012011 doi:10.1088/1742-6596/930/1/012011

Another type of resistor is the Light Dependent Resistor. The LDR resistance changes with the change in the intensity of the light that it uses. In a gela state, the LDR resistance is about 10 M ohm and in a light state, of 1 K ohm or less. LDR is made from cadmium sulphide. With this material the energy from falling light causes more discharged loads or increased current, increases.

Downloader is a tool used to insert / download programs created and compiled into the chip / ic microcontroller. The interface between the computer with the microcontroller, through downloader program that has been made on your computer can be embedded into the microcontroller, so the microcontroller can work as we expect. The downloader used is shown below:

LCD (Liquid Cristal Display) is used the LCD that can only display the characters. The LCD has a display with 16 columns width and 2 rows or often called 16x2 character LCD with 16 pin connector.

Generally, alarms can be defined as warning sounds or notifications. Alarms can also be a message of notification when a device is experiencing damage (performance degradation). This message is used to alert the operator of a problem (danger). Alarms give alarm signals, sounds, or rays. Alarms notify in case of danger or damage or unexpected events on an apparatus so as to give a clear warning in order to be anticipated. In the picture below a buzzer can be supplied DC voltage of 3 - 24 Volts.

3 Research methodology

The method used for Automatic Lock Locker With Atmega 8535 Microcontroller Based Barcode, based on data that has been collected through the laboratory.

Field method is also done, that research is done by way of directness to get good information about comparison between theory and practice in accordance with what expected, by determining the design of hardware and software that will be used in determining the classification of system. It isintended to test the concept before proceeding with the creation of the entire program as a final step is to create a user interface (User Interface).

4. Design of series and tools, materials,equipment and used it

The workings of this system is when there is more burden due to the use or short circuit so as to cause a spark of fire then the LDR sensor will detect the light and also resulted in an increase in temperature then LM 35 sensors will detect the temperature. To provide input to the microcontroller.

With the appropriate settings in accordance with the desired relay will work to decide the current to the load and issued an alarm sound to let you know that there has been excess heat or there is a spark on the control center. Also on the LCD displays the value of the degree of temperature and the presence of light from the spark.To design this sensor using a separate sensor design because the function of each sensor is different. For LM35 sensor is a single component that resembles a transistor but LM 35 is working to detect the temperature around itself, sometimes this component is given the body pads and feet so as not exposed to marinade or splashing water. For this LDR sensor is a series of kits consisting of additional components such as resistors, capacitors and potentiometers to regulate sensitivity of sensor detection.

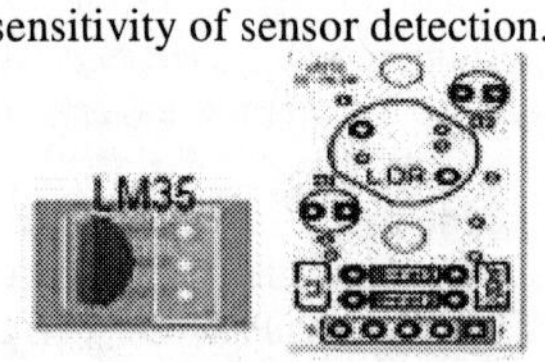

Figure 3. PCB Design Layout Sensor LM35 and LDR

The minimum system set serves to enable the microcontroller to work in accordance with the program that has been designed.

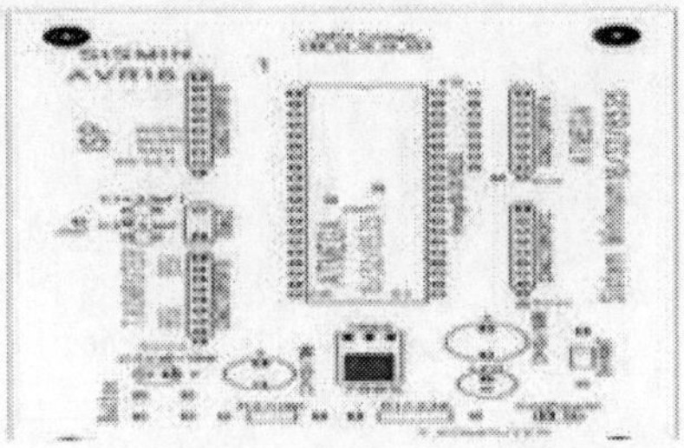

Figure 4. PCB Design Minimum System Layout Microcontroller

The series of downloaders used are systems and Atmega 8 in conjunction with other supporting components, such as: 100nF ceramic capacitors, 22pF capacitors, 4.7 K, 10K and 1K resistors, Crystal X1 12 MHz. For inputs connected to the USB through the port path of the microcontroller that is Pin D2 (INTO) and Pin B0 and Pin B1 as well as for output connected to PORT B (Miso, Mosi, SCK). From the circuit generated PCB layout then in PCB will be occupied its components. The design of the component layout can be seen in the picture below:

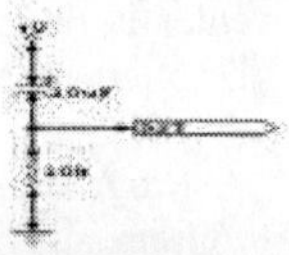

Figure 5. Reset Rows

In this circuit required R 220 ohm, 1N 4001 Diode, 6 Volt relay and C9014 transistor. Then make the PCB layout according to the circuit. Resistor here as input voltage for transistor work. The transistor acts as a switch to connect the voltage to the relay, so the relay works.

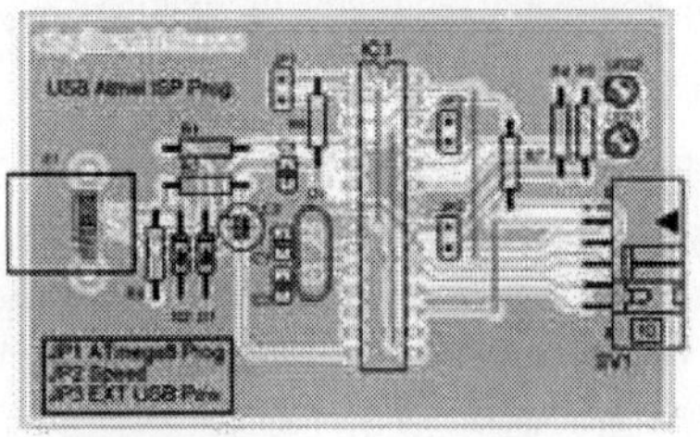

Figure 6. PCB Design Layout Relay Driver

In this circuit required R 220 ohm, 1N 4001 Diode, 6 Volt relay and C9014 transistor. Then make the PCB layout according to the circuit. Resistor here as input voltage for transistor work. The transistor acts as a switch to connect the voltage to the relay, so the relay works.

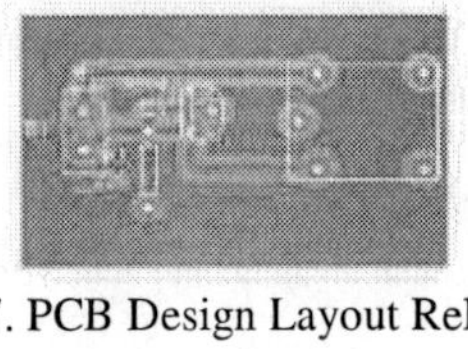

Figure 7. PCB Design Layout Relay Driver

In the simple transistor circuits below, there is an alarm that we describe as a load. One NPN transistor which will replace the working function of a switch. A potentialo meter is used so that you can perform analysis on different base current conditions by varying the conditions of the pot.

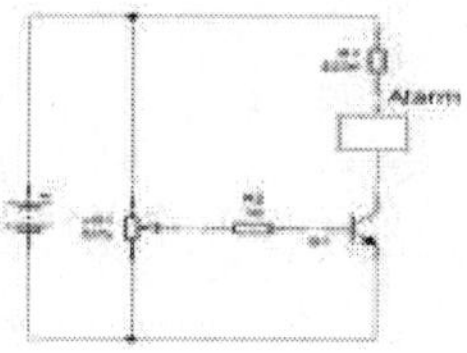

Figure 8. Transistors As Switches

5. Discussion of Design Results

The design of the fire prevention device over the mechanical load due to the short circuit current or excessive load is to detect the temperature of the cable used to flow the current and the light that is caused by sparks, consisting of a series of temperature sensors, light sensors, microcontrollers as a controller Inputs / outputs, software or programs to control that the designed device works as expected, then software and hardware testing is performed.

5.1 Software Discussion

The software used in this fire preventer is the programming language C AVR. After the program is designed then the program is inserted into the microcontroller using the downloader and then test the program entered on the microcontroller.

Testing the program is done by AV C Programming Language Language. To test the program that has been designed successfully and then the program and then compile whether an error or error occurs. Also required is the port setting of the downloader connected to the laptop / computer, by selecting Atmel AVR as a chip, Com 1 is used as the interface to the tool, and baud ratenya 115200.

And as shown in the illustration programming in the appearance of data detected the temperature and light on the LCD.

In the design of the program through Code Visioner AVR is checked again to ensure no errors in the design of the program, and if in the design of the program error occurs it will provide an error declaration at the time of checking or when compiled, as shown in the following figure:

Then the program designed by C AVR is downloaded through software by clicking on the chip program to include program designed into microcontroller IC

5.2 Hardware Discussion

To ensure that the designed hardware has functioned in accordance with its function, then the testing tool. This hardware testing starts from testing the Atmega 8535 Microcontroller circuit, sensor testing, power supply testing and LCD display testing, and relay chain testing.

5.2.1 Power Supply Discussion

Power used is full wave power supply. Power supply testing is done by providing alternating input voltage (AC) which varies by using AC voltage regulator to tap 0 and 220. For the output voltage of the travo that has been connected to the power supply circuit, a measurement is made for 12-CT-12 tap at the output of the power supply voltage. Output from a variable power supply serving a load of 1k ohm can produce a stable voltage with good voltage regulation.

Table 1. Power Supply Testing

NO	POWER*IN PUT* (AC)	Resisto r (ohm)	*POWER OUTPUT* (DC)	SIGN
1	12 V	1 k	11,75	-
2	9 V	1 k	8.85	-
3	7.5 V	1 k	7.28	-
4	4.5 V	1 k	4.35	-
5	3 V	1 k	2.91	-

Testing can be done by using a multimeter to measure the output voltage of the power supply.

5.2.2 Discussion of Microcontroller

This microcontroller testing is done by programming the microcontroller by using the C programming language AVR, and test the microcontroller used whether it is still working properly or in a state not working. Below is a picture of microcontroller test results. The voltage of the power supply (input) supplied to the microcontroller varies not passing through the voltage of the microcontroller and then measured the voltage on one of the output / input ports on the VCC and Ground pins.

5.2.3 LM 35 and LDR Sensor Testing

For testing the LM35 and LDR sensor circuit is required a multimeter to measure the output voltage of sensor test results manually. If for LM 35 by providing heat to the sensor while the LDR by providing light on the sensor so that there is a change in voltage values on the multimeter. This sensor circuit test is done by measuring the output voltage to ground.

Each sensor is given a voltage source VCC and Ground and the pin (output) is connected with a positive probe multimeter (Red wire) and black wire (negative probe) connected to the gorund, then see the change in the value of the number indicated by the needle pointer on the multimeter.

Figure 9. Sensor Testing

The LCD used in this discussion is a 16x2 character LCD. Which in this discussion LCD function as a media viewer results. The detection of the sensor is displayed on the LCD. If the temperature has reached 35o it will show "35o" and if there is a spark light it will show "Light Ada" it will be given input voltage to the microcontroller, so the microcontroller will activate the relay to decide current To the electrical load. Results LCD Display can be seen in the picture below:

Figure 10. LCD Display

So there are two logic data inputs that if there is light coming from sparks and the rise in temperature in the cable due to a short circuit or excessive load. The whole circuit image can be seen in the picture below:

Figure 11. Outomatic locker key circuit

For the whole circuit of this tool can be seen in the ganbar below:

Figure 12. Overall Circuit Tool

6. Conclusions

From the discussion in the previous chapters can be concluded that is:

1. In the current-flowing cable can be taken provisions if the burden is served excessively will lead to heat on the cable, what else happens short circuit then the heat on the cable will rise.
2. The temperature rise resulting from this overload is detected by the LM 35 temperature sensor under conditions at 35 ° C.
3. In MCB contact relationship with the output cable to the load can be worn out due to the increasing electrical load so that the cable connection is less good it will cause sparks.
4. Sparks will emit light detected by the LDR sensor so as to provide the input voltage to the relay so as to disconnect the circuit.
5. Data obtained from the detection of LM35 and LDR sensors are controlled by using Atmega 8535 Microcontroller which has been programmed first using C Programming Language C AVR as a tool for fire prevention caused by electrical overload or short circuit.

7. References

[1] WinotoArdi. (2008). *Mikrokontroler AVR ATMega8/32/16/8535 danPemrogramannyadengan Bahasa C pada Win AVR*.Yogyakarta PenerbitInformatika.

[2] WahyudiantoDariskaKukuh, Rika Rokhana, EruPuspita. (2008). *RancangBangunAlat Ukur Resistivitas Tanah Sebagai Alat Bantu Mengethaui IndikatorKualitas Tanah Untuk TanamanPadi.* Yogyakarta, Penerbit Informatika.

[3] HendawanSoebhakti. (2007). *Basic AVR Microcontroller Tutorial, PoliteknikBatam.*

[4] WardhanaLingga ,Penerbit Andi. (2006). Mikrokontroler AVR Seri ATMega 8535 Simulasi Hardware danAplikasi. Yogyakarta, PenerbitInformatika.

[5] MochChoirul Ana, S.Si. (2008). ModulElektronika 1. Yogyakarta, PenerbitInformatika.

[6] Ali Muhamad,. Chandra Ariadie N, Asmara Andik , (2013). Modul Proteus Professional Untuk Simulasi Rangkaian Digital Dan Mikrokontroler (MateriLanjutan Mikrokontroler). Yogyakarta, Penerbit Andi.

[7] Sugiri , A. 2004). Buku Elektronika Dasar dan Peripheral Komputer. Yogyakarta Penerbit Andi.

[8] Septian Taufiq Dwi,, 2010) Buku Pintar Robotika Bagaimana Merancang dan Membuat Robot Sendir Yogyakarta.Penerbit Andi.

Data Collision Prevention with Overflow Hashing Technique in Closed Hash Searching Process

Robbi Rahim[1*], Nurjamiyah[2] and Arie Rafika Dewi[2]

[1]Department of Health Information, Akademi Perekam Medik dan Infokes Imelda, Jl. Bilal Ujung Medan 20116, Indonesia
[1]School of Computer and Communication Engineering, Universiti Malaysia Perlis, Malaysia
[2]Department of Information System, Sekolah Tinggi Teknik Harapan, Jl. H.M. Joni No. 70 C, Medan 20216, Indonesia

*usurobbi85@zoho.com

Abstract. Hash search is a method that can be used for various search processes such as search engines, sorting, machine learning, neural network and so on, in the search process the possibility of collision data can happen and to prevent the occurrence of collision can be done in several ways one of them is to use Overflow technique, the use of this technique perform with varying length of data and this technique can prevent the occurrence of data collisions.

1. Introduction

The search process cannot separate from daily activities related to the information needs, the easiest to find information is to use certain keywords [1], and this search process would require an algorithm that has been tested in advance for the desired results accordingly.

Hash Search [2] is a search method that has a way of working similar to the method of direct search, which is by using certain formulas both at the time of placement and search data [3] [4]. Hash Search has better place space efficiency than Direct Search method [5]. The function (I - 1) used by the Direct Search method has a poor area of effectiveness, so the Hash Search method is added modulo operating function to cover the weakness of the direct search method. Modulo operation function is often applied to Hash Function, and the data shelter is called the Hash Table. The Hash function is not a one-on-one functionality as a function of the Direct Search method, so it is possible that some data have the same function results [5]. Collisions were occurring at the time of placement of data into tables so that a strategy is needed to overcome this collision. Strategies to overcome this there are various with advantages and disadvantages for each method.

This paper conducted search experiments with hash search method and overflow technique to avoid the occurrence of collision, the experiment is done with an extended dynamic and random data, the results are displayed with the use of overflow technique and without overflow technique, the results of the comparison shown can be an input how the search process is done and avoid collision data that could happen.

2. Methodology

Closed Hash [6] is introduced as the Hash way in which the data is directly stored in a Hash table with a specified table size that can be defined on its own [6]; in the application of the Hash function, it is possible that some data has the same result. The causes occurrence at the time of placement of data to be conducted search process [6] [7].

Strategies to overcome collisions by dividing the table into two pieces, namely the main table and overflow table with the provisions of the size of the main table is greater than the length of the table overflow [3]. The way of placement of data in the main table is done by using the Hash function. In the event of a collision, the data is placed on the overflow table. How to put data in the overflow table can be done sequentially or using the new Hash function. If the overflow table is full, then the data entered into the overflow table will be rejected, for the main table form and overflow can be seen as follows:

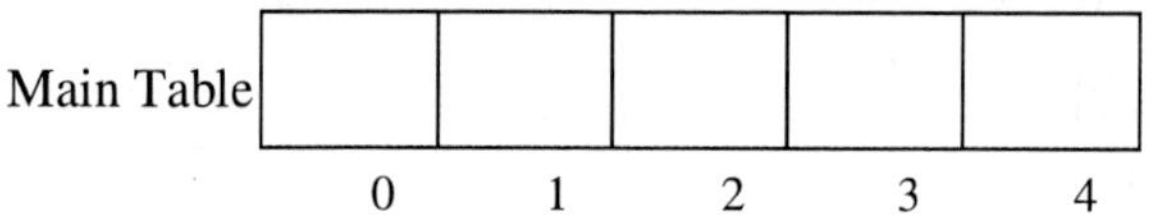

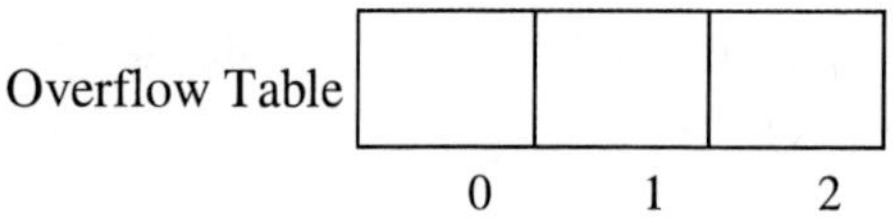

In the table above enter the number or value to be searched by using hash search algorithm and the search process starts from the new main table then to the overflow table.

3. Proposed Method

Experiments conducted with a data length of 15 pieces with the value 31,6,12,98,32,55,6,41,57,37,71,62,94,11,12 then the main table has 10 boxes and overflow table 7 boxes, to determine the position of the value in each box for the main table and overflow done calculation process as follows:

Data: 31, 6, 12, 98, 32, 55, 6, 41, 57, 37, 71, 62, 94, 11, 12

a. Data '31'
   Main table -> h(x) = 31 mod 10 = 1, Location 1 is empty so that '31' data is placed at location 1
b. Data '6'
   Main table -> h(x) = 6 mod 10 = 6, Location 6 is empty so that '6' data is placed at location 6
c. Data '12'
   Main table -> h(x) = 12 mod 10 = 2, Location 2 is empty so data '12' is placed on location 2
d. Data '98'
   Main table -> h (x) = 98 mod 10 = 8, location 8 is empty so '98' data is placed on 8 locations
e. Data '32'
   Main Table -> h (x) = 32 mod 10 = 2 -> collision with data '12' the search continues to Overflow Table
   Table overflow -> g (x) = 0, location 0 is empty so data '32' is placed at location 0

The placement of the value 32 in the 2nd box main table cannot be performed because it already contains the data, to prevent the collision then the value 32 is inserted into the overflow table box starting from location 0, so the value of 32 is in the position 0 overflow table, the process is accomplished until all the value data is stored according to the main table and the overflow table, and generates the position of the numbers as follows:

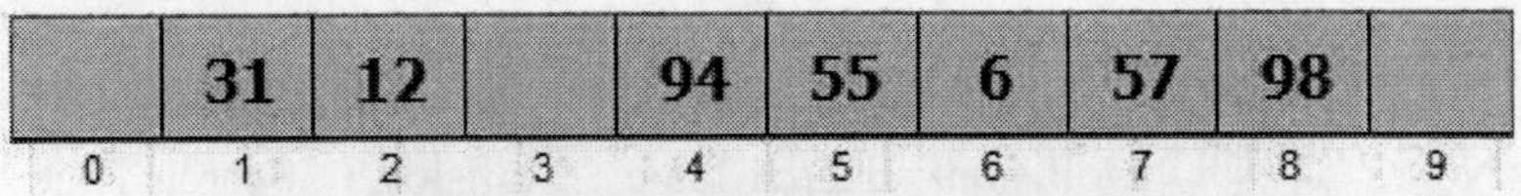

Figure 1. Value in the main table

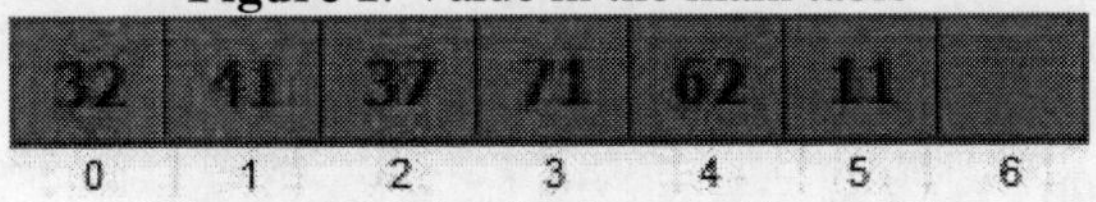

Figure 2. Value in the overflow table

After placing the values in the main table and overflow table is a technique to prevent collision data occurs, then the next process is to test to find the data contained in the main table and overflow, the process of searching the value in the overflow table is done sequentially from the initial position to end of overflow table, as an example of the value that wants to search is 37 of a set of values, then the process is to find the value into the main table with following functions:

Main Table -> h (x) = 37 mod 10 = 7 -> Data '37' not found, the search is continued into the Overflow Table

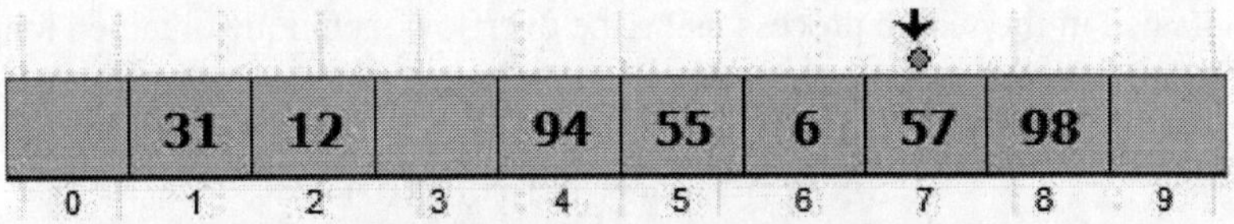

Figure 3. Value 37 not found

In the process of search overflow is done gradually with the following functions:
Overflow-> g (x) = 0 -> Data '37' not found.

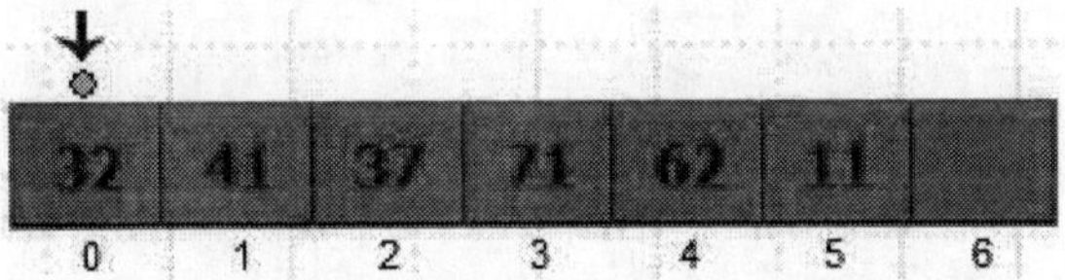

Figure 4. Value 37 in overflow table not found on location 0

Overflow-> g(x) = 1 -> Data '37' not found

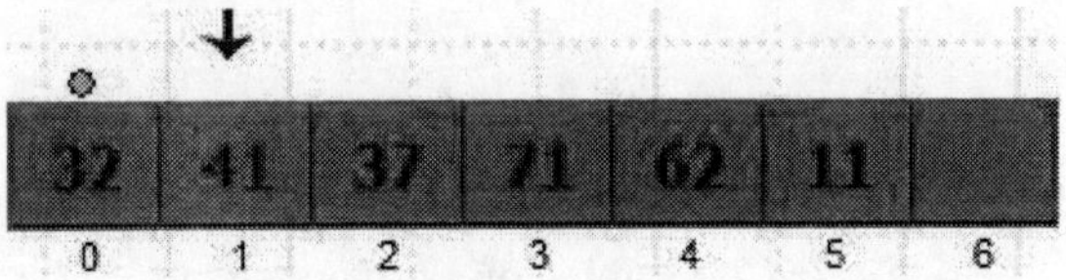

Figure 5. Value 37 in overflow table not found on location 1

Overflow-> g(x) = 1 -> Data '37' found on location 2

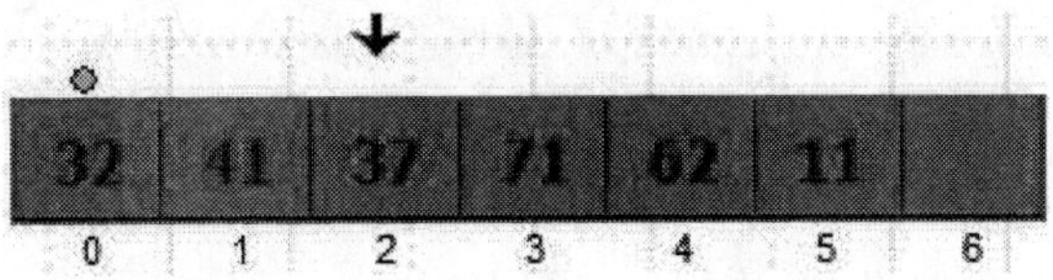

Figure 6. Value 37 in overflow table found on location 2

Experiments have been proved that collision could be bypassed and searching process with hash algorithm still can be done well, here is table test done with the amount of data which vary by using overflow technique and without technique of overflow, for data generated randomly.

Table 1. Searching Process Result

No	Amount of Data	Without Overflow		With Overflow	
		Collision	Search Time (ms)	Collision	Search Time (ms)
1	25	3	0.43	-	0.87
2	55	5	0.82	-	1.01
3	100	4	1.11	-	1.23
4	125	8	1.13	-	1.42
5	175	8	1.19	-	1.81
6	300	17	1.24	-	2.62
7	500	49	2.46	-	3.91
8	1000	67	5.76	-	8.17
9	3000	81	7.31	-	12.65
10	5000	262	8.67	-	17.32

Table 1 showed the experiment results searching process with overflow technique and without overflow technique and obtained in the search process using the overflow technique obtained longer results.

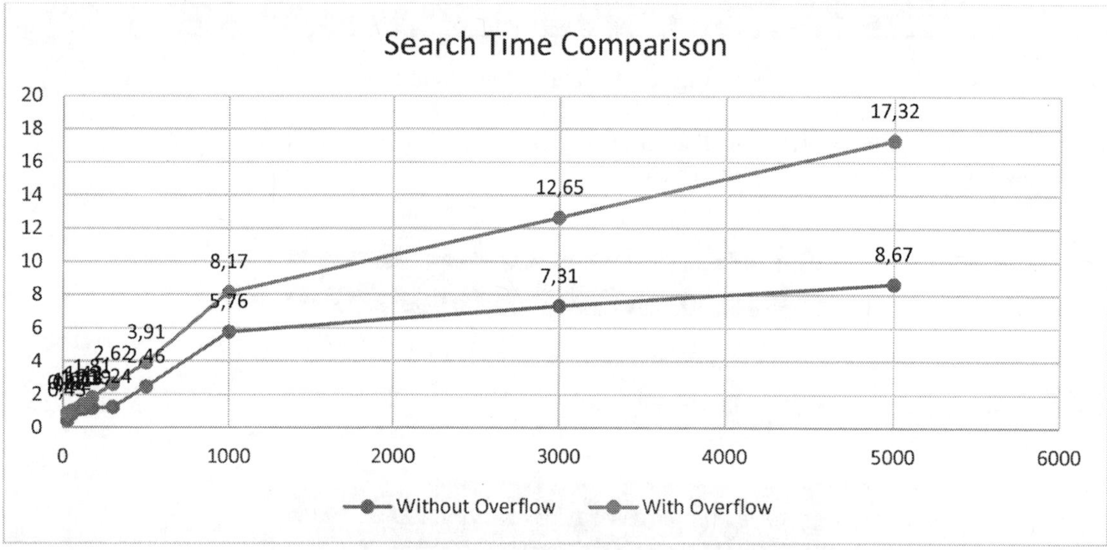

Figure 7. Graph Comparison Result

The figure above shows varying time ranges in the data search process, and the result is the greater amount of data, then the gap search time is also higher, and the searching process without overflow technique much faster, but for the prevention of collision data overflow technique successfully done with the amount of data vary.

4. Conclusion

Overflow technique used to avoid the collision on data allocation, the overflow technique is useful when the table is smaller than the main table, besides also with applied its technique of overflow search process with hash algorithm no different result only process search so long compared Without using overflow technique.

References

[1] K. K. Soni, R. Vyas, and A. Sinhal, "Importance of String Matching in Real World Problems," *International Journal Of Engineering And Computer Science,* vol. 3, no. 6, pp. 6371-6375, 2014.

[2] Y. M. K. Omar, H. Osama, and A. Badr, "Double Hashing Sort Algorithm," *Computing in Science & Engineering,* vol. 19, no. 2, pp. 63-69, 2017.

[3] L. J. Guibas and E. SZEMEREDI, "The Analysis of Double Hashing," *JOURNAL OF COMPUTER AND SYSTEM SCIENCES,* vol. 16, pp. 226-274, 1978.

[4] D. Jia, W. Xie, and Y. Jiang, "Research of database query technology based on hash function," in *IEEE 8th International Conference on Information Science and Digital Content Technology (ICIDT)*, Jeju, South Korea, 2012.

[5] K. K. Pandey and N. Pradhan, "A Comparison and Selection on Basic Type of Searching Algorithm in Data Structure," *International Journal of Computer Science and Mobile Computing,* vol. 3, no. 7, pp. 751-758, 2014.

[6] J. N, V. V, K. M and K. A, "Frequent Itemset Generation using Double Hashing Technique," in *International Co\nference on Modelling Optimization and Computing*, 2012.

[7] E. Lodi and F. Luccio, "SPLIT SEQUENCE HASH SEARCH," *Information Processing Letters,* vol. 20, pp. 131-136, 1985.

Analysis of Intergrade Variables In The Fuzzy C-Means And Improved Algorithm Cat Swarm Optimization(FCM-ISO) In Search Segmentation

Jepronel Saragih[1], Opim Salim Sitompul[2], and Zakaria Situmorang[3]

[1, 2, 3]Department of Computer Science, Faculty of Computer Science and Information Technology, Universitas Sumatera Utara, Medan, Indonesia, 20155

Email : jepoezipper@gmail.com1, opim@usu.ac.id2, zakarias65@yahoo.com3

Abstract. One of the techniques known in Data Mining namely clustering. Image segmentation process does not always represent the actual image which is caused by a combination of algorithms as long as it has not been able to obtain optimal cluster centers. In this research will search for the smallest error with the counting result of a Fuzzy C Means process optimized with Cat swam Algorithm Optimization that has been developed by adding the weight of the energy in the process of Tracing Mode.So with the parameter can be determined the most optimal cluster centers and most closely with the data will be made the cluster. Weigh inertia in this research, namely: (0.1), (0.2), (0.3), (0.4), (0.5), (0.6), (0.7), (0.8) and (0.9). Then compare the results of each variable values inersia (W) which is different and taken the smallest results. Of this weighting analysis process can acquire the right produce inertia variable cost function the smallest.

1. Introduction

The rapid development of technology makes it easier in data processing. Data that can be processed not only data in the form of text, but can be a multimedia component. In digital image processing, there is an important process that is often used is the process of segmentation. Given the importance of the segmentation process as an initial processor, a segmentation method that can perform accurate object separation [1] is required. Each pixel of a segmented result is usually unattainable that does not conform the actual image. This is due to several factors, namely when shooting is not good. So there are images that blur, lack of light and there is noise. Search for membership value on conform the actual imagesegmentation is notachieved because of the undersegmentation. In the first case, pixels be owned by to the same object are classified as be owned by different clusters. A single object can be represented by two or more clusters. Or in other cases the opposite happens, the pixels be owned by different clusters are appertainedasbe owned bythe same cluster [2]. For segmentation can be done by clustering, which is clustering part of the way in data mining that is used to enter data into the corresponding group with no deep knowledge of the group [3]Use of clustering method to explore different groupings of data, The purpose of this clustering is to optimize (the small fitness function of previous research) fitness function specified in the clustering process, which generally attempts to minimize the variation between clusters [4].Image-based clustering segmentation can be applied to any images as data set. to solve the problem of segmentation image, There are several clustering algorithms that can be used and Fuzzy C-Means one of the popular algorithms used in fuzzy clustering techniques.In his research [5] proposed a new algorithm in optimization techniques that mimic cat behavior named Cat Swarm Optimization (CSO). This algorithm has a number of advantages in solving optimization problems compared to previous techniques such as Particle-Swarm-Optimization (PSO) or PSO with Weighting Factor.

Initially this algorithm is still used to solve unconstrained minimization problem. The results of the model evaluation developed show that the model developed with CSO produces images with better quality than GA [7]. A study [8] implemented FCM algorithm for clustering with pixel-test data from digital images. The result is better than the K-Means algorithm, but in the experiment still yields a considerable cluster validity of 1.1. And in 2015 conducted research on ICSO [9] In this research the algorithm will yield index of average cluster validity which is small enough, but not optimal (small cluster validity index) according to them. Weighing the value of inertia (w) into the Cat Swarm Optimization method makes ICSO more effective for Clustering problems [10].

2. Literature Review

2.1. Algoritma Clustering (Clustering Algorithm)References

This method divides the data under clusters and data that has the same characteristics that are grouped into the same cluster and cluster data has different characteristics are congregateunder another group. Basically clustering is a method to search and classify data that have similarity between one data with other data. Clustering is one method of Data Mining which is unsupervised, meaning that this method is applied without any training (training) and without any teacher (teacher) and no need target output. Two kinds of grouping methods used in collecting data contained in Data Mining is namely Hierarchical Clustering and Non Hierarchical Clustering [14].

2.2. Clustering Based Image Segmentation

The Clustering method specifically used to compare the similarity of the value of a pixel to another. Methods in clustering based segmentation include iteration method, K-Means, Fuzzy C-means, Kohonen neural network, and various other clustering techniques. One excellent method used for image segmentation is Fuzzy C-Means clustering. Fuzzy C-Means is an improvised k-means algorithm with Fuzzy Set Theory by applying a measure of membership, in which one pixel image can be owned by multiple clusters [15].

2.3. Fuzzy C-Means Algorithm

Fuzzy C-Means (FCM) clustering is a fuzzy grouping model so that data can be a member of all clusters specified with different membership levels between 0 and 1. The probability of the existence of a data to a cluster is seen from its measure of membership or it can be said to be more likely to cluster closer to the value of the data. The basic concept of FCM, first is to establish the centre point ofcluster that will tag the around location for any cluster. In the initial conditions The centre point of the cluster is still not accurate. Each data has a membership measure for each cluster[16]. By repairing the cluster center and membership value of each data repeatedly, it can be found that the point of the cluster will go to the right location. Looping is done to minimize the objective function that describes the distance from the location of the data against the center point of a cluster is given weight by the levels of similarity data point from the set of escapes [7].

3. Research Methodology

The proposed algorithm in Figure 1 can be described in the steps below:

1) Determine the FCM-ICSOparameters Clustering the number of clusters, maximum iterations, minimum tolerance, seeking memory (SM) and seeking range (SR).The number of clusters is as much as 3 clusters.
2) Generating the initial solution of the cluster center randomly. Calculate the resultof Cost Function by using the Xie Beni Index equation.
3) Evaluate the cluster centre based on the value of the Cost Function.
4) Generatingthenewcentrepoint of the cluster based on theresultof the Cost Function is the minimum Cost Function value at the center of the previous cluster.
5) Update theresultof Cost Function.Stop Criteria if iteration has done Cost Cost Function 10 times iteration or cost function passes the minimum limit of 0.001.
6) Weighting Inertial Variables ([0.1], [0.2], [0.3], [0.4], [0.5], [0.6], [0.7], [0 , 8], [0.9]
7) Update Cluster Center with tracing mode.
8) Criteria to stop if iteration has been doing Cost Function search for 10 times iteration
9) The output of a system is the minimum Cost Function value which is the most optimum clustercenter.

4. Experimantal Results

4.1Fuzzy CMeans Performance

The result is all the V vectors that are the centre point of the cluster in the initial iteration. With m> 1 to give weighting to every measure of membership. Then, the Output generated is the value of belongingmatrix Urepresenting the membership function of the observed data.Up to a certain iteration limit or a difference of objective function that is lower than the specified error value. The hope is that every data goes in the right cluster ($| P_t - P_{t-1} | < \zeta$) or (t>maxIterasi)

4.2 Testing with Fuzzy C-Means

Testing process on both algorithm use same parameter, that is: smallest error expected $\xi = 0.001$, for weight = 2. Test image taken is horse.jpg, flower.jpg and boat.jpg. Then take a random sample of 4x4 pixels and the data set of each pixel using the pixel region image tool on Matlab R2012b as the data attribute that will be used as the data set. Furthermore, μ random and normalized random data are generated by dividing the μ value by the total μ to obtain the results according to ($| P_t - P_{t-1} | < \zeta$) or (t> maxIteration).

Table 1.The Centre point of Cluster

C	j1	j2	j3
1	150,8173073	136,970567	79,86408755
2	161,6204215	153,0618237	89,69781211
3	190,7245784	92,93978696	27,91236388

4.3 Inertial VariableWeighted Analysis on Tracing Mode Process

The optimal cluster search center search is performed with some parameters SM = 4, SR= 0.4, CD = {0,1}, Minimum tolerance = 0,0001, Number of Clusters = 3. The stopping criteria specified earlier is if the cost function≤ 0.001 or maximum 10 iteration is met. Inertial weights 0.8 produce the smallest cost functions. The explanation of the comparison of the weighting values of variables given in the FCM-ICSO algorithm can beexplainin Figure2.

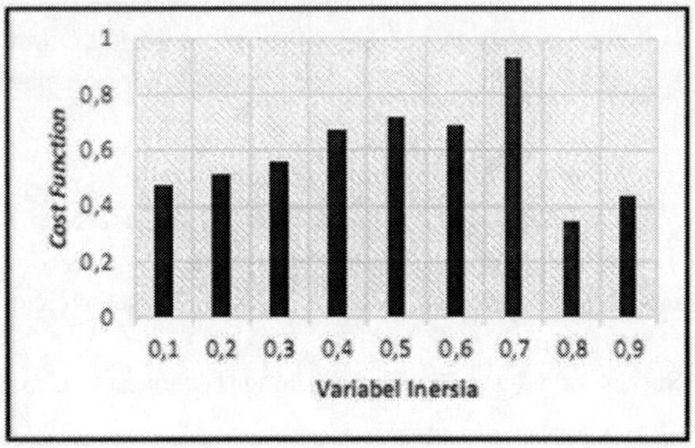

Figure 1. Cost Function comparison graph with different Inertia (W) weighting

Figure 1 shows the cost value generated at iteration 10, where iteration 10 is the maximum limit of iteration. So from the graph above can be seen that the smallest cost function on multiplication variable inertia (W) with value 0,8. Where the cost function is 0.339254076. The speed of displacement of cluster center in tracing mode is -64,27624303-55,98142603 and-27,38984305. Then each attribute test data is done update the center position of the cluster.

4.6 Assignment of inertia variables to produce smaller cost functions

By multiplying the 0.8 inertial variable in the tracing mode, the smallest error is found. But the researcher searches the lowest error point by narrowing the range of variables that will be multiplied. Because the lowest cost function graph is in the range of 0.7 to 0.8, the researcher narrows down the smallest cost function finding by determining the range of inertia variable values from 0.77 to 0.87.

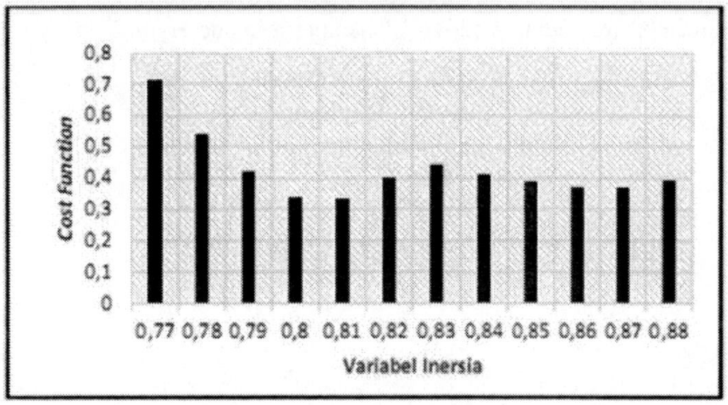

Figure 2. Cost Comparison Cost Function graph with more optimal Inersia (W) variable value

5. Conclusion

Determination of proper inertia weighting can result in a smaller Cost Function that is 0.334034524. In this study weighing 0.8 (w) on the process of tracing the Cat Swarm Optimization algorithm mode. Based on the graph of inertial multiplication results between 0.7 and 0.9 there is a chart down and rise again. Then tested again find the lowest cost function value with inertia weighting 0.77 to 0.88. There is a smaller cost value on the multiplication of inertia weight of 0.81 with cost function = 0.334034524.The result of the combinationof algorithm of FCM-ICSO was successful with both generate value cost smaller funtion so that more accurate in looking for members of the centre cluster.

6. References

[1] Gonzalez R C and Woods R E 2002. Digital Image Processing, 2nd. (New Jersey: Prentice Hall).
[2] Gonzales R C2001.Digital Image Processing. Prentice Hall: New Jersey.
[3] Santosa B and Mirsa K N 2009. ksacsoc: A coca. Scientific Research and Essays7(49), pp. 4176-4185.
[4] Berkhin, P. 2009. Survey of Clustering Data Mining Techniques. Grouping Multidimensional Data. pp 25-71.
[5] Chu S-C and Tsai P-W2007cibotboc.Ijoiciac31: 163-173.
[6] Liu Y andShenY2010.dcwcso.Jocit58.
[7] HalderA, PramanikandKarA2011Dynamic Image Segmentation using Fuzzy C-Means based Genetic Algorithm.J.Comp Appl286.
[8] Hadi I and Sabah M 2015Improvement Cat Swarm Optimization for Efficient Motion Estimation.IJhybinftech81 pp279-294.
[9] Kumal Y and Sahoo G 2015aicsoafc.CiidmVol1pp 187-197.
[10] Oliver AMunoz X, Batlle J, Pacheco L, and Freixenet J 2006icsoisusr.AQtrieeeic2:315–320.
[11] Chuang K-S, Tzeng H-L, Chen S, Wu J and Chen T-J 2006Fcmcwsiise.Cmig309–15.
[12] Oliveira J V D and Pedrycz W 2007Afc.Blcpd.
[13] Xie, X. L. & Beni, G. A. 1991. Validity Measure For Fuzzy Clustering. IEEE Trans Pattern Anal MachIntellVol. 13(8): 841-847
[14] Yang M S 1993 Asofc.Mathl. Comput. ModellingVol1811pp1-16.
[15] Shah, B., Shah, S., & Kosta. 2012. Novel Improved Fuzzy C-Means Algorithm For MR-Image Segmentation.International Journal of Soft Computing and Enggineering2(3):355-356.
[16] Herditomo., Sunaryo., & Naba, A. 2014. Penerapan Metode Hybrid Fuzzy C-Means dan PSO untuk segmentasi geografis. Jurnal Electrics Electronics Communication Controls Informatics Systems8(1): 27-32.

IOP Conf. Series: Journal of Physics: Conf. Series **930** (2017) 012014 doi:10.1088/1742-6596/930/1/012014

Application of fuzzy C-Means Algorithm for Determining Field of Interest in Information System Study STTH Medan

Edy Rahman Syahputra[1], Yulia Agustina Dalimunthe[2], Irvan[3]

[1,2] Department of Information System
Sekolah Tinggi Teknik Harapan, Medan-20217, Indonesia
[3]Department of Mathematics Education
Universitas Muhammadiyah Sumatera Utara, Medan Indonesia

[1]ydeaja@yahoo.com, [2]yuliaagustina_dalimunthe@yahoo.co.id, [3]irvan@umsu.ac.id

Abstract. Many students are confused in choosing their own field of specialization, ultimately choosing areas of specialization that are incompatible with a variety of reasons such as just following a friend or because of the area of interest of many choices without knowing whether they have Competencies in the chosen field of interest. This research aims to apply Clustering method with Fuzzy C-means algorithm to classify students in the chosen interest field. The Fuzzy C-Means algorithm is one of the easiest and often used algorithms in data grouping techniques because it makes efficient estimates and does not require many parameters. Several studies have led to the conclusion that the Fuzzy C-Means algorithm can be used to group data based on certain attributes. In this research will be used Fuzzy C-Means algorithm to classify student data based on the value of core subjects in the selection of specialization field. This study also tested the accuracy of the Fuzzy C-Means algorithm in the determination of interest area. The study was conducted on the STT-Harapan Medan Information System Study program, and the object of research is the value of all students of STT-Harapan Medan Information System Study Program 2012. From this research, it is expected to get the specialization field, according to the students' ability based on the prerequisite principal value.

1. Introduction

A change or transformation of a curriculum in the field of education continues to occur, it is intended to make a better change in the world of education in Indonesia, not least in the Information Systems Study Program at Sekolah Tinggi Teknik Harapan. In 2015 the Information Systems Study Program implements a new curriculum based on KKNI (Indonesia National Qualification Framework) which in the new curriculum requires students to choose an area of specialization that exists. There are 3 areas of specialization in the new curriculum are: (1) System Analysis, (2) Information Management, (3) Intelligent Information System. With the field of specialization that must be selected by students raises a problem of students, especially students that are in the Level III / student class of 2012. All grades that have been taken by students that are in level III have been converted into new curriculum so as to cause all the prerequisite courses of the field of specialization has been valuable, and because it is finally the Information System Study Program to make a policy by giving full rights to the students to Choose your own specialized field. From the results of the survey conducted by researchers in the field, many students that are confused must choose their own specialized area which, finally choosing a specialized area that is not appropriate with a variety of reasons such as just following a friend or because the specialized field that many choose without knowing if the self is competent in the field of selection that has been selected. Therefore, it needs recommendations of them, to recommend so many students is not easy because we have to manage the data so large that has the number of fields and the number of records that so much. Data mining technology is one of the tools for extracting data onto large databases and with complexity specifications that have been widely used in many application domains such as banking and telecommunications. In data mining there is a technique in managing large data that is clustering technique. Clustering is a grouping with unlike objects of other clusters. The principle of clustering is to maximize the homogeneity/similarity between traits within a group and to maximize the heterogeneity/inequality of traits between groups. In addition, this method is able to group data scattered irregularly. Several studies conducted research on clustering data onto various methods used, among others, research conducted by some of the following researchers and research determination of high school majors at C-Means Algorithm have higher accuracy (that is average 78,39%), if compared with method of determination majors manually which have been done only have the level of accuracy of average 56,17. Determining the ideal cluster by using variables in research that discusses the implementation of the single linkage method to determine agent performance on call centres based on asterisks for Java [1][2]. Therefore, the cluster technique with the fuzzy c-means algorithm can be used to classify the students based on the prerequisite subject value related to the specialization field so that it can give an idea of the students to choose the program of interest according to them.

2. Literature Review

2.1. Clustering Concept in Data Mining

The basic concept of data mining is to find hidden information about a database and to be part of Knowledge Discovery in Database (KDD) to find useful information and patterns in the data [3][4]. Clustering divides data onto groups or clusters based on a similarity in attributes between the data [5][6]. Characteristic of each cluster is not Determined before, but reflected by the similarity of data grouped in it. Therefore, clustering results often needs to be interpreted by the parties that really understand about the character of the data domain. In addition to being used as an independent method of data mining, clustering is also used in pre-processing of data before data is processed by other data mining methods to improve the data domain. The most important characteristic of a good clustering result is that a data instance in a cluster is more "similar" to another instance within the clustering than with an instance outside of the clustering. Similarity measures can vary and affect the calculation in determining the member of a cluster. So the data type to be clustered (quantitative or qualitative) also to determine what size is appropriately used in an algorithm. In addition to the similarity between data onto a cluster, clustering can also be done based on the distance between data or cluster with one another. Distance size (distance or dissimilarity measure) which is the opposite of this similarity size is also a lot of variety and its usage also depends on the type of data to be clustered. Both sizes are symmetrical, where if A is said to be similar to B it can be concluded that B is similar to A. There are several kinds of distance calculation formulas for clusters. For the numerical data type, a data X of X_1, $i = 1, ..., n$, each item is represented as a vector $X_1 = \{X_{i1}, X_{i2}, X_{im}\}$ with m as the number of dimensions of the item. The usual formulas used as a measure of the distance between X_i and X_j for this numerical data include.

2.2 C-Means Algorithm

In the classical clustering process (e.g. Clustering K-Means), the formation of partitions is done in such a way that each object is right on one partition, since the object is actually located between two or more other partitions. In algorithmic logic, a method that can be used to perform grouping of data is known as clustering algorithm. A clustering algorithm is said to be a clustering algorithm if the algorithm uses strategic parameters of adaptation soft competitive. Most clustering algorithms are based on the optimization of the objective function or modification of the objective function [13]. One of clustering algorithm technique is C-Means Algorithm. C-Means algorithm is a data clustering technique that the existence of each data onto a cluster is determined by the value / degree of certain membership. This technique was first introduced by Jim Bezdek in 1981 [14]. In contrast to the classical clustering technique (where an object will only be a member of several clusters. The cluster boundaries of the C-Means algorithm are soft (soft) The basic concept of C-Means algorithm, the first is to determine the cluster center that marks the flat location for each cluster of the initial conditions, the center of the cluster is still not accurate. Each data has degree of membership of each cluster by repairing the cluster center and membership value of each data repeatedly, it will be seen that the center of the cluster will move toward the right location. This recurrence is based on the minimization of objective function. The objective function used for C-Means Algorithm are [15][16].

1. The C-Means algorithm is given as follows [17][18]: 1. Determining the data to be clustered X, in the form of nxm-sized matrix (n = number of data samples, m = attribute of each data), X_{ij} = i-sample data (i = 1,2, ..., n), attributes To-j (j = 1,2, ..., mm).
2. Determine:
   - Number of clusters = c
   - Rank = w - Maximum interaction = MaxIter
   - The smallest error expected = ξ
   - The initial objective function = Po = 0
   - Initial interaction = t = 1
3. Generating random numbers μ_{ik} i = 1,2,3, ..., n: k = 1,2,3, .., c: as elements of the initial partition matrix U. Counting the number of each column:$Q_i = \sum_k^c \mu_{ik}$
   with j=1,2,..,n
4. Calculate the centre of the k-cluster: V_{kj}, with k = 1,2, ... c: and j = 1,2, m ..
5. Calculate the objective function of the interaction to-t:
6. Calculate the partition matrix changes:
7. By: i = 1,2, ..., n: and k = 1,2 „ ..., c. Checking the stop condition: - If: ($|P_t - P_{t-1}| < \xi$) or (t > Max) then stop - If not: t = t + 1, repeat step 4

3. Research Methodology

1. Identification of Problems: a preliminary study to examine the problems in the process of grouping students based on prerequisite courses in the selection process of specialization.
2. Approach in problem solving: is the stage of finding the right method in solving problems based on literature review. The Fuzzy C-Means Clustering algorithm that has been tested for several researches for the case of data grouping will be tested for classifying the preliminary subject value data onto the determination of the student interest area.
3. The tool used in modeling the sequence of activities in the using the activity diagram and for implementation system performed using software visual studio 2010.
4. Implementation and testing for the system and results analysis: is the stage of the process of testing the results of the cluster fuzzy c-means method in classifying the field of student interest based on prerequisite course value, with data training in the form of prestigious student grade 2012.
5. Drawing the conclusions of the research results: after testing the system is completed to eat a result is drawn in the form of conclusions based on the results.

4. Implementation and Testing

The data used is the data value of students of the curriculum conversion results from 2011-2015 STT-Harapan Medan Information Systems Study Program as many as 11 classes. Parameter of data to be used in research in the form of the average of the subjects of the interest groups, namely: Specialization of Information Systems Analysis (Data Systems and Information Technology, Database, Programming Language, Business Process Analysis, Analysis and Design of Information Systems and Management of Technology Services Information, Information Management, Database,

Programming Languages, Probabilistics and Statistics, Object Oriented Programming and Business Modeling and Simulation System) and Intelligent Information System (Data Systems and Information Technology, Database, Programming Language, Operating System, Design and Management of Computer Network and Intelligent System). The sample data that used in this research is Data onto Student Average Value of Special Interest Field Before Specialization and Data of student average value of Special Interest Field After Specialization.

Measurement Method

Accuracy of application of C-Means Algorithm in the specialization of its majors is the sample data onto the students and the average value of the subjects of interest before the interest grouped with the C-Means Algorithm to divide the students into the field of interest in accordance with the similarity in the average value of the field of interest. Then the outcome of the C-Means Algorithm is compared with the result of the interest in the sample data onto the average values of the subjects on the specialization that has been undertaken by the students. If the interest chosen by the student is equal to the concentration of C-Means Algorithm and the average value of the subjects acquired after specialization > = Minimum Criteria of Completeness (KKM) > (> = 75), the C-Means Algorithm is declared ACCURATE. If the interest chosen by the student is equal to the concentration of C-Means Algorithm and the average value of the specialization subject obtained after the <Minimum Exhaustiveness Criterion (> = 75), the C-Means Algorithm is NOT ACCURATE. If the interest chosen by the student is not the same as the C-Means Algorithm and the average course value obtained after the minimum > Minimum Criterion of Maximum Criterion, the C-Means Algorithm is NOT ACCURATE. If the interest obtained by the student is not the same as the C-Means Algorithm and the average value of the specialization subject obtained after the <Minimum Exhaustiveness Criterion (KKM) criterion, the C-Means Algorithm is declared ACCURATE.

Next calculated percent accuracy level Fuzzy C-Means with: % Accuracy = (Accurate Total Data / Total Samples) * 100.

After the parameter of the average field value of interest is known, then mapping / clustering the data follows the FCM algorithm with setting the initial partitions matrix in the form of n x m matrix (n is the number of data samples, ie = 81, and m is the parameter / attribute of each data, ie = 3). X_{ij} = i-th samples data (i = 1,2, ..., n), j-attribute (j = 1,2, ..., m). Then determining Parameter Value Beginning. And followed by generating random numbers μik, i = 1,2, ..., n; K = 1,2, ..., c; As the initial partition matrix element (U) formed by using Microsoft Visual Studio

```
.>>rand('state',0)
>>X=[rand(7,3);rand(7,3);rand(7,3)]
X=
```

0.9501	0.0185	0.1763	0.3529	0.2722	0.4186	0.3795	0.1934	0.86
0.2311	0.8214	0.4057	0.8132	0.1988	0.8462	0.8318	0.6822	0.8537
0.6068	0.4447	0.9355	0.0099	0.0153	0.5252	0.5028	0.3028	0.5936
0.486	0.6154	0.9169	0.1389	0.7468	0.2026	0.7095	0.5417	0.4966
0.8913	0.7919	0.4103	0.2028	0.4451	0.6721	0.4289	0.1509	0.8998
0.7621	0.9218	0.8936	0.1987	0.9318	0.8381	0.3046	0.6979	0.8216
0.4565	0.7382	0.0579	0.6038	0.466	0.0196			
			0.6813	0.1897	0.3784			

Next step is define Cluster Centre (V) In the first iteration, using the equation:

$$V = \begin{pmatrix} 7.25 & 7.66 & 5.14 \\ 7.88 & 7.79 & 6.53 \\ 8.15 & 8.08 & 6.84 \end{pmatrix}$$

And then calculating the Objective Function (P)

9.056841	0.005599	0.62408	9.68652
6.095098	76.9999	18.78412	101.8791
28.559	15.33861	67.87962	111.7772
15.47084	24.80597	55.06622	95.34303
43.0474	33.98129	9.122254	86.15094
138.9192	203.2413	190.9962	533.1567
26.91444	70.38042	0.432973	97.72783
13.84711	8.238198	19.48294	41.56825
127.5471	7.622698	138.109	273.2788
0.026695	0.06376	75.13057	75.22102
5.002971	144.6212	10.64394	160.2681

9.112387	43.89464	100.0839	153.0909
9.877825	217.2256	175.7346	402.838
94.85771	56.50128	0.099954	151.4589
137.3072	10.64514	42.35637	190.3087
38.67844	10.04519	198.6288	247.3525
178.7242	120.2178	188.2591	487.2011
65.87224	23.89041	91.81205	181.5747
44.36476	25.86134	21.73436	91.96046
36.952	4.574086	162.6364	204.1625
1030.233	1098.154	1567.618	3696.005

The value of the objective function in the first iteration (P_1) has a value of: Objective Function = 3696.005. 1. Checking Stop Conditions Because | P_1-P_0 | = | 3696.005-0 | = 3696.005 >> ξ (10^{-5}), and iteration = 1 <MaxIter (= 100), then proceed to second iteration (t = 2). 2. Calculate the partition matrix changes with: i = 1, 2, ..., n: and k = 1, 2, ..., c

Checking the stop condition:
At: (| P_t - P_{t-1} | < ξ) or (t > Max) then stop in the 48th iteration with the cluster centre value

$$V = \begin{pmatrix} 7.57 & 7.55 & 7.49 \\ 7.45 & 7.54 & 7.46 \\ 7.56 & 7.53 & 7.54 \end{pmatrix}$$

5. Conclusion

Clustering Method of Fuzzy C-Means Algorithm can classify students in the field of specialization according to the prerequisite course by utilizing UI output as determination of each student into which groups and utilize the output of Vi to determine each group is identified as specification. The Fuzzy C-Means algorithm uses repetitive iterations to specify the grouping. The aeration is repeated to the nearest optimum objective function, obtained at point (|P_t – P_{t-1}| < ξ) or (t > Max) Then it stops at 48 iterations. This research builds an application for the application of special c-means algorithm in information system studies program. Hopefully in the next research can be built an application for determinants of interest in all majors at STT-Harapan Medan. For clustering just focusing on areas of interest expected in the future can be developed for the determinant of the student majors at the final grades in school at the top level or also the value of the UN prospective students.

References

[1] Anjana Gosain, Sonika Dahiya, Performance Analysis of Various Fuzzy Clustering Algorithms: A Review, Procedia Computer Science, Volume 79, 2016
[2] Wu, Xindong, Vipin Kumar, J. Ross Quinlan, Joydeep Ghosh, Qiang Yang, Hiroshi Motoda, Geoffrey J. McLachlan et al. "Top 10 algorithms in data mining." Knowledge and Information Systems 14, no. 1 (2008).
[3] Kaur, Prabhjot, and Anjana Gosain. "Improving the performance of fuzzy clustering algorithms through outlier identification." In Fuzzy Systems, 2009. FUZZ-IEEE 2009. IEEE International Conference on, pp. 373-378. IEEE, 2009.
[4] Baoju Zhang, Shan Qin, Wei Wang, Dan Wang, Lei Xue, Data stream clustering based on Fuzzy C-Mean algorithm and entropy theory, Signal Processing, Volume 126, 2016
[5] Zhang BO, Fuzzy clustering based on the data flow model, Comput. Eng. Apl. (2010)

[6] Yinghua Shen, Witold Pedrycz, Collaborative fuzzy clustering algorithm: Some refinements, International Journal of Approximate Reasoning, Volume 86, 2017
[7] M. Zarinbal, M. H. Fazel Zarandi, I. B. Turksen, Relative entropy collaborative fuzzy clustering method, Pattern Recognit. 48 (2015) 933-940.
[8] Y. Jiang, F. Chung, S. Wang, Z. Deng, J. Wang, P. Qian, Collaborative fuzzy clustering from multiple weighted views, IEEE Trans. Cybern. 45 (2015)
[9] P. Rastin, G. Cabanes, N. Grozavu, Y. Bennani, Collaborative clustering: How to select the optimal collaborators? in: Proceedings of the 2015 IEEE Symposium Series on Computational Intelligence, 2016, pp. 787-794.
[10] C. T. Chu, H. K. Chiang and Y. S. Chang, "Clustering algorithms applied Gaussian basis function neural network compensator with fuzzy control for magnetic bearing system," 2016 5th International Symposium on Next-Generation Electronics (ISNE), Hsinchu, 2016, pp. 1-2.
[11] J. I. Sanchez-Gomez, L. Morales-Velazquez, R. A. Osornio-Rios and E. Guillen-Garcia, "Fuzzy c-means clustering for steady state events classification of electrical signals," *2016 12th Congreso Internacional de Ingenier¿¿a (CONIIN)*, Santiago de Queretaro, 2016, pp. 1-5.
[12] José Angel BARRIOS, César VILLANUEVA, Alberto CAVAZOS, Rafael COLAS, Fuzzy C-means Rule Generation for Fuzzy Entry Temperature Prediction in a Hot Strip Mill, Journal of Iron and Steel Research, International, Volume 23, Issue 2, 2016
[13] Engin Esme, Bekir Karlik, Fuzzy c-means based support vector machines classifier for perfume recognition, Applied Soft Computing, Volume 46, 2016
[14] Hanuman Verma, R.K. Agrawal, Aditi Sharan, An improved intuitionistic fuzzy c-means clustering algorithm incorporating local information for brain image segmentation, Applied Soft Computing, Volume 46, 2016
[15] Hassan Fathabadi, Power distribution network reconfiguration for power loss minimization using novel dynamic fuzzy c-means (dFCM) clustering based ANN approach, International Journal of Electrical Power & Energy Systems, Volume 78, 2016
[16] Jeetashree Aparajeeta, Pradipta Kumar Nanda, Niva Das, Modified possibilistic fuzzy C-means algorithms for segmentation of magnetic resonance image, Applied Soft Computing, Volume 41, 2016
[17] J. Aparajeeta, P.K. Nanda, N. Das, Bias field estimation and segmentation of MR image using modified fuzzy-C means algorithms, in: Proceedings of Eighth International Conference on Advances in Pattern Recognition (ICAPR), IEEE, 2015, pp. 1–6.
[18] Yi Ding, Xian Fu, Kernel-based fuzzy c-means clustering algorithm based on genetic algorithm, Neurocomputing, Volume 188, 2016.

Research of Simple Multi-Attribute Rating Technique for Decision Support

Dodi Siregar[1], Diki Arisandi[2], Ari Usman[1], Dedy Irwan[3] and Robbi Rahim[4*]

[1]Department of Computer Engineering, Sekolah Tinggi Teknik Harapan, Jln. H.M. Joni No. 70 C Medan, Medan, Indonesia
[2]Department of Computer Engineering, Universitas Abdurrab, Jl. Riau Ujung No. 73, Pekanbaru 28291, Indonesia
[3]Department of Information System, Sekolah Tinggi Teknik Harapan, Jln. H.M. Joni No. 70 C Medan, Medan, Indonesia
[4]School of Computer and Communication Engineering, Universiti Malaysia Perlis, Malaysia

*usurobbi85@zoho.com

Abstract. One of the roles of decision support system is that it can assist the decision maker in obtaining the appropriate alternative with the desired criteria, one of the methods that could apply for the decision maker is SMART method with multicriteria decision making. This multi-criteria decision-making theory has meaning where every alternative has criteria and has value and weight, and the author uses this approach to facilitate decision making with a compelling case. The problems discussed in this paper are classified into problems of a variety Multiobjective (multiple goals to be accomplished) and multicriteria (many of the decisive criteria in reaching such decisions).

1. Introduction

Decision Support Systems is represented as a system capable of providing the ability to solve problems and communications capabilities for semi-structured problems. In particular [1] [2] [3] [4], Decision Support Systems are defined as a system that supports the work of a leader or decision maker in solving semi-structured problems by providing information or suggestions to a particular decision [4] [5] [6].

Decision support systems utilize private resources in a manner with computer skills to improve decision results, so this is a computer-based support system for decision-making that deals with semi-issues structured [1] [2]. The unstructured problem contains relationships between elements that are not understood by the troubleshooter. While the semi-structured problem is an issue that includes some elements recognized by the problem solvers [1]. Decision-making is always correlated with the uncertainty of the results of decisions taken, to reduce this risk factor, the decision requires valid information about the conditions that have been, and may occur, then processed the information into several alternative problems solving as a material consideration in deciding the steps to be implemented, so that the decision taken is expected to provide maximum benefits [4].

The decision support system that will introduce in this research use SMART method which aims to gather information about all data related to multiple attributes and multiple-criteria [7]. The data in the parameters is the data of the past and the future, from the data, will be generated classification and the relationship between data one and other data so that the final result will get the best results solution,

IOP Conf. Series: Journal of Physics: Conf. Series **930** (2017) 012015 doi:10.1088/1742-6596/930/1/012015

choosing SMART (Simple Multi Attribute Rating Technique) method because this method can make multi-decision decisions, so it is expected to create an accurate decision-making system and can solve the problem of decision making the best [7] [8] [9] [10].

In previous research conducted Risawandi [1], SMART method completion process is not elaborated in detail and there are no results that can explain how the results of using SMART method with alternatives and dynamic criteria, in this study the author tries to display the calculation of SMART procedure in detail toward alternatives based on criteria and weights; and dynamically differentiate process time with alternatives and criteria.

2. Methodology

SMART (Simple Multi Attribute Rating Technique) is a multi-criteria decision method developed by Edward in 1977 [1] [7]. This multi-criteria decision-making technique is proposed on the theory that each alternative consists of some criteria that have values and each criterion have weights that describe how important compared to other criteria. This weighting is used to assess each alternative to obtain the best choice. SMART is a linear additive model to predict the value of each option. SMART Method is more useful because of its simplicity in responding to the needs of decision makers and how it responds. The analysis involved is transparent so that this approach provides a great understanding of the problem and is acceptable to the decision maker [1] [10]. The multi-linear function model used by SMART is as follows:

$$\text{Maximize} \sum_{j=1}^{k} w_j \cdot u_{ij}$$

Explanation:

1. W_j is the criteria weighted value to j of k criteria.
2. U_{ij} is an alternative utility value i on criterion j.
3. The decision selection is to identify which of the n alternatives have the greatest practical value.

The value of this function can also be used to rank alternatives.

3. Result and Discussion

Below is a test of decision support system using the SMART method, the case used with multi-criteria and multi-attribute concept so that it can be performed in different cases.

1. Known weights of assessment as follows:

Table 1. Selection of Weights

No	Criteria	Weight
1	C1	100
2	C2	80
3	C3	50
	j	230

2. From the above weights are made into relative weights, as follows:

Table 2. Determination of Relative Weights

No	Criteria	Weight	Relative Weights
1	C1	100/230	0.434
2	C2	80/230	0.347
3	C3	50/230	0.217

3. Alternative and weight value

Table 3. Alternative and weight value

Alternative	Criteria		
	C1	C2	C3
A1	77	79	78
A2	80	85	80
A3	85	85	74
A4	77	85	88
A5	77	88	75
A6	72	89	90
A7	72	70	89
A8	90	90	88
A9	74	90	78
A10	77	89	89

4. The next stage is to determine the alternative value based on each criterion, in this case, it is estimated the alternative value on the scale between 0-100, for C1 value 77 as the minimum value and 90 as the maximum value, as example below:
   a. Weight C1 for A1

$$\left(\frac{\text{Maximum C1 - A1 Value}}{\text{Max - Min}}\right)$$

Weight the value of C1.A1 $= \left(\frac{90 - 77}{90 - 77}\right) = 100$

Do this process until C1 for A4 and every procedure get the result from the function. Next is to count each weight for criterion C2, from table the maximum value is 90 and minimum 80

   a. Weight C2 for A1

$$\left(\frac{\text{Maximum C2 - A1 Value}}{\text{Max - Min}}\right)$$

Weight the value of C2.A1 $= \left(\frac{90 - 85}{90 - 80}\right) = 50$

Do this process until C1 for A4 and every procedure get the result from the function. Next is to count each weight for criterion C3, from table the maximum value is 90 and minimum 74

   a. Weight C3 for A1

$$\left(\frac{\text{Maximum C3 - A1 Value}}{\text{Max - Min}}\right)$$

Weight the value of C3.A1 $= \left(\frac{90 - 78}{90 - 74}\right) = 75$

Based on the above process obtained value evaluation factor as follows:

Table 4. Evaluation Factor

Criteria	Alternative			
	A1	A2	A3	A4
C1	100	62.5	38.4	100
C2	55	100	50	50
C3	75	93	100	31,25

IOP Conf. Series: Journal of Physics: Conf. Series **930** (2017) 012015 doi:10.1088/1742-6596/930/1/012015

In the next stage determined the total evaluation value for each alternative, the total value of its assessment as follows:

Table 5. Evaluation Factor for A1

Criteria	Evaluation Factor A1	Factor Weight	Weight Evaluation
C1	100	0,434	43,4
C2	55	0,347	19,085
C3	75	0,217	16,275
Total	234	0,998	78,76

Table 6. Evaluation Factor for A2

Criteria	Evaluation Factor A2	Factor Weight	Weight Evaluation
C1	62,5	0,434	27,25
C2	100	0,347	34,7
C3	93	0,217	20,181
Total	230	0,998	82,006

Table 7. Evaluation Factor for A3

Criteria	Evaluation Factor A3	Factor Weight	Weight Evaluation
C1	38,4	0,434	16,66
C2	50	0,347	17,35
C3	100	0,217	21,7
Total	188,4	0,998	55,71

Table 8.Evaluation Factor for A4

Criteria	Evaluation Factor A4	Factor Weight	Weight Evaluation
C1	100	0,434	43,4
C2	50	0,347	17,35
C3	31,25	0,217	6,78
Total	218	0,998	67,53

Based on the above table, the highest total weight value is an A3 alternative; and the following is a table of calculation results using the SMART method:

Table 9. Evaluation Factor for A4

Criteria	A1	A2	A3	A4
C1	100	62,5	38,4	100
W_{C1}	**0,434**	**0,434**	**0,434**	**0,434**
C2	55	100	50	50
W_{C2}	**0,347**	**0,347**	**0,347**	**0,347**
C3	75	93	100	31,25
W_{C3}	**0,217**	**0,217**	**0,217**	**0,217**
Total Value	**78,76**	**75,06**	**55.71**	**67,53**

The results of the experiment based the number of alternatives to the number of predefined criteria (3 criteria) get a processing time of not more than 20 seconds, where the addition of the number of alternatives does not increase the processing time significantly, with 50 alternatives and 3 criteria only takes 12 seconds, The next process tests 50 alternatives to the number of criteria at random, see table below:

Table 10. Testing with many criteria

Alternative	50	50	50	50	50
Criteria	3	6	9	12	15
Time(Second)	13	40	65	90	197

From the value of the table above can be seen in the graph below:

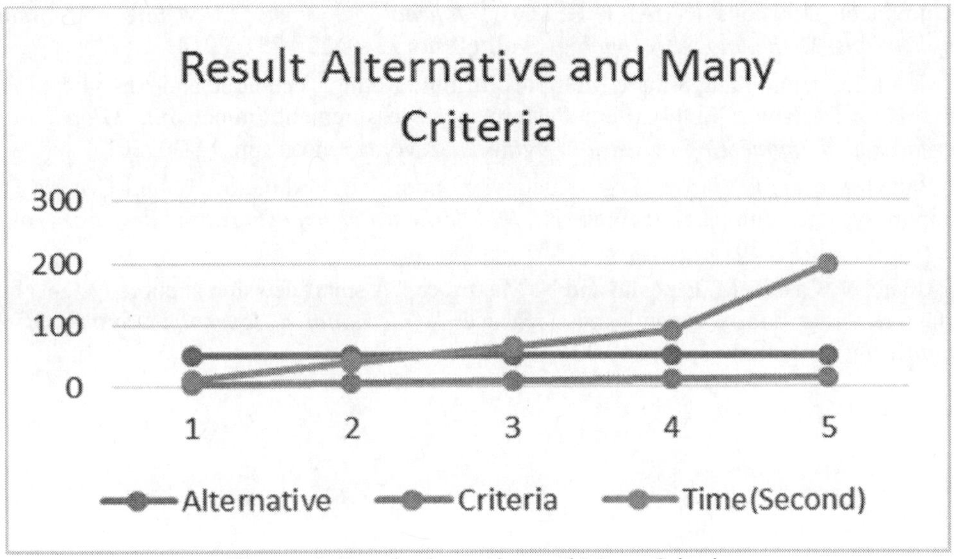

Figure 1. Result alternative and Many Criteria

Testing 50 alternatives with the varying number of criteria get a significant processing time and take a long time.

4. Conclusion

Based on the results of tests conducted with the number of dynamic alternatives and use three criteria then the process of calculation using the SMART method does not require a long time, but different if the alternatively added dynamically with the number of constant alternatives then the process of calculating the SMART method will require process time which is longer.

References

[1] Risawandi and R. Rahim, "Study of the Simple Multi-Attribute Rating Technique For Decision Support," *International Journal of Scientific Research in Science and Technology (IJSRST),* vol. 2, no. 6, pp. 491-494, 2016.

[2] Syamsudin and R. Rahim, "Study Approach Technique for Order of Preference by Similarity to Ideal Solution (TOPSIS)," *International Journal of Recent Trends in Engineering & Research,* vol. 3, no. 3, pp. 268-285, 2017.

[3] Nurmalini and R. Rahim, "Study Approach of Simple Additive Weighting For Decision Support System," *International Journal of Scientific Research in Science and Technology (IJSRST),* vol. 3, no. 3, pp. 541-544, 2017.

[4] Mesran, G. Ginting, R. Rahim and Suginam, "Implementation of Elimination and Choice Expressing Reality (ELECTRE) Method in Selecting the Best Lecturer (Case Study STMIK BUDI DARMA)," *International Journal of Engineering Research & Technology (IJERT),* vol. 6, no. 2, pp. 141-144, 2017.

[5] R. Rahim, Mesran, A. P. U. Siahaan and S. Aryza, "COMPOSITE PERFORMANCE INDEX FOR STUDENT ADMISSION," *International Journal of Research In Science & Engineering,* vol. 3, no. 3, pp. 68-74, 2017.

[6] Jasri, D. Siregar and R. Rahim, "Decision Support System Best Employee Assessments with Technique for Order of Preference by Similarity to Ideal Solution," *INTERNATIONAL JOURNAL OF RECENT TRENDS IN ENGINEERING & RESEARCH,* vol. 3, no. 3, pp. 6-17, 2017.

[7] J. M. Taylor and B. Love, "Simple Multi-Attribute Rating Technique for Renewable Energy Deployment Decisions (SMART REDD)," *Journal of Defense Modeling and Simulation: Applications, Methodology, Technology,* vol. 11, no. 3, pp. 227-232, 2014.

[8] F. M. Kasie, "Combining Simple Multiple Attribute Rating Technique and Analytical Hierarchy Process for Designing Multi-Criteria Performance Measurement Framework," *Global Journal of Researches in Engineering Industrial Engineering,* vol. 13, no. 1, pp. 14-30, 2013.

[9] U. Baizyldayeva, O. Vlasov, A. A. Kuandykov and T. B. Akhmetov, "Multi-Criteria Decision Support Systems. Comparative Analysis," *Middle-East Journal of Scientific Research,* vol. 16, no. 12, pp. 1725-1730, 2013.

[10] F. Amato, V. Casola, M. Esposito and N. Mazzocca, "A smart decision support systems based on a fast classifier and a semantic post reasoner," *International Journal of System of Systems Engineering,* vol. 4, no. 3, pp. 317-336, 2013.

IOP Conf. Series: Journal of Physics: Conf. Series **930** (2017) 012016 doi:10.1088/1742-6596/930/1/012016

K-Means Algorithm Performance Analysis With Determining The Value Of Starting Centroid With Random And KD-Tree Method

Kamson Sirait[1], Tulus[2], Erna Budhiarti Nababan[3]

[1,2,3]Department of Computer Science, Faculty of Computer Science and Information Technology, Universitas Sumatera Utara, Medan, Indonesia, 20155

Email: kamsonsirait@gmail.com[1], tulus_jp@yahoo.com[2], ernastp@yahoo.com[3]

Abstract. Clustering methods that have high accuracy and time efficiency are necessary for the filtering process. One method that has been known and applied in clustering is K-Means Clustering. In its application, the determination of the begining value of the cluster center greatly affects the results of the K-Means algorithm. This research discusses the results of K-Means Clustering with starting centroid determination with a random and KD-Tree method. The initial determination of random centroid on the data set of 1000 student academic data to classify the potentially dropout has a sse value of 952972 for the quality variable and 232.48 for the GPA, whereas the initial centroid determination by KD-Tree has a sse value of 504302 for the quality variable and 214,37 for the GPA variable. The smaller sse values indicate that the result of K-Means Clustering with initial KD-Tree centroid selection have better accuracy than K-Means Clustering method with random initial centorid selection.
Keyword: *Clustering , K-Means, KD-Tree.*

1. Introduction

One of the known and applied clustering methods is K-Means Clustering. Clustering is one of the data mining techniques used to obtain groups of objects that have common characteristics in large enough data. K-Means Clustering algorithm can be used to predict student academic performance [1]. In terms of time and accuracy, K-Means produces better results [2]. K-Means grouping is a partitioning grouping technique in which groups are formed with the base of the centroid [3].

K-Means is a known grouping technique based on partitions that try to find centroid that represent the number of clusters. But the output is enough sensitive to the beginning of the cluster center position. Finding the starting point of centroid can make the algorithm more effective and efficient so that the results obtained are more accurate and the time required to process the data is substantially reduced [4]. The K-Means algorithm always selects a different starting center for each run of algorithm for the same dataset and k value, then the algorithm is optimized by modifying it with the initial center selected, resulting in fewer execution times for large datasets [5].

Selection of the starting position from the middle point of a bad cluster will result in the algorithm of K-Means stuck in an optimal local solution [6]. Therefore, the determination of the starting point value of the clustering center will greatly determine the results obtained by the K-Means Clustering algorithm [7]. To determine the correlation between two objects by using the following Euclidean Distance formula:

$$d_{Euclidean}(x,y) = \sqrt{\sum_{i=1}^{n}(x_i - y_i)^2} \quad ....(1)$$

Dimana :

$d(x,y)$ = The distance of data to x to the center of the cluster y.

x_i = data i in n data.

y_j = data j in n data.

Clustering using K-Means with initial centroids is determined randomly.

a. Determine the value of K desired, where the value of K is the number of clusters to be formed.
b. Determine the initial centroids. The initial centroid is randomly assigned from the existing data and the number of cluster is equal to the number of initial centroid.
c. Find the nearest centroid of each data by calculating the distance to each centroid using the Euclidean Distance formula.
d. Grouping data by the minimum distance. A data will be a part of a cluster that has the closest (smallest) distance from its cluster center.
e. Based on these groupings, then look for new centroids based on the average of the data of each cluster.
f. Go back to stage 3.
g. The recurrence stops if no data is migrated.

KD-Tree is a data tree structure that which each node can accommodate many data sets and is used as a tool to accommodate the evaluation data elements, by devide the the data set into two subsets in which each evaluation stage. Each subset is stored in a node tree to form a left and right subtree. The same evaluation on both subtrees is done as long as the node is found with the number of members more than the specified maximum limit. If each node is assumed as "h" and many desired maximum data elements for each node are "w", then each leaf node must meet the following conditions:

$$1 \leq h \leq w \quad \text{...(2)}$$

Simply the properties of the k-d tree data structure can be described as follows [8]:

a. KD-tree is a binary tree.
b. The root node holds all elements of the data set.
c. The data elements at each node are divided into two parts based on a particular variant, such that one part becomes the left subtree elements and one part becomes the right subtree elements.
d. The subtrees on the left and right will be recursively reconstructed, as long as each node "h" does not meet $1 \leq h \leq w$.

KD-Tree K-Means Clustering is clustering method that the initial centroid determined by KD-Tree.

a. The establishment of KD-Tree from the data set. The KD-Tree created has a leaf bucket.\
b. For each leaf bucket (L1, L2, ... Lj), calculate the density value (Pj) of each leaf bucket Lj, and calculate the middle point value of leaf bucket (Mj) by finding the average value of all points on the leaf bucket Lj .
c. Select the middle point of the first cluster C1 = Mz, where z = argmax (Pj).
d. For t = 2, ..., K:
   For j = 1, ..., q calculate the leaf bucket ranking (Gj).
e. Ct = Mz, where z = arg max (Gj)
f. Remove 20% leaf bucket with the lowest density value. Go back to step 3 and calculate the position of the new middle point of the cluster K (c1, c2, ... ck)
g. Run the K-Means Clustering algorithm with the initialization of the middle point (C1, ... Ck).

2. Computational Analysis

2.1 *Clustering Using K-Means With Starting point of Centroid Specified by Random.*

In this section will be discussed about the results obtained in this study, to the analysis that has been done is whether by determining the initial centroid using KD-Tree on the grouping of data with K-Means can give better results with a smaller sum square error Compared to the initial determination of

centroid randomly in determining the potential dropout of students' academic data. GPA data is a cumulative prastasi index of students from the semester that has been lived, while the quality data is the achievement of SKS that has been lived times with the GPA obtained by students.
The first stage is initialization of the begining of cluster where the value of K to be formed is 3. Conducted selection K = 3 data as the initial centroid randomly, that is selected 100th, 500th, and 900th data. Table 1 presents the centroid values Named C1, C2 and C3. Data to 100 (N100) becomes first Centroid (C1), data to 500 (N500) becomes second Centroid (C2) and data to 100 (N100) becomes Centroid third (C3). Next each data is calculated the distance to the nearest centroid. The nearest Centroid will be the cluster followed by the data. Next is calculated the new centroid for each cluster ie the average of all the data that joins on each cluster as shown in table 2. The average value on each cluster will be the new centroid value for the next iteration.

Table 1. Initialization of initial Centroid Random Value

Centroid	Data	Quality	GPA
1	N100	424	3,59
2	N500	362	3,07
3	N900	399	3,33

Table 2. Result Clustering New Cluster Iteration 1

Cluster	Member	Quality Average	GPA Average
1	253 point	430	3,64
2	356 point	240	2,51
3	391 point	396	3,34

Table 3. Cluster Center

Centroid	Quality	GPA	Member
C1	432	3,65	237 point
C2	83	1,76	162 point
C3	388	3,28	601 point

In this 3rd iteration, the average value of each cluster is equal to the initial cluster value, so that on the 2nd and 3rd iterations this cluster position does not change anymore and no more data moves from one cluster to another , Then the iteration is stopped and the final result obtained by 3 clusters with 10 iterations as seen in table 3. It is known that the more potential drop out students are in cluster 2 due to the highest Quality and GPA of the lowest value compared to cluster 1 and cluster 3. On Table 3 can be seen that there are 162 students who joined in cluster 2 which means most potentially dropout. To see the accuracy of clustering process by using K-Means clustering method then calculated the square error value (SE) of each data in cluster 2. The value of square error is calculated by squaring the difference of the quality score or GPA of each student with the value of centroid cluster 2.
The value of square error is obtained by calculating the difference in the value of the data with the average of all data in the Cluster 2 group. The value of sum square error (SSE) obtained is 952972 for the quality variable and 232.48 for the GPA which is the sum of all square error values. This SSE value will be compared with the SSE value of the groupings obtained by using K-Means Clustering K-Means Clustering method which determines the initial centroid in KD-Tree.

2.2 *K-Means Clustering with Early Centroid Determination by KD-Tree*
Student's GPA and Quality Data will be clustered using K-Means KD-Tree Clustering to predict students with the potential to drop out. The initial determination of the KD-Tree cluster center is done

by dividing the data into 32 leaf bucket (Lj) by limiting each leaf bucket to a maximum of 35 data. Further calculated the middle point value of each leaf bucket (Mj) and its density (Pj). The values of M1 and M2 are respectively the average values of data incorporated in L1 and L2.
Next is calculated the range of each leaf bucket (Lj) that is by taking the value of data range (Vj) is significant (in this case is taken the value of Quality range, the value of the GPA range can be ignored because it is small enough than the value of Quality) is the highest data difference with the data Lowest on the leaf bucket. For data at L1 has the highest data (Quality value) 8 and the lowest data 0, so that obtained value V1 = 8-0 = 8. For data at L2 has the highest data (Quality value) 55 and the lowest data 10, so obtained value V2 = 55 - 10 = 45. Next we calculate the density value (Pj) of each leaf bucket with equation (3):

$$P_j = \frac{N_j}{V_j} \qquad (3)$$

The highest density (Pj) leaf bucket (Lj) density obtained is L20 with value P20 = 15,00 and is directly selected to be the first initial C1 centroids. Furthermore, throw 20% leaf bucket which has the lowest density value to avoid the election of outlier data becomes the middle point candidate. 20% of 32 Leaf bucket is 6. The next cluster is determined by calculating the rank of leaf bucket (Gj) which is the multiplication of distance (dj) between the middle point candidate with centroid C2 with the density value (Pj) as equation (4):

$$G_j = \left\{ \min_{k=0 \ldots t} \left[d(C_k . M_j) \right] \right\} \times P_j \qquad (4)$$

Calculation of Gj value for leaf bucket L1, L3 and L7 can be shown as follows:

$$G_1 = d(C_1 . M_1) \times P_1 = \sqrt{(1{,}4 - 401)^2 + (0{,}01 - 3{,}40)^2} \times 3{,}75 = 400 \times 3{,}75 = 1499{,}30$$

$$G_3 = d(C_1 . M_3) \times P_3 = \sqrt{(55 - 401)^2 + (1{,}02 - 3{,}40)^2} \times 0{,}70 = 346 \times 0{,}70 = 241{,}40$$

$$G_7 = d(C_1 . M_7) \times P_7 = \sqrt{(367 - 401)^2 + (3{,}11 - 3{,}40)^2} \times 3{,}33 = 34 \times 3{,}33 = 114{,}45$$

Leaf bucket (Lj) which has the highest Gj is L1 with the value G1 = 1499.30 and immediately become the second initial centroid C2 with the value of Quality 1.4 and the value of GPA 0.01. Furthermore, throw 20% leaf bucket which has the lowest density value to avoid the election of outlier data becomes the middle point candidate. 20% of 25 leaf bucket is 5. The next cluster is determined by calculating the rank of leaf bucket (Gj) which is the multiplication of distance between candidate middle point with centroid C2 (dj) with the density value (Pj) as equation (3).
Obtained leaf bucket which has the highest Gj leaf bucket L26 and directly elected to the third initial centroid C3 with the value of Quality 425 and the value of GPA 3,59. Then obtained the initial 3 centroid C1, C2, and C3 as table 4. Next, the initial centroid processed with K-Means Clustering. In the 8th and 9th iterations the cluster position does not change anymore and no more data moves from one cluster to another, the iteration is stopped and the end result is 3 clusters with 9 iterations.

Table 4. Selection of KD-Tree Cluster Center

Centoid	Quality	GPA
C1	401	3,40
C2	1,40	0,01
C3	425	3,59

Table 5. End Cluster Center KD-Tree K-Means Clustering

Centoid	**Quality**	**GPA**	**Member**
C1	377	3,21	444
C2	68	1,67	149
C3	421	3,56	407

Table 6. SSE Value Method K-Means Clustering and KD-Tree K-Means Clustering

Method	SSE		Member
	Quality	GPA	
K Means	952972	232,48	162
KD Tree K Means	504302	214,37	149

It can be seen that more potential drop out students are in cluster 2 due to the highest Quality and GPA of the lowest score compared to cluster 1 and cluster 3. Table 5 shows that there are 149 members in cluster 2. The value of square error obtained by calculating the difference in the value of the data with the average of all data in the cluster group 2. The sum of square error (SSE) obtained is 504302,2 for the Quality variable and 214,37 for the GPA which is the sum of all square error values. Students with the most potential to drop out are in clusters that have the highest Quality and GPA of the lowest value compared to other clusters. In table 6 can see the results of grouping with two methods that have been tested and then compared the value of sum square error (sse) between K-Means Clustering results with KD-Tree K-Means Clustering. The method with the smallest sse value is the best.

3 Conclution

Value of sum square error (SSE) obtained from K-Means Clustering method with initial centroid determination at random is 952972 for Quality variable and 232,48 for GPA. Value of sum square error (SSE) obtained from KD-Tree K-Means Clustering Method with initial centroid determination by KD-Tree is 504302 for Quality and 214,37 for IPK. It can be seen that the SSE value of the K-Means Clustering KD-Tree Method is smaller than the K-Means Method which selects the centroid initially at random. This suggests that early KD-Tree centroid selection in the K-Means method provides better accuracy than the K-Means Clustering method that selects the initially random centroid.

4 References

[1] Oyelade, O. J., Oladipupo, O. O. & Obagbuwa, I. C. 2010. Application Of K-Means Clustering Algorithm For Prediction Of Students Academic Performance. *International Journal of Computer Science and Information Security*. Vol. 7, No. 1.

[2] Godara, S. & Yadav, R. 2013. Performance Analysis Of Clustering Algorithms For Character Recognition Using Weka Tool. *International Journal of Advanced Computer and Mathematical Sciences*. Vol 4, pp 119-123.

[3] Chaturvedi, S. K., Richariya, V. & Tiwari, N. 2012. Anomaly Detection in Network using Data mining Techniques. *International Journal of Emerging Technology and Advanced Engineering*. Vol. 2, No. 5.

[4] Dash, R., Mishra, D., Rath, A. K. & Acharya, M. 2010. A Hybridized K-Means Clustering Approach For High Dimensional Dataset. *International Journal of Engineering, Science and Technology*. Vol. 2, No. 2, pp. 59-66.

[5] Bhatia, M. P. S. & Khurana, D. 2013. Experimental study of Data clustering using k Means and modified algorithms.*International Journal of Data Mining & Knowledge Management Process*. Vol.3, No. 3.

[6] Jain, A. K. 2009. Data Clustering: 50 Years Beyond K-Means. *Pattern Recognition Letters.* Vol 31, pp 651–666.
[7] Sahu, S. K. & Jena, S. K. 2014. A Study of K-Means and C-Means Clustering Algorithms for Intrusion Detection Product Development. *International Journal of Innovation, Management and Technology.* Vol. 5, No. 3.
[8] Likas, A., Vlassis, N. & Verbeek J. J. 2003. The Global K-Means Clustering Algorithm. *The Journal of the Pattern Recognition Society*. Vol 36, pp 451 – 461.

ROBOT TRACER WITH VISUAL CAMERA

Abdul Jabbar Lubis[1], Yuyun Dwi Lestari[2], Haida Dafitri[3], Azanuddin[4]

[123] Department of Informatics Technic, Sekolah Tinggi Teknik Harapan, Jl. H.M. Jhoni No. 70 C, Medan 20217, Indonesia

[4]STMIK Triguna Dharma, Jl Jend. A.H. Nasution no. 73, Sumatera Utara Telp: (061) 8224051 website: trigunadharma.ac.id

Email: abduljabbarlbs@gmail.com

Abstract - Robot is a versatile tool that can function replace human work function. The robot is a device that can be reprogrammed according to user needs. The use of wireless networks for remote monitoring needs can be utilized to build a robot that can be monitored movement and can be monitored using blueprints and he can track the path chosen robot. This process is sent using a wireless network. For visual robot using high resolution cameras to facilitate the operator to control the robot and see the surrounding circumstances.

1. Introduction

The development of robotics technology has brought many changes to all areas, such as industry, medicine, manufacturing, space exploration and construction. Self-propelled technology has long been the goal of researchers and producers. The application of autonomous vehicle technology that is able to move it without being driven by human beings has long been a dream. In addition to providing convenience it also reduces the number of errors due to human factors. One of the revolutionary developments is the unmanned aircraft technology. Which can fly in a controlled way from a distance without having to have a pilot inside. The work presented in this study is the use of robots that are given the ability to back and forth independently and avoid all obstacles that exist around. The modeling of the environmental structure of this study is (ignored). It is expected that this robot can overcome the problems that occur when experiencing various obstacles that occur.
Utilization of autonomous navigation means to make the robot move in the surrounding environment without colliding with objects around it. This mobile robot is controlled by a computer that communicates wirelessly from an existing computer monitoring center. The moving robot consists of several classifications. Among them wheeled robots are usually used for field investigations with terrestrial or soil surfaces. Flying robots are usually used for unmanned flight purposes. Underwater robots are usually used for underwater investigations, or ice surfaces and others. Robot navigation aims to guide the direction of movement of the robot and represent the circumstances surrounding it and send monitored of data center information. And expected with zero error or minimum error.

2. Robot Working *System*

In the Central monitoring console there are two parts that serve to capture the image around the robot and send to the console database. High-speed control systems are sent using 4G or *WiFI* for data to be transferred properly.

A. Data sheet will be accepted by the robot transceiver system.

As a microcontroller issued a control signal that serves as a dual motor drive input connected which robot scent. The high-resolution camera is designed in such a way as to take real-time images from around the robot and send them via 4G or *Wi-Fi*

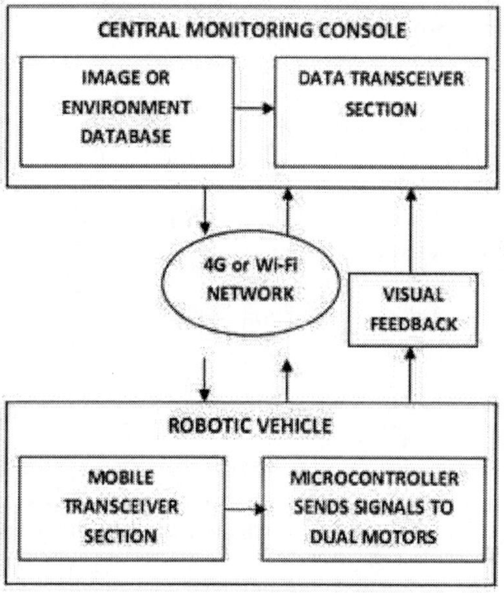

Fig 1. Robot diagram system

B. Design of Robotic Vehicle Unit

The chassis is designed such a way by design using many holes to facilitate the installation of components needed for making the robot. Two the rear wheel and front wheel are designed for normal speed and to reduce complexity when moving in circuits. If both motors rotate in the same direction, the robot moves straight ahead or backward. If one motor is running faster than the other, the robot takes turns to the left or right and if both motors run at the same velocity the opposite, the robot will rotate in a circle. The wheels are connected to DC motors due to their high efficiency, high quality and low cost. The batteries required for the operation of this robot vehicle are stored under a metal chassis and should be placed right in the center of the vehicle mass to improve the stability to make the weight more balanced.

C. Robotic Vehicle Design Formula

The design of robotic vehicles is discussed in this section:
Torque = Style x Radius That is,
T = k x F x R The k value is the scaling factor and is taken as 1/20.
Motor torque, T = 5 Kg cm,
Wheel radius, R = 3.5 cm 5 = 1/20 x F x 3.5 That is,
F = (5 x 20) / 3.5 F = 28.6Kg

This design allows weight to move the robot to the desired direction according to the control of the control center

3. Software and Hardware.

Utilization of software in helping the robot work here has an important role. The hardware used also has a very important role in data processing. The software used include: Windows *XP* platform or higher. Then the *Math Lab* Application to simulate the data received in the form of graphical display. Then 4G rival for high speed data communication. While hardware such as processors and *microcontrollers* also have an important role. The processor used should have a high specification. Ag Core Processors i5 and i7, *Microcontroller* AT89C52 which has a flash memory of 8K. C programming language is coded embedded in microcontroller according to user needs. Transceiver part will send data via Bluetooth to the microcontroller. The control system depends on the code it receives. The code will be changed by the microcontroller into a signal which then controls the signal to the motor driver *IC* L293D. Then the driver *IC* will drive the motor according to the given signal.

4. Detail Implementation

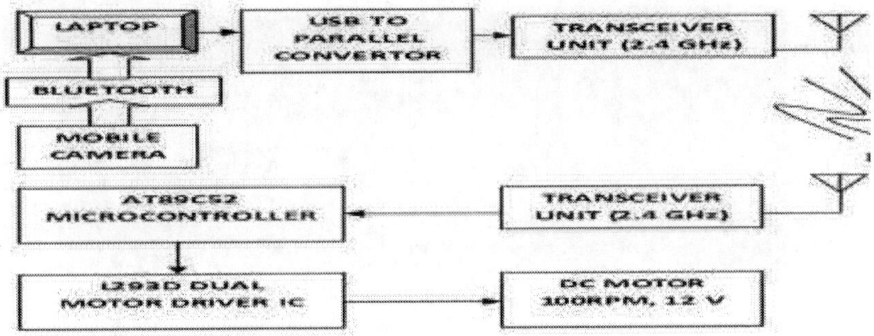

Fig 2 Diagram Implementation

Laptop as a control center apply the concept of *real time* in order to *transit* data continuously. Images taken by the camera will be converted into *JPG*, or *JPEG* form. The *MATLAB* application converts pixel changes that represent a change of direction. There are 4 codes used: F for forward movement, H for backward movement, L for left movement and R for right movement. The image image given to it will be converted into a set of codes. This code is sent *wirelessly* to the robot. Communication of laptop data to the robot using Bluetooth module. This transmitted code will be read by microcontroller and converted into a signal. Then forwarded to Driver *IC* L293D which has 2 outputs to drive a wheeled motor owned robot that serves to drive the robot.

The differential drive system will work in accordance with the given current. If the positive polarity of the name is given, then the motor will move in the same direction. But if the opposite, then the motor will move in the opposite direction. The camera phone is installed using a visual feedback system designed to capture images around the robot wherever it moves and is sent back to the laptop via Bluetooth. The *MATLAB* program is connected to the computer using the serial port (COM). The User Interface (GUI) graphic shown in *MATLAB* is used as a software to control the movement of robots. The GUI uses the *CLI* command (Command Line Interface) where the command must use the keyboard. START, RESUME and STOP are the Hyper terminal commands in the GUI window. The picture below is a screenshot of the GUI when input is given. The black line is a clumsy input.

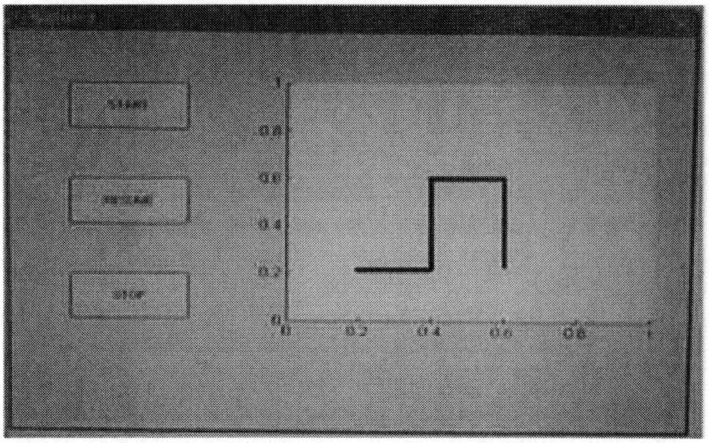

Fig 3 The state before the screenshot START button is pressed

When the start button is pressed, the robot starts moving. This is visible from the black path. Then a red *line* will appear, indicating the state that the code has been sent to the microcontroller and will be converted into a signal to fine-tune the two motors according to the state of the given signal.

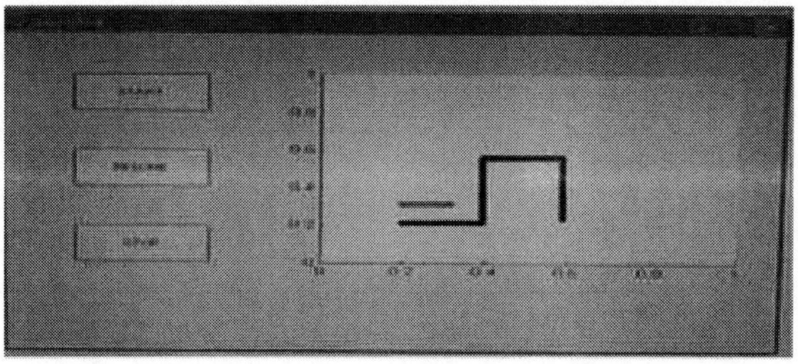

Fig 3 The state After the screenshot START button is pressed

If there are obstacles that exist in front of the robot, then the user pressing the STOP button that causes the movement of the robot to stop. At the same time also the red line will stop. When the obstacle pulls away from the front of the robot, the user presses the RESUME button, and the robot will move to the last stop point and the red line will be traced back from the last point stalled. And so on until the entire track is tracked.

5. Conclusions

Control of robots can be done centrally. The monitoring system designed to allow users to perform remote control. The visual feedback the operator receives serves to see if the robot is moving or not. The concept is executed in real time

6. References

1. Celeste W.C., Bastos Filho T.F., and Sarcinelli Filho M., "Implementing strategies for cautious navigation of robotic wheelchairs", IEEE International Symposium Industrial Electronics, 2010.

2. Cain F.A., Carelli R., De la Cruz C., Muller S., and Bastos Filho T.F., "Autonomous assistance navigation for robotic wheelchairs in confined spaces", IEEE Engineering in Medicine and Biology Society International Conference, 2010.

3. Fletcher J., Fountain C., Kerley C., Torres J., Barott W.C., Selma S., and Stansbury R.,"Immersive audio-visual system for an Unmanned Robotic Vehicle in a Senior Design Project", IEEE Southeastcon proceedings, 2012.

4. Leonard J.J, "Challenges for Autonomous Mobile Robots", IEEE Machine Vision and Image Processing Conference, 2007.

5. Chih-Lyang, Hwang Han-Chen Wu, Nien-Wen Lu, Ming-Lung Lin, Chun-Hao Huang, Ting-Chia Hsu, I-Chiao Luo, and Hsien-Ju Chou, "Realization of humanoid service robotic system", IEEE Advanced robotics and Its Social Impacts, 2008.

Analysis of Artificial Neural Network Backpropagation Using Conjugate Gradient Fletcher Reeves In The Predicting Process

Anjar Wanto[1], Muhammad Zarlis[2], Sawaluddin[3], Dedy Hartama[4], Jaya Tata Hardinata[5], Herlan F Silaban[6]

[145] STIKOM Tunas Bangsa, Jl. Sudirman Pematangsiantar, North Sumatera, 21127
[12] University of North Sumatera, Faculty of Computer Science and Information Technology, University of Sumatera, Medan, Indonesia, 20155
[3] Faculty of MIPA, University of North Sumatera, Jl. Dr. Mansur 9 Medan Indonesia
[6] AMIK Parbina Nusantara, Jl. Sutomo Pematangsiantar, North Sumatera 21136
Email : anjarwanto@amiktunasbangsa.ac.id, m.zarlis@yahoo.com, sawal@usu.ac.id, dedyhartama@amiktunasbangsa.ac.id,jayatatahardinata@gmail.com,herlan76ban@gmail.com

Abstract. Backpropagation is a good artificial neural network algorithm used to predict, one of which is to predict the rate of Consumer Price Index (CPI) based on the foodstuff sector. While conjugate gradient fletcher reeves is a suitable optimization method when juxtaposed with backpropagation method, because this method can shorten iteration without reducing the quality of training and testing result. Consumer Price Index (CPI) data that will be predicted to come from the Central Statistics Agency (BPS) Pematangsiantar. The results of this study will be expected to contribute to the government in making policies to improve economic growth. In this study, the data obtained will be processed by conducting training and testing with artificial neural network backpropagation by using parameter learning rate 0,01 and target error minimum that is 0.001-0,09. The training network is built with binary and bipolar sigmoid activation functions. After the results with backpropagation are obtained, it will then be optimized using the conjugate gradient fletcher reeves method by conducting the same training and testing based on 5 predefined network architectures. The result, the method used can increase the speed and accuracy result.

1. Introduction

Artificial Intelligence is one area that is quite reliable in solving problems such as prediction (forecasting) . Artificial Neural Network (ANN) is one of the artificial representations of the human brain that always tries to simulate the learning process in the human brain [1][2]. ANN approach can imitate any complex and non-linear relationship through non-linear units (neurons) and has been widely used in the forecasting area [3]. Prediction (forecasting) is basically a presumption about the occurrence of an event or event in the future. Prediction (forecasting) is very helpful in planning and decisionmaking activities of a policy [4]. One of the sub areas of Artificial Intelligence that can be relied upon in doing a prediction is the *artificial neural network backpropagation*. To produce a good Backpropagation, the selected parameters must be precise. Therefore, an algorithm that can help accelerate Backpropagation training, one of the reliable algorithms is Conjugate Gradient Fletcher Reeves. This algorithm is expected to improve the performance of the system, because the algorithm is able to optimize so that it can minimize a function, where the search is based on the direction of conjugation orthogonal value [5]. Due to its orthogonal searching nature, so this algorithm can quickly reach convergence on the solution sought. These methods will be used to predict. The data to be predicted is the data of Consumer Price Index of Pematangsiantar that come from the Central Bureau of Statistics. The Consumer Price Index is one of the common economic indicators used to measure the rate of price change (inflation / deflation) at the consumer level, especially in urban areas. In Indonesia, the inflation rate is measured from the percentage change of the Consumer Price Index and announced to the public at the beginning of each month (first business day) [6]. The expenditure group from the Consumer Price Index can be seen in Table 1 below:

Table 1. Consumer Price Index Expenditure Groups

No	Group
1	Foodstuff
2	Food So, Drink, Cigarette And Tobacco
3	Housing, Water, Electricity, Gas and Fuel
4	Clothing
5	Health
6	Education, Recreation and Sports
7	Transport, Communication and Financial Services

Source: Central Bureau of Statistics Pematangsiantar

Based on table 1, each group consists of several subgroups. The subgroup data from each sector of Consumer Price Index can be seen in table 2 below :

Table 2. Subgroups of the Consumer Price Index Sector

No	Group	Subgroups
1	Foodstuff	Grains (Bulbs), Meat, Fresh Fish, Fish Preserved, Eggs and Milk, Vegetables, Nuts, Fruits, Seasonings, Fats and Oils, Other Foodstuffs
2	Food So, Drink, Cigarette And Tobacco	Food, Non-Alcoholic Beverages, Tobacco and Alcoholic Beverages
3	Housing, Water, Electricity, Gas and Fuel	Housing, Fuel (Illumination) and Water Supply, Household Supplies, Household Implementation
4	Clothing	Men's Clothing, Women's Clothing, Kids clothing, Personal Goods and other Clothing
5	Health	Medical Services, Pharmaceuticals, Physical Services, Physical and Health Care
6	Education, Recreation and Sports	Educational Services, Courses / Training, Educational Equipment / Equipment, Sports Recreation
7	Transport,Communication and Financial Services	Transport, Communication and Shipping, Transport Support Facilities, Financial Services

Source: Central Bureau of Statistics Pematangsiantar

Table 3. Previously Research

No	Author	Title	Results	Deficiency
1	Ramanda, K [7]	Prediction of Premature Birth Using Neural Network and PSO	95%	Learning rate is too big ie 0.4

2	Amrin [8]	Using Backpropagation and Multiple Linear Regression For Inflation Prediction	Backpropagation : MAD = 0.0577, MSE = 0.0069, RMSE = 0.0830 Multiple Linier : MAD = 0.0380, MSE = 0.0023, RMSE = 0.0481	Comparison

2. Material and Methods

2.1. Data Used

The data used in this paper is the Consumer Price Index (CPI) data based on the Foodstuffs of Pematangsiantar from 2014 to 2016, from January to December. Consumer Price Index Data (CPI) is derived from the Central Bureau of Statistics Pematangsiantar.

Table 4. Input Data Used Before Normalization

Consumer Price Index 2014-2016												
Sector: Foodstuff												
Year	**Month**											
	Jan	Feb	Mar	Apr	May	Jun	Jul	Aug	Sep	Oct	Nov	Dec
2014	116,22	116,03	117,54	113,89	120,12	119,79	120,05	119,22	119,98	123,53	126,17	127,07
2015	125,95	119,60	118,37	117,91	122,16	127,04	125,53	124,15	122,17	122,85	123,72	128,40
2016	130,65	128,53	130,70	128,30	130,83	131,66	131,90	134,01	135,67	138,00	141.85	144,06

Based on table 3. It can be explained that, the Consumer Price Index (CPI) dataset based on Foodstuff Sector on 2014-2015 is used as training data, while dataset on 2015-2016 is used as testing data. The data presented are January to December data each year.

2.2. Stage Data Processing

At this writing, it was Created a pattern recognition system and predicted the Consumer Price Index (CPI). This process has two stages where the first stage is to do pattern recognition by finding the best architecture of the Artificial Neural Network model created. The process of training and testing the data to get the best model is obtained from the Consumer Price Index (CPI) data at the Central Bureau of Statistics Pematangsiantar as much as 24 Patterns counted from 2014 to 2016, with the criteria Pattern 1 to Pattern 12 is training data (training) , While Pattern 13 through Pattern 24 is the Test data. The second stage is to make predictions with the best architectural patterns obtained in the first stage.

2.3. Data Normalization

Normalization of the data is done so that the output of the network appropriateto the activation function used. The activation function that the author uses in this writing is the sigmoid activation function. Sigmoid function is asymptotic function (never reach 0 or 1), then data transformation is done at smaller interval [0,1; 0.9], shown by the equation (following) [9].

$$x' = \frac{0.8(x-a)}{b-a} + 0.1 \quad ....(1)$$

Based on table 4, Then we will get the result of transformation as follows:

Table 5. Training Data After Normalization

Data	Input												Target
	Jan	Feb	Mar	Apr	May	Jun	Jul	Aug	Sep	Oct	Nov	Dec	
1	0,2285	0,2180	0,3012	0,1000	0,4435	0,4253	0,4396	0,3939	0,4358	0,6315	0,7771	0,8267	0,7649
2	0,2180	0,3012	0,1000	0,4435	0,4253	0,4396	0,3939	0,4358	0,6315	0,7771	0,8267	0,7649	0,4148
3	0,3012	0,1000	0,4435	0,4253	0,4396	0,3939	0,4358	0,6315	0,7771	0,8267	0,7649	0,4148	0,3470
4	0,1000	0,4435	0,4253	0,4396	0,3939	0,4358	0,6315	0,7771	0,8267	0,7649	0,4148	0,3470	0,3216
5	0,4435	0,4253	0,4396	0,3939	0,4358	0,6315	0,7771	0,8267	0,7649	0,4148	0,3470	0,3216	0,5560
6	0,4253	0,4396	0,3939	0,4358	0,6315	0,7771	0,8267	0,7649	0,4148	0,3470	0,3216	0,5560	0,8250
7	0,4396	0,3939	0,4358	0,6315	0,7771	0,8267	0,7649	0,4148	0,3470	0,3216	0,5560	0,8250	0,7418
8	0,3939	0,4358	0,6315	0,7771	0,8267	0,7649	0,4148	0,3470	0,3216	0,5560	0,8250	0,7418	0,6657
9	0,4358	0,6315	0,7771	0,8267	0,7649	0,4148	0,3470	0,3216	0,5560	0,8250	0,7418	0,6657	0,5565
10	0,6315	0,7771	0,8267	0,7649	0,4148	0,3470	0,3216	0,5560	0,8250	0,7418	0,6657	0,5565	0,5940
11	0,7771	0,8267	0,7649	0,4148	0,3470	0,3216	0,5560	0,8250	0,7418	0,6657	0,5565	0,5940	0,6420
12	0,8267	0,7649	0,4148	0,3470	0,3216	0,5560	0,8250	0,7418	0,6657	0,5565	0,5940	0,6420	0,9000

Training Data on 2014-2015 is done using rotary rotation, it means each dataset has the same rights to achieve the target. The data value in data 1 is taken from the CPI of the foodstuff sector in 2014. While the Target value is taken from the CPI of the foodstuff sector in January 2015. The data values in data 2 are taken from the CPI of the foodstuff sector in 2014 from February to December and the month dataset January 2015. Target value in data 2 was taken from the CPI of the foodstuff sector in February 2015. Data values in data 3 were taken from the CPI of the food sector in 2014 March-December and the dataset from January to February 2015. Target value on data 3 is taken from the CPI food sector in March of 2015. So on until all values are completed in play. The maximum value (b) of the dataset is 128.40. While the minimum value (a) is 113.89.

By using the normalization formula as follows : $x' = \frac{0,8\,(116,22 - 113,89)}{128,40 - 113,89} + 0,1$
Then will get result Normalization data 1 for January 0,22885. So on for all data, normalized by using the same formula in equation (1).

Table 6. Testing Data After Normalization

Data	Input												Target
	Jan	Feb	Mar	Apr	May	Jun	Jul	Aug	Sep	Oct	Nov	Dec	
1	0,3460	0,1517	0,1141	0,1000	0,2300	0,3793	0,3331	0,2909	0,2303	0,2511	0,2777	0,4209	0,4898
2	0,1517	0,1141	0,1000	0,2300	0,3793	0,3331	0,2909	0,2303	0,2511	0,2777	0,4209	0,4898	0,4249
3	0,1141	0,1000	0,2300	0,3793	0,3331	0,2909	0,2303	0,2511	0,2777	0,4209	0,4898	0,4249	0,4913
4	0,1000	0,2300	0,3793	0,3331	0,2909	0,2303	0,2511	0,2777	0,4209	0,4898	0,4249	0,4913	0,4179
5	0,2300	0,3793	0,3331	0,2909	0,2303	0,2511	0,2777	0,4209	0,4898	0,4249	0,4913	0,4179	0,4953
6	0,3793	0,3331	0,2909	0,2303	0,2511	0,2777	0,4209	0,4898	0,4249	0,4913	0,4179	0,4953	0,5207
7	0,3331	0,2909	0,2303	0,2511	0,2777	0,4209	0,4898	0,4249	0,4913	0,4179	0,4953	0,5207	0,5280
8	0,2909	0,2303	0,2511	0,2777	0,4209	0,4898	0,4249	0,4913	0,4179	0,4953	0,5207	0,5280	0,5925
9	0,2303	0,2511	0,2777	0,4209	0,4898	0,4249	0,4913	0,4179	0,4953	0,5207	0,5280	0,5925	0,6433
10	0,2511	0,2777	0,4209	0,4898	0,4249	0,4913	0,4179	0,4953	0,5207	0,5280	0,5925	0,6433	0,7146
11	0,2777	0,4209	0,4898	0,4249	0,4913	0,4179	0,4953	0,5207	0,5280	0,5925	0,6433	0,7146	0,8324
12	0,4209	0,4898	0,4249	0,4913	0,4179	0,4953	0,5207	0,5280	0,5925	0,6433	0,7146	0,8324	0,9000

3. Analysis And Results

3.1. Analysis

Previously, the data to be tested should be divided into two (2) sections, where the first part is for training data and the second part is for the test data. Standard Backpropagation method uses gradient decrease algorithm (descent gradient). Variations on the standard model are done by replacing the algorithm with another algorithm. The parameters used in this training and testing : Architecture = 1 hidden layer, Input neurons = 12, Activation Function = Sigmoid, Initialization weights = Random, Target Error Minimum= 0,001 – 0,09, Maximum Epoch = 10000, Learning Rate = 0,01.

3.2. Results

This research uses 5 architecture. Among others are 12-6-1, 12-15-1, 12-24-1, 12-33-1, and 12-34-1. From the 5 architectures, this is the best architecture is 12-15-1. Whether it is using standard backpropagation or merging between backpropagation with conjugate gradient fletcher reeves. The architecture is better because of the high accuracy and stable results, the lesser Epoch and the smaller MSE (Mean Square Error).

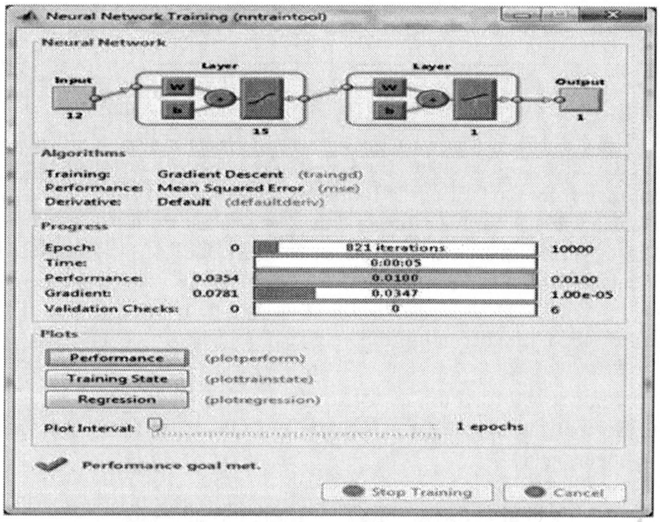

Figure 1. Training and testing with standard backpropagation

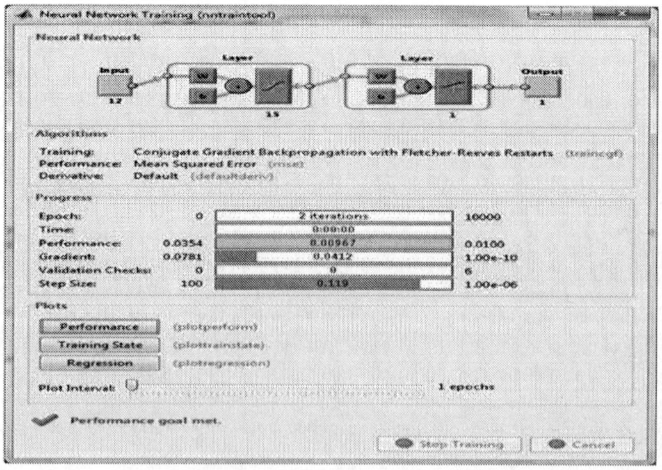

Figure 2. Training and testing with backpropagation with Fletcher Reeves

Table 7. Architectural Results Backpropagation and Backpropagation + CGFR

Method	Architecture	Epoch (Iteration)	MSE (Mean Square Error)	Accuracy
Backpropagation	12-6-1	5308	0,0228736153	50%
	12-15-1	821	0,0142803691	75%
	12-24-1	4999	0,0205544065	58%
	12-33-1	961	0,0366150738	25%
	12-34-1	1491	0,0175775766	25%
Backpropagation + Conjugate Gradient Fletcher Reeves	12-6-1	6	0,0185594035	58%
	12-15-1	2	0,0090116088	67%
	12-24-1	15	0,0261824790	50%
	12-33-1	16	0,0162138928	42%
	12-34-1	149	0,0356825278	33%

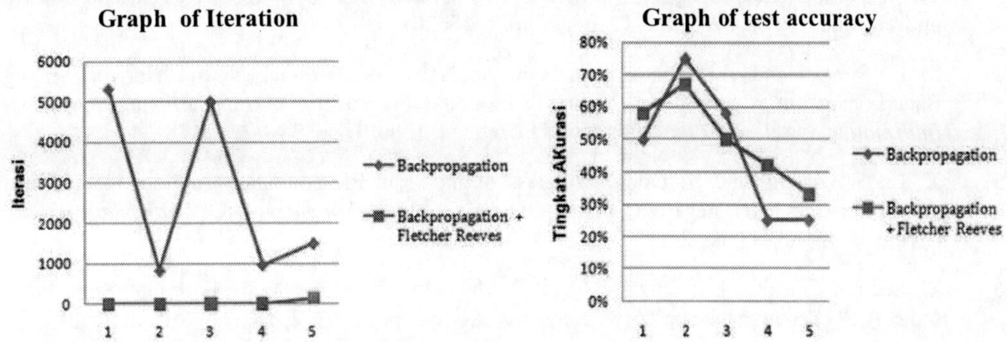

Figure 3. Graph of Iteration and Accuracy between standard backpropagation with BP + CGFR

The results of standard backpropagation methods have been optimized using Conjugate Gradient Fletcher Reeves. Optimization is done by using the Traincgf function. Traincgf is a network function that updates the weights and biases according to the backpropagation gradient convergence with the Fletcher-Reeves update. Based on the results of tests conducted with 5 experiments on backpropagation with different architectures, there was an average of 2716 iterations. While with the same architecture using Backpropagation with Conjugate Gradient Fletcher Reeves there was an increase of iteration with an average of 38. For accuracy of training conducted with the same 5 experiments, there was increased accuracy of training in the first (1st) experiment with the network architecture 12-6-1, the fourth (4th) experiment with 12-33-1 network architecture and fifth (5th) experiment with 12-34-1 network architecture. The decline in the level of accuracy of training in second (2nd) experiment with network architecture 12-15-1 and third (3th) experiment with the network architecture 12-24-1. For accuracy testing conducted with the same 5 experiments, increased accuracy testing in first (1st) experiment with network architecture 12-6-1, fourth (4th) experiment with network architecture 12-33-1 and fourth (5th) experiment with network architecture 12-34- 1. The decline in the level of accuracy testing in second (2nd) experiment with the network architecture 12-15-1 and third (3th) experiment with the network architecture 12-24-1.

4. Conclusion

From the above discussion can be drawn conclusion with the same architectural parameters on the backpropagation that the backpropagation network architecture affect the ups and downs of the rate of learning and the level of accuracy of training and testing.

References

[1] A. Wanto, A. P. Windarto, D. Hartama, and I. Parlina, "Use of Binary Sigmoid Function And Linear Identity In Artificial Neural Networks For Forecasting Population Density," *International Journal Of Information System & Technology (IJISTECH)*, vol. 1, no. 1, pp. 43–54, 2017.

[2] S. F. Weng, J. Reps, J. Kai, J. M. Garibaldi, and N. Qureshi, "Can Machine-learning improve cardiovascular risk prediction using routine clinical data?," *PLoS ONE*, vol. 12, no. 4, pp. 1–15, 2017.

[3] D. Huang and Z. Wu, "Forecasting outpatient visits using empirical mode decomposition coupled with backpropagation artificial neural networks optimized by particle swarm optimization," *PLoS ONE*, vol. 12, no. 2, pp. 1–18, 2017.

[4] S. P. Siregar and A. Wanto, "Analysis Accuracy of Artificial Neural Network Using Backpropagation Algorithm In Predicting Process (Forecasting)," *International Journal Of Information System & Technology (IJISTECH)*, vol. 1, no. 1, pp. 34–42, 2017.

[5] X. Li, W. Zhang, and X. Dong, "A class of modified FR conjugate gradient method and applications to non-negative matrix factorization," *Computers & Mathematics with Applications*, vol. 73, no. 2, pp. 270–276, 2017.

[6] M. S. Frits Fahridws Damanik, SST and S. Magdalena Sinaga, "Analysis of Consumer Price Index (CPI) Pematangsiantar City," *Economic Census*, pp. 1–63, 2015.

[7] K. Ramanda, "Performance Improvement of Neural Network Based Particle Swarm Optimization for Predicting Premature Birth Case Study of Cipto Mangunkusumo General Hospital Jakarta," in *National Seminar on Innovation and Trend (SNIT) 2015*, 2015, pp. 203–208.

[8] Amrin, "Analysis of Consumer Price Index (CPI) Pematangsiantar City Comparative Analysis of Neural Network Backpropagation And Multiple Linear Regression For Forecasting Inflation Rate," *Journal of Computer Engineering*, vol. II, no. 2, pp. 1–6, 2016.

[9] Sumijan, A. P. Windarto, A. Muhammad, and Budiharjo, "Implementation of Neural Networks in Predicting the Understanding Level of Students Subject," *International Journal of Software Engineering and Its Applications*, vol. 10, no. 10, pp. 189–204, 2016.

IOP Conf. Series: Journal of Physics: Conf. Series **930** (2017) 012019 doi:10.1088/1742-6596/930/1/012019

Design And Control Of Agricultural Robot For Tomato Plants Treatment And Harvesting

Arnes Sembiring[1], Arif Budiman[2] and Yuyun D Lestari[3]

[1,2,3]Teknik Informatika, Sekolah Tinggi Teknik Harapan, Medan, Indonesia

arnessembiring@gmail.com

Abstract. Although Indonesia is one of the biggest agricultural country in the world, implementation of robotic technology, otomation and efficiency enhancement in agriculture process hasn't extensive yet. This research proposed a low cost agricultural robot architecture. The robot could help farmer to survey their farm area, treat the tomato plants and harvest the ripe tomatoes. Communication between farmer and robot was facilitated by wireless line using radio wave to reach wide area (120m radius). The radio wave was combinated with Bluetooth to simplify the communication between robot and farmer's Android smartphone. The robot was equipped with a camera, so the farmers could survey the farm situation through 7 inch monitor display real time. The farmers controlled the robot and arm movement through an user interface in Android smartphone. The user interface contains control icons that allow farmers to control the robot movement (formard, reverse, turn right and turn left) and cut the spotty leaves or harvest the ripe tomatoes.

1. Introduction

Although Indonesia is one of the biggest agricultural country in the world, implementation of robotic technology, otomation and efficiency enhancement in agriculture process hasn't extensive yet. Indonesian's tomato farm in the modern channel which adopt agricultural technology about 25% Indonesian's tomato farm **[1].** One component of modern farm is involving automation technology in farming such agricultural robot. Research in agricultural robot has begun since 1980 when citrus harvesting robot designed in University of Florida. Subsequently, numerous agricultural research (fruit and vegetable) were initiated and studied **[2]**. While many prototypes of agricultural robots have been design and developed for various plants, commercial deployment still inhibited by many challenges. The technical challenges include the unstructured environment of the field and the constantly changing field conditions. Obstruction of fruit by leaves or other fruit is one an other technical challenges because it limit fruit identification and robot effector reach. After that, the fragileness of fruit caused by pressure and bad handling also the important challenges **[3]**.

The economic infeasibility and low return of agricultural robot also prevent commercial development **[4]**. One approach to improve the performance of agricultural robots is combining human workers (farmer) and agricultural robot synergistically. Human workers (farmer) collaborate with robot in target detection dan control of end effector. This approach has increased average target detection from 75-85% (robot only) to average of 94% (collaboration human and robot) and reduced the time required for detection by 20% **[5]**. An other approach is separating the system for environment sensoring from agricultural robot (robotic harvester). The robot operate separately (independently) from the sensing system so the speed of operation is no longer limited by the sensing method but by its menchanical capabilities and the fruits susceptibility **[4]**.

In this paper we introduce the new architecture of tomato harvesting robot. This robot contains wireless communication unit via radio wave combined with Bluetooth for communication to Android smartphone, low cost wireless video surveillance system unit, Arduino main controller unit to control arm, end effector and wheels and action unit which consist wheels, arm and end effector.

2. Work Environment

The work environment for the agricultural robot is shown in Fig.1. The space between rows about 70 cm. The field of robot working is uneven surface farm and sometime contain soft and moist soil. For harvesting, robot need to reach and cut the ripe fruit. In common tomato, maximum plant height is 2.5m but fruit to be harvested about 0.15m to 1m. For treatment, robot need to reach spotty leaves which contain blight. Spotty leaf must be removed from plant.

Fig. 1 Tomato field condition

3. Treatment And Harvesting Robot System

The proposed model of tomato harvesting robot is shown in Fig. 2. This robot has a vision unit for image acquisition, a jointed manipulator of 4 freedom degrees, 2 driven wheels and one omni wheel, the system controller and container. The vision unit consists a surveillance camera, video transmitter and receiver using 5.8 GHz radio frequency, a 7 inch monitor for real time displaying of tomato image and environment for user. The manipulator built from thick acrylic and end effector contain scissors for ripe fruit and spotty leaf cutting. The manipulator is jointed by 5 Servo and controlled by Arduino Uno. The motor servo has 10 kgf.cm torsion and 4.8 to 7.2 V operating voltage. The controller unit consists Arduino Uno, the 433 MHz radio frequency transmitter and receiver modul, the bluetooth modul for connecting data to Android smartphone. A user interface is built in Android smartphone and gives farmer full accessibility to control robot movment and action.

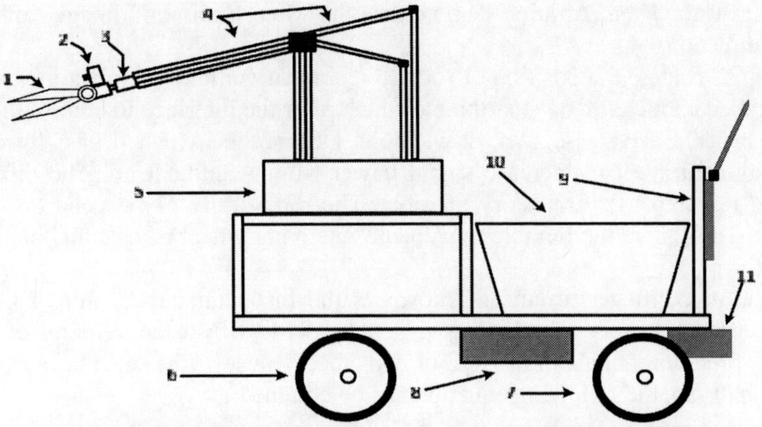

Fig. 2 Tomato harvesting robot model

The structure of the agricultural robot is shown in Fig. 3. Main controller of the robot system is Arduino Uno which consists ATmega328 microcontroller. Arduino Uno controls 4 system consist velocity controlling of DC motor to drive the wheels and robot positioning, five servo motors to control arm and end effector position, environment surveillance system activation and wireless communication system between farmer and robot. A surveillance camera is mounted on the end of arm in order to give the farmer wide area and flexible surveillance.

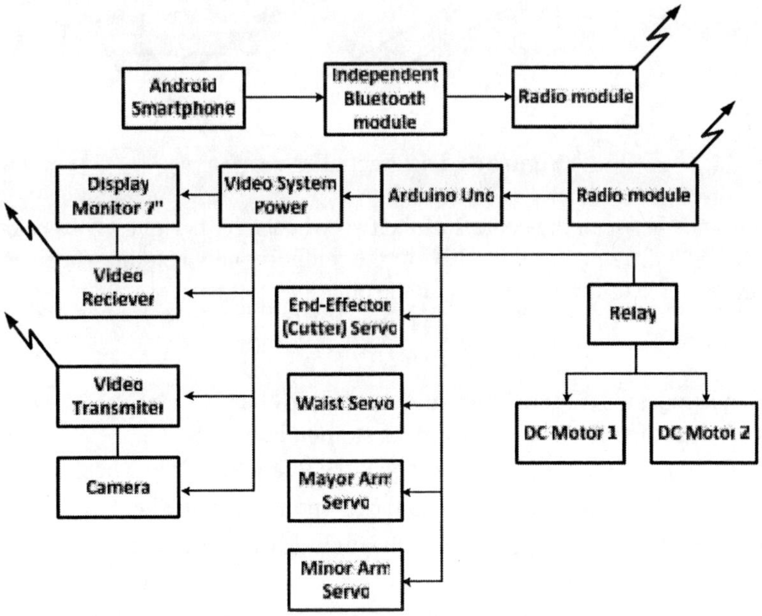

Figure 3. Structure of agricultural robot system

Communication between farmer and robot is facilitated by radio wave. Farmer Android smartphone send farmer direction via Bluetooth to independent Bluetooth module. The independent Bluetooth module communicate serially to radio transmitter and the radio transmitter send the data via radio wave to radio receiver in robot side. The radio receiver communicate the data of farmer direction serially to Arduino Uno. Then, Arduino Uno control the wheels, arm and image capturing activation based on farmer direction.

If the farmer want to change the position of robot for farm surveillance or plant treatment, the farmer just open the installed application on smartphone which offer the interface to control the DC motor and each servo. The wheels driver are two 12-24 Volts DC motors which drived by 4 channel relay module. The relay continue and increase signal power from Arduino Uno. The differential driving technique is used to control the trajectory of robot. The robot trajectory depend on variation of two wheels velocities. To achieve the flexible movement, one omni wheel is used and placed in the center front of the robot.

The point that the robot rotates around is known as the Instantaneous Center of Curvature (ICC), shown in Fig. 4. The trajectory of robot movement is defined by each velocity of the two driven wheels. The rate of rotation (ω) about the ICC of two wheels must be the same because the two wheels have a common shaft. So, the following equation can be obtained:

$$Vr = \omega * \left(R + \frac{d}{2}\right) \quad (1)$$

$$Vl = \omega * \left(R - \frac{d}{2}\right) \quad (2)$$

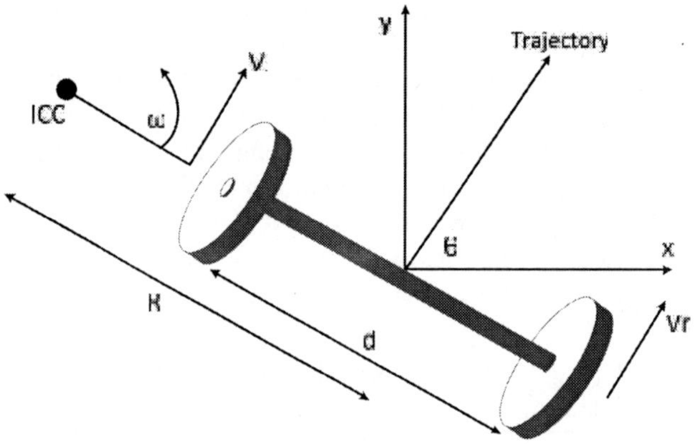

Figure 4. Differential drive system

Where d is the distance between the centers of the two wheels. Vr is right wheel velocity and Vl is left wheel velocity. R is the distance from the ICC to the midpoint between the wheels [**6**]. From equation (1) and (2), at any instance of time ω and R can be obtain as follows:

$$\omega = (Vr - Vl)/d \quad (3)$$

$$R = 0.5 * (Vr + Vl)/(Vr - Vl) \quad (4)$$

From the above equations, the 4 following situation of control are valid:

1. If $Vr = Vl$, then the robot move forward in straight line.
2. If $Vr = -Vl$, then the robot rotate 360° about the midpoint of the wheel shaft.
3. If $Vr = 0$, then the robot rotate 360° and the midpoint is right wheel.
4. If $Vl = 0$, then the robot rotate 360° and the midpoint is left wheel.

4. Results

The main purpose of this agricultural robot is farm surveillance, treatment and harvesting tomatoes. Differential drive technique is implemented for robot navigation. Navigating between crop rows is enough fluent and the robot pass through the uneven land. The robot run in flat land easily but still face with a little difficult in steep grade land. The dimension of robot is 60 x 45 cm, maximum vertical distance from ground that the robot can reach is 80 cm and maximum horizontal distance from outer body of robot is 95 cm. Experiments is implemented at a tomato field to test the robot performance. Test speed of the robot is 0.65 m/s and test load is 10 kg. Maximum distance from robot to Android smartphone and monitor (farmer side) in line of sight is 120 m. There is no significant delay time between user action in Android smartphone and robot responses. Average single harvest cycle from fruit cutting till fruit placing to container is 35 seconds. The captured image by camera and transmitted by radio wave still contain noise. The test field shown in Fig. 5.

5. Conclusion

In this study, agricultural robot, which could navigate between tomatoes rows, cut the ripe fruit and spotty leaves based on farmer order, was developed. The control interface in Android smartphone gave the farmer full control to robot movement and action. Although the image still contain noise, the farmer could survey the land and plants via monitor. The further work is mechanical structure enhacement, image restoration and processing, autonomous fruit identification and web based plant treatment (IoT).

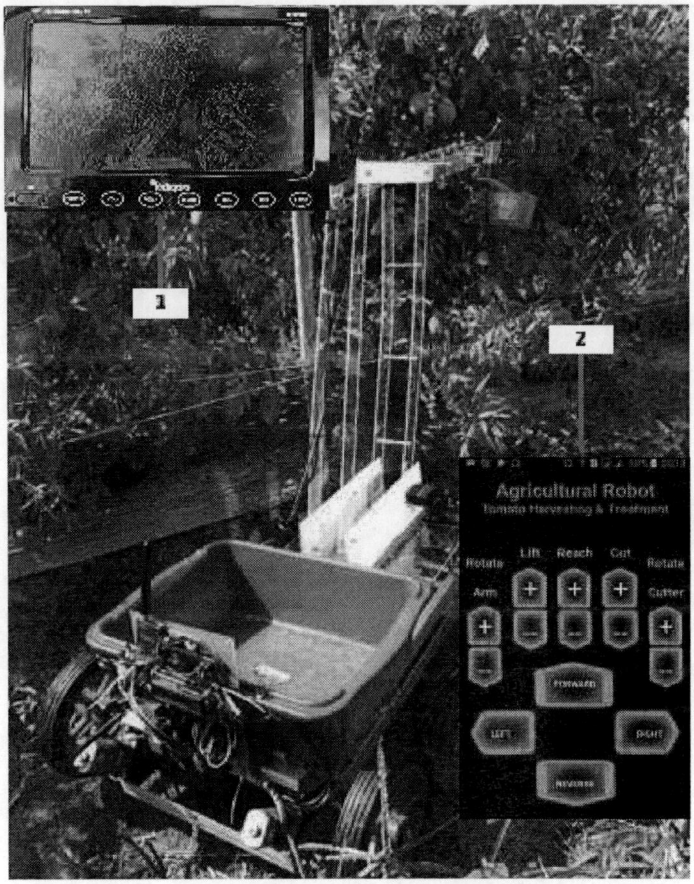

1. Monitor 7 inch in farmer side
2. Android interface fo controlling in farmer side

Figure 5. Field test

Acknowledgments

This research was granted by Ministry of Research, Technology and Higher Education, Republic of Indonesia (Kementerian Riset Teknologi dan Pendidikan Tinggi Republik Indonesia)

IOP Conf. Series: Journal of Physics: Conf. Series **930** (2017) 012019 doi:10.1088/1742-6596/930/1/012019

References

[1] Hernandez R, Reardon T, Natawidjaja R and Shetty S 2015 Tomato Farmers and Modernising Value Chains in Indonesia *Bulletin of Indonesian Economic Studies* **Vol. 51** *No.3, 2015: 425-44*

[2] Mehta S.S. and Burks T.F. 2014 Vision-based control of robotic manipulator for citrus harvesting *Computer and Electronic in Agriculture* **102** *(2014) 146-158*

[3] Edan Y., S. Han and N. Kondo 2008 Automation in Agriculture *Handbook of Automation Chap. 63, 1095-1128*

[4] Mann M., Zion B. Shmulevich I. and Rubistein D. 2015 Determination of robotic melon harvesting efficiency: a probabilistic approach *International Journal of Production Research*

[5] Bechar A. and Edan Y. 2003 Human-robot Collaboration for Improved Target Recognition of Agricultural Robots *Industrial Robot: An International Journal* **30***: 432–436.*

[6] Celen I.H. and Onler E. 2015 A Design of an Autonomous Agricultural Robot to Navigate Between Rows *International Conference of Electrical, Automation and Mechanical Engineering (EAME 2015)*

The Effect Of Atributes Product's Analysis, Halal Certification, And Product Innovation To The Interest Of Consumer Buying-Back Through The Advantage Competitive Of Micro Small And Medium Business (MSMB) In Medan

Bunga Aditi

Lecturer of Sekolah Tinggi Ilmu Ekonomi Harapan Medan, Indonesia

Email:bunga.aditi16@gmail.com

Abstract. This research is done to know and analyzing the effect of attribute, halal certification product, and product innovation on the competitiveness of Micro Small and Medium Business in Medan, knowing and analyzing the effect of product attribute, halal certification, and product innovation to consumer buying surge. The type of research used is explanatory research with quantitative approach. The population in this study is the community as consumers who are in the area of Medan province of North Sumatera. Total sample of 150 people. Sampling method which use an accidental sampling is the technique of determining samples by chance, example: consumers whomeets with researcher coincidentally can be sampled if the consumer is appropriate or suitable as a source of data. This research uses the method of Structural Equation Modeling analysis. The mainting of this research is showed that product attributes, halal certification, and product innovation had a positive and significant impact on competitive advantage which impact on consumer buy-back (surge) interest. The advantage competitive has affects consumer buying surge interest positively. Keywords: Product attributes, halal certification, product innovation, competitive advantage, consumer buying interest

1. Introduction

The Law No. 20 of 2008, "A company classified as a SMMB is a small company owned and managed by a person or owned by a small group of people with a limit amount of wealth and incomeSmall and Medium Business (SMB) has a fairly strong role in the growth up development of Indonesia and it is considered as the driving force of the economy of a region. This is indicated when the economic crisis in 1998 and small business is the most sustained group when not a few big businesses have gone bankrupt.

Data from the Central Bureau of Statistics shows that after the economic crisis of 1997-1998 the number of MSMB has not decreased, it has been increasing, even absorbing 85 million to 107 million manpower (worker) until 2012. In that year, the number of entrepreneurs in Indonesia is 56,539,560 units. Of this amount, Micro Small and Medium Business (MSMB) of 56,534,592 units or 99.99%. The rest, around 0.01% or 4,968 units is a major undertaking. The data proves that MSMB (UMKM) is a potential market for the financial services industry.

According to the Head of Cooperatives and Micro Small Medium Business (MSMB) Medan, Arjuna

Sembiring in Medan said that," today the development of MSMB in the capital of North Sumatra province continues to show significant growth. In 2009 there were about 22 thousand perpetrators of MSMB in the city of Medan and the number continues to increase until 2015 amounted to approximately 300 thousand. The high growth of MSMB is certainly very encouraging because it also can be minimilize the unemployment (economy.okezone.com).

The increasing number of MSMB, it is also increasing competition among MSMB (UMKM) entrepreneurs, they jointly compete in producing, gaining competitive advantage and provide value for customers, as far as possible in according with the value expected by the Consumer.

Moslem is the largest in Southeast Asia of 207,176,162 (www.bps.go.id). Indonesia with a predominantly Moslem population is a large potential market for many producers of goods and services. Moslem consumers in Indonesia seek authentic halal certification issued by Indonesian Ulama Council (MUI =Majelis Ulama Indonesia). This certification authorizes companies to use halal logos to be printed on product packaging or to be displayed on the company premise (www.halalmui.org).

According to Yunus, Ariffin and Rashid (2014) the research which related to the interest of buying halal products shows that the composition of ingredients contained in the products have a significant positive effect on the buying interest of Moslem consumers. On the other hand, the research was conducted by Hussin, Hashim, Yusuf, and Alias (2013) related to the interest of buying Moslem consumers on halal products showed the different things. The study proves that the composition or ingredients contained in the product have a significant negative effect on consumer buying interest. The above research shows inconsistent results leading to a research gap.

Based on the description of the background then the researcher is interested to conduct research with the title " The effect of attributes product's analysis, halal certification, and product innovation to the interst of consumer buying-back through the advantage competitive of Micro Small and Medium Business (MSMB) in Medan.

2. Research Objectives

The purpose of this research is knowing and analyzing the effect of attribute, halal certification product, and product innovation on the competitiveness of UMKM in Medan, and knowing and analyzing the effect of product attribute, halal certification, and product innovation to consumer buying surge interest.

3. Literature Review

3.1. Product Attribute

According to Kotler & Armstrong (2010: 254), product attributes are characteristics of a product or service that results in an ability to satisfy that is expressed or implied in the consumer's needComponents:

1. Product Brand

   Brands may be defined as names, terms, signs, designs, or combinations there of, intend to identify goods and services from a seller or group and distinguish them from its competitors (Tjiptono, 2008).
2. Product Quality

According Tjiptono (2008), quality is the level of performance of an item, the quality of a product can be shown from the level of customer satisfaction of the results and processes.

1. Product Design

   Understanding product design according to David (2009: 251) is striving to increase sales through sales through product improvement or product development today.
2. Product Labels

   According to Kotler (2012: 478) "Labels can be just a simple patch on products or drawings are designed with a complex that is a unity with the packaging.

3. Packaging Products
   According to Tjiptono (2008), product packaging is a physical wrapper to make a protecting the product and also create a unique identity.

3.2. Halal certification

Halal certification is a guarantee of security for a Moslem consumer to choose good food for him and accordancing with the rules of religion. Food productshalal certification are products in their processing that meet the standards of safety and hygiene (Lada et al., 2009).

3.3. Product Innovation
The definition of product innovation according to Myers and Marquis in Kotler (2007: 36) states that product innovation is a combination of various processes that affect each other between the one with the other. So innovation is not the concept of a new invention or the development of market, but innovation is a picture of all those processes.

3.4. Competitive Advantages
The competitive advantage according to Michael E Porter (2008) is a company's ability to achieve economic benefits above the profits that can be achieved by competitors in the market in the same industry. Companies that have it always be ability to understand the changing market structure and are able to choose an effective marketing strategy.

3.5.Consumer buying-back interest
Mital (in Sutantio 2004, p.253) says that one indication of success or failure of a product is the amount of consumer buying-back interest in the product. Meanwhile, repurchase interest is basically customer behavior in which the customer responds positively to the service quality of a company and intends to re-visit or re-consume the company's products.
Based on the previous explanation it can formulate the conceptual framework as the flow of thought in this study as follows:

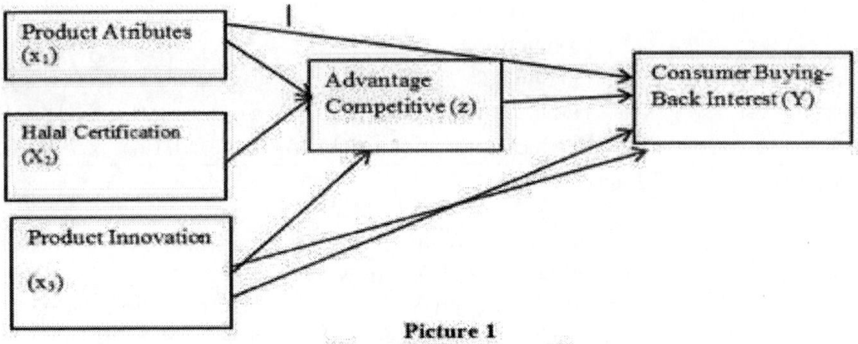

Picture 1
Conceptual Framework

4. Data Analysis Method
The model data analysis in this research is path analysis. According Sarwono (2007) path analysis a technique for analyzing the causal relationships that occur in multiple regression if independent variables affect the dependent variable directly and indirectly.

5. Research Results
Further analysis is a Structural Equation Modeling (SEM) analysis in Full Model which is intended to test the model and hypothesis developed in this research. Testing model is done by two test, that are

model compatibility test and significance test of causality through regression coefficient test. Results of data processing for SEM analysis seen in table 1:

Table 1. Model Feasibility Test Results

Goodness of Fit Indeks	Cut off Value	Results	Evaluation Model
λ2 (Chi-square)	Expected of small	8,928	Good
Significanced Probability	≥0,05	1	Good
RMSEA	≤0,08	0.00	Good
GFI	≥0,90	0,994	Good
AGFI	≥0,90	0,991	Good
TLI	≥0,95	1,027	Good
CFI	≥0,95	1	Good
CMIN/DF	≤2.00	0,059	Good

Source: Research Results, 2017 (processed data)

In table 1 the resuts of testing full model has met the fit criteria. It based on all criteria already met well based on cut off value.

5.1. Direct Effect

The direct effect of the independent variables on the dependent variable can be seen in Table 2:

Table 2. Results of Hypothesis Testing Direct Effect

			Estimate	S.E.	C.R.	P	Evaluation
Competitive Advantages	<---	Product Atributes	.253	.075	3.353	***	Significant
Competitive Advantages	<---	Halal Certification	.293	.078	3.752	***	Significant
Competitive Advantages	<---	Product Innovation	.290	.079	3.682	***	Significant
Buying-Back Interest	<---	Product Atributes	.196	.076	2.593	.010	Significant
Buying-Back Interest	<---	Competitive Advantages	.226	.082	2.744	.006	Significant
Buying-Back Interest	<---	Halal Sertification	.225	.080	2.818	.005	Significant
Buying-Back Interest	<---	Product innovation	0,244	0.081	2.799	0.005	Significant

Source: Research Results, 2017 (processed data)

In Table 2 the interpretation of the direct effectin each path coefficient or the direction of causal relation, the result of hypothesis testing is stated as follows:

1. Testing the first hypothesis shows that product attribute positively influence to competitive advantage. This is visible from path coefficient marked positive 0.253 with value of C.R equal to 3.353 and obtained significant probability (p) as small as *** smaller than significant level (α) specified equal to 0.05.
2. Second hypothesis testing shows that halal certification positively influences competitive advantage.It is from the coefficient of the path marked positive by 0.293 with the value of C.R of 3.752 and obtained significant probalitas (p) *** smaller than the significant level (α) determined by 0.05.
3. Testing of the third hypothesis shows that product innovation positively effect the competitive advantages. This resultscoefficient of the path marked positive by 0.290 with the value of C.R amounted to 3.682 and obtained significant probability (p) of *** smaller than the significant level (α) determined by 0.05.

4. Hypothesis testing can show that product attribute has a good effect to buying-back interest. The data iscoefficient of the path marked positive by 0.196 with the value of C.R of 2.593 and obtained significant probalitas (p) of 0.010 smaller than the significant level (α) which has been determined by 0.05.
5. Testing of the fifth hypothesis shows that thepositive value and significant on buying-back Interest. The positive path coefficient of 0.226 with the value of C.R of 2.744 and obtained a significant probability (p) of 0.006 smaller than the significance level (α) which has been determined by 0.05.
6. The testing of the sixth hypothesis shows that halal certification has a positive effect on buying-back Interest. The coefficient of the path marked positive by 0.225 with the value of C.R of 2.818 and obtained a significant probability (p) of 0.005 smaller than the significant level (α) determined by 0.05.
7. The testing of the seventh hypothesis shows that product innovation in buying and purchase interest, is coefficient of the line marked positive by 0.244 with the value of C.R of 2.799 and obtained a significant probability (p) of 0.005 smaller than the significant level (α) which has been determined by 0.05.

5.2.Indirect Effect

The effect of in indirect the independent variables on the dependent variable in table 3:

Tabel 3. Results of Indirect Effect Hypothesis Testing

			Estimate	S.E.	C.R.	P	Keterangan
Buying-Back Interest	<---	Product Atributes	.248	.059	4.193	***	Significant
Buying-Back Intterest	<---	Halal Sertification	.286	.064	4.471	.002	Significant
Buying-Back Interest	<---	Product Innovation	.292	.063	4.627	***	Significant

Source: Research Results, 2017 (processed data)

In Table 3 the interpretation shows that each path coefficient or the direction of causal relation, the result of hypothesis testing is stated as follows:

1. Testing the first hypothesis shows that productattribute positively effect the buying-back interest. This is visible from path coefficient marked positive equal to 0.248 with value of C.R equal to 4,193 and obtained significant probability (p) as small as *** smaller than significant level (α) specified equal to 0.05.
2. Testing the second hypothesis shows that the halal certification positively affects the buying-back Interest. This is coefficient of the path marked positive by 0.286 with the value of C.R of 4.471 and obtained significant probalitas (p) of 0.002 smaller than the significant level (α) determined by 0.05.
3. Testing of the third hypothesis shows that product innovation positively affects the buying-back Interest. It is coefficient of the path marked positive by 0.292 with the value of C.R of 4.627 and obtained significant probalities (p) of *** smaller than the significant level (α) determined by 0.05.

6. Conclusion

1. Product attributes, halal certification, and product Innovation have a positive and significant impact on competitive advantage.
2. Product attributes, halal certification and product innovation have right effect on buying-Back Interest

IOP Conf. Series: Journal of Physics: Conf. Series **930** (2017) 012020 doi:10.1088/1742-6596/930/1/012020

References

[1] Arikunto S 2005. *Prosedur Penelitian Suatu Pendekatan Praktek* (Jakarta: Rineka Cipta)

[2] Buchari A 2004*Manajemen Pemasaran dan Pemasaran Jasa* (Bandung: Alfabeta)

[3] Kevin, L.K.*Manajemen Pemasaran Edisi Kedua Belas* Cetakan Ketiga(Jakarta: Indeks)

[4] Koncoro. M. 2003.*Metode Riset Bisnis Dan Ekonomi* : Bagaimana Meneliti (Jakarta: Ghalia)

[5] Kotler, Philip & Kevin Lanew Keller 2007*Manajemen Pemasaran*, Indeks

[6] Lada*et al*2009. *Predicting intention to choose halal products using theory of reasoned action. International Journal Islamic and Middle Eastern Finance and Management***2**(1)**66-76**.

[7] Porter Michael E 2008. Keunggulan Bersaing – *Menciptakan dan mepertahankan Kinerja Unggul.* (Jakarta:Erlangga)

[8] Saladin D Oesman Y 2002*Intisari Pemasaran dan Unsur Pemasaran* (Bandung : Lindakarya)

[9] Sugiyono 2010*Statistika untuk Penelitian*(Bandung: Alfabeta)

[10] Tjiptono Fandy dan Teguh Budiarto 2002*Pemasaran Internasional*(Yogyakarta: BPEE)

[11] Tjiptono F 2004. *Strategi Pemasaran(Yogyakarta*: andi).

[12] Yunuz *et a*l 2014*Muslim"s Purchase Intention Towards Non-Muslim"s Halal Packaged Food Manufacturer Procedia - Social And Behavioral Sciences.pp*

[13] http://www.bi.go.id/id

[14] http://www.bps.go.id

[15] http: www.halalmui.org

[16] http://www.kemenpar.go.id

IMPLEMENTATION OF K-MEANS CLUSTERING METHOD FOR ELECTRONIC LEARNING MODEL

Herlina Latipa Sari[1], Dewi Suranti[2], Leni Natalia Zulita[3]

[1 2 3]Teknik Infromatika Fakultas Ilmu Komputer Universitas Dehasen Bengkulu Jalan Meranti Raya No.32 Sawah Lebar Kota Bengkulu

Email:[1]herlinalatipasari@ymail.com, [2]Mrs.dewisuranti@gmail.com, [3]Natalia.af25@gmail.com

Abstract - Teaching and Learning process at SMK Negeri 2 Bengkulu Tengah has applied e-learning system for teachers and students. The e-learning was based on the classification of normative, productive, and adaptive subjects. SMK Negeri 2 Bengkulu Tengah consisted of 394 students and 60 teachers with 16 subjects. The record of e-learning database was used in this research to observe students' activity pattern in attending class.KMeans algorithm in this research was used to classify students' learning activities using e-learning, so that it was obtained cluster of students' activity and improvement of student's ability. Implementation of K-Means Clustering method for electronic learning model at SMK Negeri 2 Bengkulu Tengah was conducted by observing 10 students' activities, namely participation of students in the classroom, submit assignment, view assignment, add discussion, view discussion, add comment, download course materials, view article, view test, and submit test. In the e-learning model, the testing was conducted toward 10 students that yielded 2 clusters of membership data (C1 and C2). Cluster 1: with membership percentage of 70% and it consisted of 6 members, namely 1112438 Anggi Julian, 1112439 Anis Maulita, 1112441 Ardi Febriansyah, 1112452 Berlian Sinurat, 1112460 Dewi Anugrah Anwar and 1112467 Eka Tri Oktavia Sari. Cluster 2:with membership percentage of 30% and it consisted of 4 members, namely 1112463 Dosita Afriyani, 1112471 Erda Novita, 1112474 Eskardi and 1112477 Fachrur Rozi.
Keywords: Clustering, Learning Model, K-Means Algorithm

1. Introduction

National education goal is to educate nations. It can't be separated from the role of teacher. In order to improve the quality of education, the development of curriculum, learning innovation, and education facilities contentmentare needed. To improve students' learning achievement, teacher should make innovative learning that support the students learn optimally, either in self-learning and in class learning. In an attempt to improve the quality of education, teachers are required to make learning more effective, innovative and fun [2]. In learning process, the role of teachers is asmotivator and facilitator. In consequence of learning paradigm change from instructional based learning into constructional based learning, the teachers should capable to design learning that activate students.To make more effective, innovative, and fun, the teachers can utilize all school resources, either human resources, facilities and infrastructuresor other resources [9]. The survey result of this research showed that Sekolah Menengah Kejuruan (SMK) Negeri 2 Bengkulu Tengahhas had computer network and already utilize Information and communication Technology (ICT) by utilizee-learning in implementing learning into teaching and learning process. It was a step forwardthat capable to improve students' learning motivation and also to meet the shortage of meeting in classroom doe to the vocational curriculum that require field study, so that the students can conduct self learning without depend on teacher and books [8]. E-learning is learning media technology that utilize either electronic media or software [9]. E-learning can be developed by using Learning Management System (LMS)called Moodle. Some facilities

provided by Moodle includereading module, assignment module, chat module, forum module, option module, and quize module[2]. Once the pattern obtained,analyzing each pattern of subject taken by students was performed, so that it can describe the subject taken by students based on students' participation interest in following the lessons. Fuzzy K-Means algorithm in this research was used to cluster teaching and learning activities between teacher and students so that it resulted informations about activities group attended by students in e-learning.Beside that, k-means algorithm is also versatile which means that it easy to modify steps in the algorithm, such as in initialization of function to calculate distance and also criterion on stoping iteration [7].

2. Literature Review

Clustering is about grouping of records, observation and forming a class of objects that have similarities.Cluster is a set of records which have similarities to records in another cluster.Clustering is different from the classification that the absence of the target variable in clustering[7]. Clustering plays an important role in data mining applications, such as the exploration of scientific data, accessing information and text mining, spatial database applications, and web analysis[4]. Clustering has two methods, namely hierarchical clustering method and non hierarchical clustering method.Hierarchical method is used when desired number of groups is not yet known, while the nonhierarchical method is used when the number of groups desired predetermined. Fuzzy K-Means algorithm belongs to the group of hierarchical clustering methods [3]. Fuzzy K-means is a popular cluster algorithm which is used in the process of clustering data sets because of its simplicity[6]. However, this algorithm can also be used for the establishment of a database cluster that its attributes derived from different types, by altering these attributes into the index of similarity or dissimilarity [10].In the K-Means algorithm cluster number K are already predetermined,object with particular atribute are grouped within cluster K. Each cluster contains cluster center (centroid) point and members of a cluster is sellected in accorandce with distance from cluster center point [1]. K-Means method belongs to partitioning clusteringalgorithm [11].Steps for kmeans clustering are as follow.

a. Decide the number of cluster K

b. Inisialization of the cluster center (centroid) can be conducted by using various ways.However, the most frequently used is by using random way. Clusters centers are assigned by random numbers.

c. Allocate all data/objects to the closest cluster.Determination of closeness of two objects are determined based on the distance of two objects. To calculate the distance of all data to each centroid point, Euclidean Distance theory is used that is formulated as follow.

$$D(i,j)=\sqrt{(X_{1i}-X_{1j})^2+(X_{2i}-X_{2j})^2+\ldots+(X_{ki}-X_{kj})^2}$$

Where: D (i,j) =distance of ithdata to cluster center j Xki = the ith on the kthdata atribut Xkj = the jth center pointon the kthdata atribut

d. Recalculate centroid with current cluster membership. Centroid is an average (mean) of all data/objects within particular cluster. If desired, the median of this cluster can also be used. Thus the meanis not the only one parameter that can be used.

e. Reassign each object by using new clustercenter, if the cluster does't change, then clusetringprocess finished otherwise repeat step 3 until there is no change for each cluster.

3. Research Methodology

The methodology used in this research illustrates the steps taken to perform the analysis of the implementation of K-Means Clustering Algorithm for electronic based learning in the form of Elearning application at SMK Negeri 2 Bengkulu Tengah. Figure 2 shows research framework started by definingscope of the problems to the testing of K-Means Clustering algorithm. Once directobservation done, literature study was conducted by comparing some previous research. Partitioning data aims to classify data that has same characteristics into one same clusterand data that has different characteristics into one other group [1]. After defining scope of the problem, then analyzing the problem and determine the expected goals was performed followed by studying the literature about KMeans Clustering Algorithm and the results of calculations and grouping so that it yielded grouping of student activity in learning activities in the E-Learning application.
The data in this research were drawn from the database that is contained in the elearning application in the form of teacher login data and student login data that contains student activity in participating lessons given in the teacher [12].
There are three group of subject in the elearning, namely normative, adaptive, and productive. Activities type that exist in elearning are shown in Table 1.

Table1. Students Activities in E-Learning

No	Students Activities	E-Learning Module
1	Students participation in classroom	course
2	Submit Assignment	Assignment
3	View Assignment	Assignment
4	Add Discussion	Forum
5	View Discussion	Forum
6	Add Comment	Forum
7	Download Course Material	Resource
8	View Article	Blog
9	View Text	Quiz
10	Attempt Test	Quiz

4. Discussion

K-Means algorithm is used in clustering dataset doe to simplicity of its algorithm to process big dataset, in addition the K-means algorithm is a partitioning clustering that separates data into different groups. Through partitioning iteratively it can minimiza the average distance of each data into clusters [12]. The implementation of K-Means Clustering algorithm in this research was conducted by simulating students activities with consist of 396 students. In this study, 10 students were taken as samples by observing the activities of 10 students in applying E-Learning with 10 activities that students activities with NIS 1112438 named Anggi Julian are 60 times entering classroom, 18 times submiting assignments, 20 times viewing assignments, never adding discussion, never viewing discusion, never adding comments, 25 times downloading course materials, never viewing articles, 25 times viewing test, and 25 submiting tests.
From the total of students activities data shown in Table 2, the steps in implementeing KMeans algorithm ar as follow.

IOP Conf. Series: Journal of Physics: Conf. Series **930** (2017) 012021 doi:10.1088/1742-6596/930/1/012021

a. Number of desired clusters are 2 clusters(C1 and C2)
b. Determiniation of Centroid (cluster center point) In this research, second data is used to determine centroid for cluster 1 and fifth data is used for determination centroid for cluster 2. These two data are as folow:
   Cluster 1 : 79, 11, 19, 0, 0, 0, 29, 0, 35, 30
   Cluster 2 : 45, 0, 21, 0, 0, 8, 26, 1, 26, 22
c. Calculate the distance for every data to its centroid Iteration 1 : Calculation of distance for 1st data: 60, 18, 20, 0, 0, 0, 25, 0, 30, 25 to centroid 1 and 2

$$D_{1,1} = \sqrt{\begin{aligned}&(79-60)^2+(11-18)^2+(19-20)^2+(0-0)^2\\&+(0-0)^2+(0-0)^2+(29-25)^2+(0-0)^2\\&+(35-30)^2+(30-25)^2\end{aligned}} = 21{,}84$$

$$D_{2,1} = \sqrt{\begin{aligned}&(45-60)^2+(9-18)^2+(21-20)^2+(0-0)^2\\&+(0-0)^2+(8-0)^2+(26-25)^2+(1-0)^2\\&+(26-30)^2+(22-25)^2\end{aligned}} = 19{,}95$$

d. The same calculation is then conducted toward 2nd to 10th data, once calculating the distance of centroid for all data, the distance matrix is then obtained as follow. D1 :

Table 2. The distance matrix

Cluster	**Student**									
	1	2	3	4	5	6	7	8	9	10
C1	21,84	0	16,46	41,06	0	33,60	21,93	32,97	49,44	61,36
C2	19,93	37,19	32,47	24,10	40,78	12,47	19,39	13,19	30,66	40,71

Every colomn in the matrix represents the distance value of data to centroid. First row in the matrix represents distance value of data to the first centroid (C1) : 79, 11, 19, 0, 0, 0, 29, 0, 35, 30 and the second row in the matrix represents distance value of data to second cluster (C2) : 45, 0, 21, 0, 0, 8, 26, 1, 26, 22. In the cluster 1 (C1), the distance for the first activity (participation of students in classroom) to the centroid point is 21.84 and in the cluster 2 (C2) the distance value for the first activity (participation of students in classroom) to the centroid point is 19.95, In the cluster 1 (C1), the distance value for the second activity (submitting task) to the centroid point is 0, and in the second cluster (C2) the distance value for the second activity (submitting task) to the centroid point is 37.19.

e. Data Grouping The calculated distances are then compared to sellect the closest distance between data and centroid, this distance show that the data are in a cluster with closest centroid. Here will be shown matrix data of cluster, value 1 means that the data are in the cluster dan value 0 means that the data are not in the cluster. G1 =

Table 3. Matrix data

Cluster	**Activity**									
	1	2	3	4	5	6	7	8	9	10
C1	0	1	1	0	1	0	0	0	0	0
C2	1	0	0	1	0	1	1	1	1	1

In the cluster 1 the value of activity 1 (A1) is 21.84, in the cluster 2 is 19,95, it means that C1A1 <C2A1 so that the smaller value close to centroid is asigned value 1 whreareas C1A1 is assigned

value 0. The same process was also conducted to C1A2 and C2A2 to C1A10 and C2A10.In the data grouping in G1 for the number of cluster 1 (C1) contains 3 members (2,3,5) and the number of cluster 2 (C2) contains 7 members (1,4,6,7,8,9,10).

f. Determination New Centroid Once the members of each cluster obtained, then a new centroid are calculated based on data from each cluster member in accordance with the formula of the center of the cluster members. So that we get the following calculation: Cluster 1 (C1) obtained 3 members with calculation of new centroid is as follow.

g. Data Grouping As in the first iteration, the calculated distances are then compared to sellect the closest distance between data and centroid, this distance show that the data are in a cluster with closest centroid. Here will be shown matrix data of cluster, value 1 means that the data are in the cluster dan value 0 means that the data are not in the cluster. G2 =In the cluster 1 the value of activity 1 (A1) is 13.35, in the cluster 2 is 18.84, it means that C1A1 < C2A1 so that the smaller value close to centroid is asigned value 1 whreareas C1A1 is assigned value 0. The same process was also conducted to C1A2 and C2A2 to C1A10 and C2A10. In the data grouping in G1 for the number of cluster 1 (C1) contains 5 members (1,2,3,6,7) and the number of cluster 2 (C2) contains 5 members (4,5,8,9,10). Based on the calculation in the first and second iterationwith G1and G2 grouping, the equal number of cluster 1 (C1) and cluster 2 (C2)has not been reached so that it needs to repeat the calculation in determining new centroid. In this research, the stabil clustering result and convergent was reached at iteration 5 with the matrix of students activities G5 as follow.

h. Simulation Result of Clustering Analysis Based on the the calculation K-Means clustering algorithm using simulation can be interpreted as follow. Cluster 1: with membership percentage of 70% and it consisted of 6 members Cluster 2: with membership percentage of 30% and it consisted of 4 members h. Implementation Process of K-Means Clustering Algorithm using RapidMiner In implementing K-Means algorithm using Rapid Minerthe first step is by importing excel data "Aktifitas Siswa"

Table 4. Descriptive Statistics

Activity	Min	Max	Average	Deviation
Entering Class	48	79	53.400	14.238
Submit Assignment	6	18	11.300	3.917
Adding Discussion	19	58	27.600	11.316
Viewing Discussion	0	0	0	0
Adding Comment	0	8	0.900	2.514
Downloading Course Material	7	40	24.600	9.228
Viewing Article	0	1	0.400	0.516
Viewing Test	19	42	30.200	7.524
Attempting Test	15	40	25.700	7.072

After running K-Means Clustering, it will appear the graphic of data clustering using RapidMiner as shown in Figure 7.

5. Conclusion

K-Means algorithm in this research was used to classify students' learning activities using e-learning, so that it was obtained cluster of students' activity and improvement of student's ability. The testing was conducted toward 10 students that yielded 2 clusters of membership data (C1 and C2). Cluster 1: with membership percentage of 70% and it consisted of 6 members, namely 1112438 Anggi Julian, 1112439 Anis Maulita, 1112441 Ardi Febriansyah, 1112452 Berlian Sinurat, 1112460 Dewi Anugrah Anwar and 1112467 Eka Tri Oktavia Sari. Cluster 2: with membership percentage of 30% and it consisted of 4 members, namely 1112463 Dosita Afriyani, 1112471 Erda Novita, 1112474 Eskardi and 1112477 Fachrur Rozi.

REFERENCES

[1] K J. Kim and H Alin"A Recommended System Using GA K-Means Clustering in an Online Shopping Market ", Expert System with Application Vol. 34, PP.1200-1290, 2008

[2] Miyazaki Y & Kurasige K, "Use of Reward Independent Knowledge on Reinforcement Learning for Dynamic Environment", ICACSIS 2010 Faculty of Computer Science Universitas Indonesia.

[3] Handoyo R, Rumani R, Nasution SM," Perbandingan Metode Clustering menggunakan Metode Single Linkage dan K-Means pada pengelompokan dokumen" JSM STMIK Mikroskil Vol 15, No.2 Oktober 2014 ISSN 1412-0100

[4] J. Santoso, G.O, Van Albada, B. Nazief, P.M.A. Sloot, "Hierarchical Job Schedulling for Cluster of Workstation", (Published Conference Proceeding Style in Proc the 6th ASCI Delft Netherland, June 2010.

[5] Celebi, M.E, Kingraul, HA & Vela P.A, "Comparative Study of Efficient Intralization Methods for the K-Means Clustering Algorithm", Expert System with Application, 2013.

[6] Grossman, David A dan Ophir Frieder, "Information Retrieval Algorithm and Heuristics", Second Edition Springer, The Netherland, 2004.

[7] Suhendi, "Analisa Penggunaan E-learning untuk meningkatkan kemudahan Mahasiswa dalam pembelajaran", Seminar Nasional Teknologi Informasi dan Multimedia 2015, STMIK Amikom Yogyakarta, ISSN: 2302-3805, 2015

[8] Surjono HD, "Membangun Course E-Learning Berbasis Moodle" I, UNY press Yogyakarta, 2011

[9] Rokach L, "A Survey of Clustering Algorithm", in O Malman &L. Rokach (eds). Data Mining and Knowledge Discovery Handbook (2nd ed) springer science + Business Media LLC.

[10] Rivani Admira, "Aplikasi K-Means Cluster untuk Pengelompokkan Provinsi Berdasarkan Produksi Padi, Jagung, Kedelai, dan Kacang Hijau tahun 2009", Jurnal Mat Stat, Vol.10 No.2 Juli 2010 :122-134. [11] Wijaya AK, "Implementasi Data Mining dengan Algoritma Fuzzy C-Means Studi Kasus Penjualan di UD. Subur Baru", Fasilkom Udinus, 2014

The Accuracy Of Fuzzy Sugeno Method With Antropometry On Determination Natural Patient Status

Dinur Syahputra[1], Tulus[2], Sawaluddin[3]

[1] Master of Informatics Engineering, University of North Sumatra, Indonesia, didinsyahdin231@gmail.com

[2] Lecturer of Computer Science and Information Technology, University of North Sumatra, Indonesia,tulus_jp@yahoo.com

[3] Lecturer of Computer Science and Information Technology, University of North Sumatra, Indonesia,sawal@usu.ac.id

Abstract. Anthropometry is one of the processes that can be used to assess nutritional status. In general anthropometry is defined as body size in terms of nutrition, then anthropometry is reviewed from various age levels and nutritional levels. Nutritional status is a description of the balance between nutritional intake with the needs of the body individually. Fuzzy logic is a logic that has a vagueness between right and wrong or between 0 and 1. Sugeno method is used because in the process of calculating nutritional status so far is still done by anthropometry. Currently information technology is growing in any aspect, one of them in the aspect of calculation with data taken from anthropometry. In this case the calculation can use the Fuzzy Sugeno Method, in order to know the great accuracy obtained. Then the results obtained using fuzzy sugeno integrated with anthropometry has an accuracy of 81.48%.

Keywords: Anthropometry, Nutrition Status, Fuzzy Sugeno

1. Introduction

Anthropometry is one of the processes that can be used to assess nutritional status. In general anthropometry is defined as body size in terms of nutrition, then anthropometry is reviewed from various age levels and nutritional levels. Anthropometry is very commonly used to measure nutritional status for various imbalances between protein-generated energy [20]. From a nutritional point of view has been much expressed by experts, one of them is [1] revealed that:

"Nutritional anthropometry is the measurement of the variations of the physical dimensions and the gross composition of the human body at different age levels and degree of nutrition".

Fuzzy Inference System (FIS) is one part of fuzzy logic. This FIS can make reasoning like a human being. FIS requires numbers as inputs and then translates them into linguistic terms such as small, medium, and large. There are several types of FIS namely Mamdani, Sugeno and Tsukamoto. Mamdani and Sugeno methods are two methods that researchers often use in their research, but Mamdani and Sugeno methods have their respective advantages and disadvantages. The advantages of the Mamdani method is that this method is more intuitive and the way it works is more like a human. However FIS Mamdani has a problem in terms of computing. While the Sugeno method is more adaptive in terms of optimization, computing is more efficient especially in terms of nonlinear dynamic systems [5,4,3].Sugeno method is used because in the process of calculating nutritional status so far is still done in anthropometry, so there is still a level of error in the calculation. Currently information technology is growing in any aspect, one of them in the aspect of calculation with data taken from anthropometry. In this case the calculation can use the Sugeno Method, which can anticipate errors with the rules - rules that will input.

The research is done by the calculation using Fuzzy Sugeno method with data taken from anthropometry in the determination of nutritional status of the patient, so it can give accurate result in determining nutritional status.

2. Fuzzy Logic

In his journal [2] said that fuzzy logic is a powerful concept for nonlinear handling, different timing, and adaptive systems. This allows the use of linguistic values of variables and improper relationships for system modeling behavior.

The seminal ideas of fuzzy logic applied to systems modelling and control can be found in the early papers of Zadeh in the 1960s and 1970s [15,16]. There are many interpretations of fuzzy modelling. For instance, a fuzzy set is a fuzzy model of human concept. In other words, fuzzy models consist of linguistic explanations about the system behaviour and deals with fuzzy modelling of a plant for control [9,16,17,18,19]. Just as modern control theory, a fuzzy controller can be designed based on a fuzzy model of a plant if the fuzzy model can be identified [12].

In this research is to assess the nutritional status of patients using fuzzy sugeno method which is generally done with anthropometry. With the aim of how much accuracy Fuzzy Sugeno with Anthropometry on the determination of nutritional status of patients, so it can provide benefits that on fuzzy sugeno also can be done assessment of nutritional status of patients.

3. Sugeno Fuzzy Inference

The Sugeno fuzzy model, also known as the TSK fuzzy model, was proposed by [9,10]in an effort to develop asystematic approach to generate fuzzy rules from a given input–output data set.A typicalfuzzy rule in Sugeno fuzzy model has the form

$$\text{If } x \text{ is } A \text{ and } y \text{ is } B \text{ then } z = f(x,y)$$

where A and B are MFs in the antecedent part, while $z = f(x,y)$ is a linear functionin the consequent part. Usually $f(x,y)$ is polynomial in the input variables x andy but it can be any function as long as it can appropriately describe the output ofthe model within the fuzzy region specified by the antecedent of the rule. When $f(x,y)$ is a first-order polynomial, the resulting fuzzy inference system is called afirst-order Sugeno fuzzy model which was proposed in [12,10,9]. When $f(x,y)$ is aconstant, it is zero-order Sugeno fuzzy model, which can be considered as a specialcase of Mamdani fuzzy model, in which the consequent of each rule is specified bya fuzzy singleton or by a pre-defuzzified consequent or a special case of Tsukamotofuzzy model in which the consequent of each rule is specified by an MF ofa step function.

Moreover, a zero-order Sugeno fuzzy model is functionally equivalent to aradial basis function network under certain minor constraints [13,14]. The output of zero-order Sugeno model is a smooth function of itsinput variables as long as neighbouring MFs in the antecedent have enoughoverlap.

4. Anthropometry

The term anthropometry [8] in his book comes from "anthro" which means man and "metri" which means size. Anthropometry by [6,7] is a collection of numerical data related to the physical characteristics of the human body, its contribution, shape and strength and the application of the data to design problem handling.

Body Mass Index (BMI) is a simple tool to monitor the nutritional status of adults, especially those related to deficiency and overweight. BMI use only applies to adults over 18 years of age and is not applicable to infants, children, adolescents, pregnant women, and athletes.

The formula for calculating BMI is as follows:

$$\text{BMI} = \frac{\text{Weight (kg)}}{\text{Height (m)x Height (m)}}$$

Based on the formula issued by [11], that category of Body Mass Index threshold for Indonesia as follows:

Table 1. BMI limit threshold category for indonesia

Nutritional Status	BMI (kg/m^2)
Very Thin	< 17.0
Thin	17.0-18.4
Normal	18.5-25.0
Fat	25.1-27.0
Very Fat	> 27.0

Body Mass Index released in the table above, then the nutritional status of a person based on the information below:
1) If the nutritional value <17.0, then the nutritional status is very thin
2) If the nutritional value is 17.0-18.4, then the nutritional status is thin
3) If the nutritional value is 18.5-25.0, then the nutritional status is normal
4) If nutritional value 25.1-27.0, then the nutritional status is fat
5) If the nutritional value> 27.0, then the nutritional status is very fat

5. Fuzzy Membership Function
The fuzzy set and the membership function of the variable weight, height, and nutritional value are represented as follows:
a. The Fuzzy Set of Weight Variables
In the weight variable is defined three fuzzy set, namely LIGHT, NORMAL, and WEIGHT. The fuzzy set image for the weight variable is shown in Figure 1.

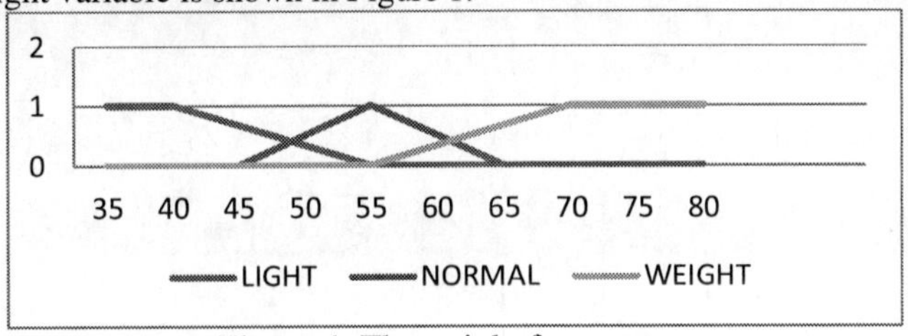

Figure 1. The weight fuzzy set

Where the horizontal axis is the input value of the variable weight, while the vertical axis is the membership level of the input value.
With membership function is as follows:

$$\mu_{Light} = \begin{cases} 1; & x \leq 40 \\ \frac{55-x}{55-40}; & 40 \leq x \leq 55 \\ 0; & x \geq 55 \end{cases} \quad (1)$$

$$\mu_{Normal} = \begin{cases} 0; & x \leq 45 \\ & or\ x \geq 65 \\ \frac{x-45}{55-45}; & 45 \leq x \leq 55 \\ \frac{65-x}{65-55}; & 55 \leq x \leq 65 \\ 1; & x \geq 55 \end{cases} \quad (2)$$

$$\mu_{Weight} = \begin{cases} 1; & 70 \leq x \leq 80 \\ \frac{x-55}{70-55}; & 55 \leq x \leq 70 \\ 0; & x \leq 55 \end{cases} \quad (3)$$

b. TheFuzzy Variable Height Set of Bodies

In the variable height is defined three fuzzy set, namely LOW, NORMAL, and HIGH. The representation of the fuzzy set for the height variables is shown in Figure 2.

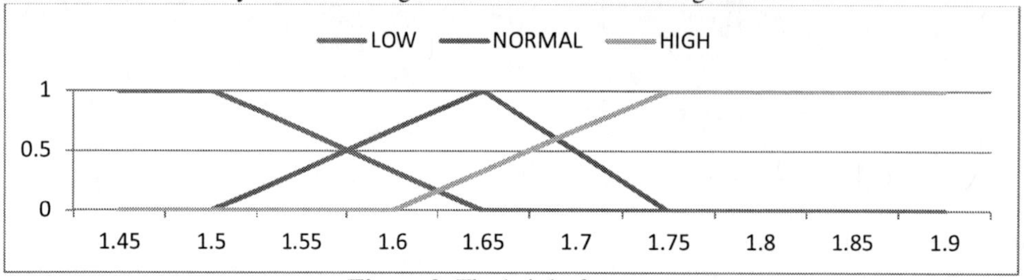

Figure 2. The height fuzzy set

Where the horizontal axis is the input value of the variable height, while the vertical axis is the membership level of the input value.

With membership function is as follows:

$$\mu_{Low} = \begin{cases} 1; & x \leq 1.5 \\ \frac{1.65-x}{0.15}; & 1.5 \leq x \leq 1.65 \\ 0; & x \geq 1.65 \end{cases} \quad (4)$$

$$\mu_{Normal} = \begin{cases} 0; & x \leq 1.5 \\ & or\ x \geq 1.75 \\ \frac{x-1.5}{0.15}; & 1.5 \leq x \leq 1.65 \\ \frac{1.75-x}{0.1}; & 1.65 \leq x \leq 1.75 \end{cases} \quad (5)$$

$$\mu_{High} = \begin{cases} 0; & x \leq 1.6 \\ \frac{x-1.6}{0.15}; & 1.6 \leq x \leq 1.75 \\ 1; & x \geq 1.75 \end{cases} \quad (6)$$

c. The Fuzzy Set of Nutritional Value Variables

The fuzzy set of nutritional values is obtained by the classification of the Body Mass Index (BMI), which is represented by the fuzzy set. In the nutritional value variables are defined five fuzzy set, namely VERY THIN, THIN, NORMAL, FAT, VERY FAT. The fuzzy set of images for the nutritional value variable is shown in Figure 3.

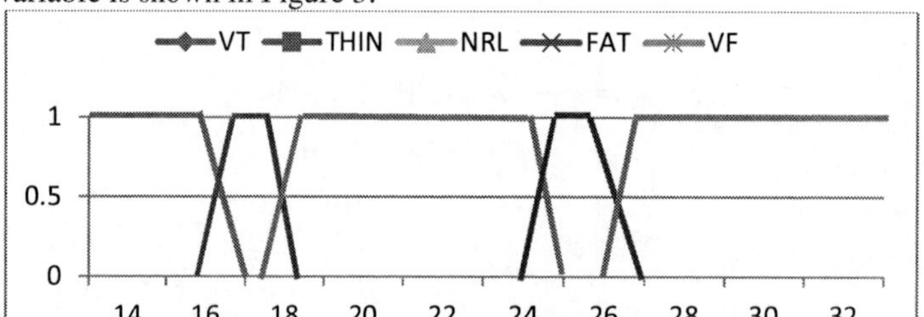

Figure 3. The nutrition fuzzy set

Information:
VT : Verry Thin
THIN
NRL: Normal
FAT
VF : Verry Fat

Based on Figure 3, the horizontal axis is the input value of the nutritional value variable, while the vertical axis represents the membership level of the input value.
With membership function is as follows:

$$\mu_{Very\ Thin} = \begin{cases} 1; & x \leq 16 \\ 17 - x; & 16 \leq x \leq 17 \\ 0; & x \geq 17 \end{cases} \quad (7)$$

$$\mu_{Thin} = \begin{cases} 0; & x \leq 16 \\ & or\ x \geq 18.5 \\ x - 16; & 16 \leq x \leq 17 \\ 1; & 17 \leq x \leq 17.5 \\ 18.5 - x; & 17.5 \leq x \leq 18.5 \end{cases} \quad (8)$$

$$\mu_{Normal} = \begin{cases} 0; & x \leq 17.5 \\ & or\ x \geq 25 \\ x - 17.5; & 17.5 \leq x \leq 18.5 \\ 1; & 18.5 \leq x \leq 24 \\ 25 - x; & 24 \leq x \leq 25 \end{cases} \quad (9)$$

$$\mu_{Fat} = \begin{cases} 0; & x \leq 24 \\ & or\ x \geq 27 \\ x - 24; & 24 \leq x \leq 25 \\ 1; & 25 \leq x \leq 26 \\ 27 - x; & 26 \leq x \leq 27 \end{cases} \quad (10)$$

$$\mu_{Very\ Fat} = \begin{cases} 0; & x \leq 26 \\ x - 26; & 26 \leq x \leq 27 \\ 1; & x \geq 27 \end{cases} \quad (11)$$

6. Fuzzy Rules

By category in BMI, the following rules may be established:

Table 2. Fuzzy rules in determining nutritional status

No	Weight	Height	Value	Status
1	40	1.5	17.77777778	Thin
2	40	1.65	14.69237833	Very Thin
3	40	1.75	13.06122449	Very Thin
4	55	1.5	24.44444444	Normal
5	55	1.65	20.2020202	Normal
6	55	1.75	17.95918367	Thin
7	70	1.5	31.11111111	Very Fat
8	70	1.65	25.71166208	Fat
9	70	1.75	22.85714286	Normal

7. Rule Evaluation and Accuracy

After the rule is formed, then evaluated the rule to see the results obtained from the fuzzy sugeno and anthropometry methods. And after the evaluation rule is done, then the next to evaluate the level of accuracy taken from anthropometry data.

Table 3. Rule evaluation results and accuracy

Data	Patient Status		Nutritional Value	Nutritional Status	Sugeno	Sugeno Status	Statement
	Weight (Kg)	Height (M)					
P001	62	1.52	26.83518	Fat	27.0713	Very Fat	False
P002	90	1.65	33.057851	Very Fat	24.99803	Normal	False
P003	50	1.49	22.521508	Normal	21.77778	Normal	True
P004	89	1.6	34.765625	Very Fat	27.51148	Very Fat	True
P005	55	1.64	20.449137	Normal	19.95313	Normal	True
P006	52	1.5	23.111111	Normal	22.96296	Normal	True
P007	63	1.55	26.222685	Fat	26.91508	Fat	True
P008	65	1.53	27.767098	Very Fat	29.86508	Very Fat	True
P009	60	1.51	26.314635	Fat	26.53811	Fat	True
P010	48	1.65	17.630854	Thin	16.18468	Very Thin	False
P011	67	1.5	29.777778	Very Fat	31.11111	Very Fat	True
P012	63	1.5	28	Very Fat	29.29293	Very Fat	True
P013	72	1.5	32	Very Fat	31.11111	Very Fat	True
P014	58	1.48	26.479182	Fat	25.92593	Fat	True
P015	62	1.54	26.142688	Fat	26.22731	Fat	True
P016	58	1.65	21.303949	Normal	20.8197	Normal	True
P017	74	1.6	28.90625	Very Fat	27.51148	Very Fat	True
P018	62	1.49	27.92667	Very Fat	28.50242	Very Fat	True
P019	75	1.5	33.333333	Very Fat	31.11111	Very Fat	True
P020	68	1.55	28.30385	Very Fat	29.31129	Very Fat	True
P021	70	1.65	25.711662	Fat	24.99803	Normal	False
P022	68	1.68	24.092971	Normal	24.47728	Normal	True
P023	62	1.65	22.773186	Normal	22.19013	Normal	True
P024	45	1.5	20	Normal	17.77778	Thin	False
P025	51	1.6	19.921875	Normal	19.72368	Normal	True
P026	87	1.6	33.984375	Very Fat	27.51148	Very Fat	True
P027	52	1.55	21.644121	Normal	21.08881	Normal	True

Based on the above table, the overall data were 27 patients where the correct statement was 22 patients and the wrong statement was 5 patients.

8. Performance Evaluation Method

From the data table 4.2. Shows that the results obtained using fuzzy sugeno integrated with anthropometry have the following levels of accuracy:

$$Accuracy\ rate\ = \frac{\sum correct}{\sum data} x100\%$$

$$Accuracy\ rate = \frac{22}{27} x100\% = 81{,}48\% \qquad (12)$$

So the accuracy level analysis in determining nutritional status based on Body Mass Index (BMI) produced by Sugeno fuzzy method has percentage of 81,48%.

Conclusions

Sugeno fuzzy method can be an alternative in determining nutritional status.Differences in the calculation of nutritional value using fuzzy sugeno and anthropometry methods. The accuracy level of fuzzy sugeno method with data taken from anthropometry of 81.48%.Procedure to obtain nutritional status is by adjusting the nutritional value that has been obtained into the function of the fuzzy set membership on the variable of nutritional value. The fuzzy set of nutritional values is obtained by the classification of the Body Mass Index (BMI), which is represented by the fuzzy set.

References

[1] Jellife, Derrick B.1996. *Child Nutrition in Developing Countries.* Washinton DC

[2] Hosseinzadeh, B., Zareiforoush, H., Adabi, M.E. & Motevali, A. 2011. Development of a Fuzzy Model to Determine the Optimum Shear Strength of Wheat Stem. *International Journal of Computer Science and Telecommunications*2(4): 56-60.

[3] Kaur, A., & Kaur, A. 2012. Comparison of fuzzy logic and neuro-fuzzy algorithms for air conditioning system. *International journal of soft computing and engineering*, 2(1), 417-20.

[4] Pang, L. M., Tay, K. M., & Lim, C. P. 2016. Monotone Fuzzy Rule Relabeling for the Zero-Order TSK Fuzzy Inference System. *IEEE Transactions on Fuzzy Systems,* 24(6), 1455 – 1463.

[5] Singla, J. 2015, March. Comparative study of Mamdani-type and Sugeno-type fuzzy inference systems for diagnosis of diabetes. In *Computer Engineering and Applications (ICACEA), 2015 International Conference on Advances in* (pp. 517-522). IEEE.

[6] Stevenson, M.G. 1989. *Principles Of Ergonomics, Center For Safety Science-University Of NSW*, Australia

[7] Nurmianto, E. 1991. *Ergonomics, Basic Concepts and Application.* ITSN , Surabaya

[8] Wignjosoebroto, Sritomo. 2000. *Ergonomics, Motion and Time Studies.* Jakarta: Guna Widya

[9] Takagi T, Sugeno M. 1985. *Fuzzy identification of systems and its application to modeling and control.* IEEE Trans Syst Man Cybern 15:116–132

[10] Sugeno M, Kang GT. 1988. *Structure identification of fuzzy model.* Fuzzy Sets Syst 28:15–33

[11] Directorate of Community Nutrition. 2003. *Technical Guidelines for Adult Male Nutritional Status Monitoring with Body Mass Index (BMI).* Director General of Binkesmas. Ministry of Health RI: Jakarta.

[12] Sugeno M, Yasukawa T. 1993. *A fuzzy-logic-based approach to qualitative modelling.* IEEE Trans Fuzzy Syst 1(1):7–31

[13] Jang J-SR, Sun C-T. 1993. *Functional equivalence between radial basis function networks and fuzzy inference systems.* IEEE Trans Neural Netw 4(1):156–159

[14] Jang JSR. 1994. Structure determination in fuzzy modelling: a fuzzy CART approach. In: Proceedings of international conference on fuzzy systems, Orlando, FL, pp 480–485

[15] Zadeh LA. 1968. *Fuzzy algorithms.* Inf Control 12:94–102

[16] Zadeh LA. 1973. *Outline of a new approach to the analysis of complex systems and decision process. I*EEE Trans Syst Man Cybern 3:28–44

[17] Zadeh LA. 1994. *The role of fuzzy logic in modelling.* Ident Control Model Ident Control 15(3):191–203

[18] Kovacic Z, Bogdan S. 2006. *Fuzzy controller design: theory and application.* CRC Press, Taylor and Francis Group

[19] Shin YC, Xu C. 2009. *Intelligent systems: modelling, optimistion and control.* CRC Press, Taylor and Francis Group

[20] Gibson, 2005. LILA Library Review. (Online) http://www.scribd.com/doc/46253718/Tinjauan-pustaka-Lila-Antropo-Dsb (Retrieved on July 28, 2017)

Optimization Of Feature Weight TheVoting Feature Intervals 5 Algorithm Using Partical Swarm Optimization Algorithm

Eka Hayana Hasibuan[1], Herman Mawengkang[2],and Syahril Efendi[3]

[1]Faculty of Computer Science and Informatics Technology, University Of Sumatera Utara, Indonesia,
[2]Faculty of Mathematic Science, University of Sumatera Utara, Indonesia
[3]Faculty of Computer Science and Informatics Technology, University Of Sumatera Utara, Indonesia
eka_hayana@yahoo.com, hayana_eka@students.usu.ac.id

Abstract. The use of Partical Swarm Optimization Algorithm in this research is to optimize the feature weights on the Voting Feature Interval 5 algorithm so that we can find the model of using PSO algorithm with VFI 5. Optimization of feature weight on Diabetes or Dyspesia data is considered important because it is very closely related to the livelihood of many people, so if there is any inaccuracy in determining the most dominant feature weight in the data will cause death. Increased accuracy by using PSO Algorithm ie fold 1 from 92.31% to 96.15% increase accuracy of 3.8%, accuracy of fold 2 on Algorithm VFI5 of 92.52% as well as generated on PSO Algorithm means accuracy fixed, then in fold 3 increase accuracy of 85.19% Increased to 96.29% Accuracy increased by 11%. The total accuracy of all three trials increased by 14%.In general the Partical Swarm Optimization algorithm has succeeded in increasing the accuracy to several fold, therefore it can be concluded the PSO algorithm is well used in optimizing the VFI5 Classification Algorithm. Keywords: Partical Swarm Optimation, Voting Feature Intervals, fold, Accuracy

1. Introduction

Research conducted [1] using the classification algorithm Voting Feature Intervals 5 (VFI5) the average accuracy of Pulmonary TB data model is 83%. The feature weight used in the study is one for each feature. The feature weight used in the study is one for each feature. Accuracy of the study is still predicting many wrong data. In a previous study [2] using a genetic algorithm to find the weight of the K-NN feature and the result was applied to the VFI5 classification algorithm, and the accuracy obtained from 95% to 99%, therefore in this study the accuracy of the previous study [1] Can increase.Pre-existing research [3] using the Voting Feature Intervals 5 (VFI5) Classification algorithm on feature data of DM Disease (Diabetes Mellitus) versus Dyspepsia (Dys). This research uses 3-fold cross validation that is divided into first iteration, second iteration and third iteration, each with combination of different training and test data. The first iteration yielded an accuracy of 92.31% with two predictive errors in the 41st DM patient and 80th dyspepsia patient, a second iteration accuracy of 92.59% with 2 predictive errors in the 21st DM patient and the dyspeptic patient to- 67 and a third iteration accuracy of 85.19% with 4 predictive errors in 14th DM patients, 51st dyspepsia patients, 52nd dyspepsia patients, 57th dyspepsiapatients. Theresults of the feature hose on the training process and the final normalization in this study are similar. Subsequent research [4] is Optimization of Voting Feature Intervals 5 Algorithm with Genetic Algorithm on Tubercolysis of Lung Data results in improved accuracy by using genetic algorithm ie in fold 1 of 76% -95% accuracy increase of 19%. Improved accuracy in fold 2 from 84% -95% increase in accuracy 11%. Improved accuracy in fold 3 from 89% -95% increase in accuracy of 6%. In general, genetic algorithms have improved accuracy for each fold. Therefore it can be concluded that a good genetic algorithm is used in optimizing the VFI5 Classification Algorithm.

Particle Swarm Optimization (PSO) is a frequently applied method for solving optimization problems in nonlinear cases. PSO produces more optimal weight as an alternative Back Propagation (BP) in the ANN training [5]. PSO provides optimal results in clusters center as an alternative to KMeans on the problem of pattern classification or pattern classification [6] The combination of PSO with Support Vector Machine (SVM) on project scheduling optimization results in more performance [7]. The combination of PSO with Statistical Clustering (SC) on the feature classification problem shows that both algorithms can select features with much smaller numbers and achieve better classification performance [8]. The PSO algorithm is also used to optimize the Power of Communication Network with excellent results because the PSO combines three advantages of having good search performance, continuous variable control and discrete [9].

The formulation of the problem in this research is to find the accuracy of the optimal feature weight. The purpose of this research is to get the most optimum feature weight accuracy on VFI5 classification algorithm using PSO algorithm so as to find PSO making model with VFI5. Limitation of the problem in this research is weight data feature used is data of DM symptom symptom that is dizziness, nausea, vomiting, weakness, fever, heartburn, shortness of breath, chest pain, diarrhea, cough [3].

2.Voting Feature Intervals 5

The VFI5 classification algorithm represents a concept that describes the concept of interfacial hose. This algorithm was developed by GülşenDemiröz and Halil Altay Güvenir in 1997. It has been argued that the VFI5 algorithm is included in a supervised algorithm, ie an algorithm that has a target of data classes and is non-incremental meaning that all training data are processed simultaneously [10]. VFI algorithm continues to grow until the final version of the VFI5 algorithm. The process of classifying new instances is based on the feature vote. Each of these features gives its value to the votes among the classes. The class that receives the highest vote value is then specified as the predicted class. The advantage of the VFI5 algorithm is that the algorithm is robust against the irrelevant features but is able to deliver good results on the existing real world dataset. VFI5 is also capable of eliminating the unfavorable effects of features that are irrelevant to its voting mechanism [2].

2.1 Training

This training process aims to find the model that will be used for the classification process. In this training process will be generated hose on each feature. A hose represents the set of values of a given feature.The end point on the hose should be known to produce a specific feature hose. The process of finding different end points for linear features and nominal features. For linear features the values have a sequence and are comparable in rank. To determine the end point of this feature by finding the maximum and minimum values for that feature for each class.

2.2 Prediction

If ef is known then the interval can be found. The interval can store training instances of some classes. The classes in an interval are represented by the votes of the 2 classes at that interval. For each class c, feature f gives the same vote with interval_class_vote [f, I, c]. The notation represents the vote feature f given for class c. Each feature f gathers its votes in a vector <feature_vote [f, C1], ..., feature_vote [f, Cj], ..., feature_vote [f, Ck]>, where feature_vote [f, Cj] is a vote feature f For class Cj and k is the number of classes.Then d vector vote, where d is the number of features, summed to get the total

vector vote <vote [C1], ..., vote [Ck]>. The class with the biggest vote is predicted as the class from the test instance e. Pseudocode prediction algorithm VFI5 can be seen in figure 1

3. Partical Swarm Optimization

Algorithm Particle Swarm Optimization (PSO) is widely used to solve optimization problems as well as feature selection issues [11]. Optimization is the process of adjusting to input or device characteristics, mathematical processes, or experiments to find the minimum or maximum output of results. The input consists of a variable, a process or function known as a cost function, an objective function or a function and an output capability is a cost or a goal, if the process is an experiment, then a variable is the physical input for the experiment [12]. PSO is a population-based method like GA, but its basic concept is cooperation, not competition [13] [14]. A mathematical formula that describes the position and velocity of the particles on A certain dimension of space:
Xi(t) = xi1

$$Xi\ (t) = xi1\ (t), Xi2\ (t), \ldots xiN\ (t) \ldots\ldots\ldots (1)$$

$$Vi\ (t) = vi1\ (t), vi2\ (t), \ldots viN\ (t) \ldots\ldots\ldots\ldots (2)$$

Where : X = particle position V = particle velocity I = particle index T = the iteration of the t N = size of the space dimension Here is a mathematical model that describes the mechanism Updating the particle status of Kennedy and Eberhart 1995:

$$Vi\ (t) = Vi\ (t-1) + c1r1\ (XiL - Xi\ (t-1)) + c2r2(XG - Xi\ (t-1)) \ldots\ldots (3)$$

$$Xi\ (t) = Vi\ (t) + Xi\ (t-1) \ldots\ldots\ldots (4)$$

Where = 1, 2 , ..., represents the local best of the i-particles. Whereas = 1, 2 , ..., represent the global best of the whole herd. While c1 and c2 are a constant constant positive constantsReferred to as learning factor. Then r1 and r2 are a random numberWhich ranges from 0 to 1. Equation (1) is used to calculate the velocity of a new particle based on its previous velocity, the distance between the current position with the best particle position (local best), and the distance between the current position and the best position of the flock (global best) . Then the particle flies toward the new position based on equation (2). After the PSO algorithm is run with a certain number of iterations to reach the criteria of dismissal, it will get a solution located in the global best.

4. Research methodology

The research methodology used can be seen in Figure 2.

4.1. Data Sharing and Training Data sharing using 3 fold cross validation, same with previous training. The data will be divided into 2 for training data and 1 for many test data. The training was conducted using the VFI5 algorithm, where the specified trainer data was used as input of the VFI5 algorithm in the training process. Symptoms of disease as a feature of each patient, while DM and Dyspepsia is a class of data.

4.2. Normalization / Hose and Features

The lapse of each feature is obtained from the training process. Before determining the lapse do normalization on the training process. Hose contains a vote value for each class on each symptom.

4.3. Classification Classification is divided into 3 stages: first stage determination of vote value in test data, second stage summing value of each instances, and determine prediction class.

4.4. PSO optimization The process that will be run to optimize the feature weight with several stages as below:

a. Determining the number of particles (particle size), the range of particles is 10 to 40 particles.
b. Determines maximum speed (MaxVelocity), Maximum velocity (v) is set for particle displacement. If the velocity V is between -10 s / d 10, then the maximum speed is 20.
c. Learning factor Learning factors (c1 and c2) generally have a value of 2. Different The problem will also be different in value and the range is 0-2.
d. Stop condition Includes maximum number of iterations of PSO Algorithm and error conditions Reach a minimum. The stop condition depends on the optimized problem.
e. Inertia weight Inertia weight used is w = 1

4.5. Accuracy

The final stage of this research method is to calculate the accuracy of the results obtained at the data processing stage using VFI5 algorithm. Accuracy can be calculated by:

$$\text{Accuracy} = \frac{\sum \text{data classification}}{\sum \text{total data}} \times 100\%$$

5. Result and Discussion

5.1. 3- Fold Cross Validation The data is processed using Voting Feature Intervals (VFI5) algorithm. The first step is the training process, where each data sought the minimum and maximum value up to get the value of his vote weight.

Table 1. Distribution of training and testing data

Data Training	Data Testing
Subset I	Subset III
Subset II	
Subset I	Subset II
Subset III	
Subset I	Subset I
Subset III	

Table 2. Accuracy of each fold Using VFI5

Accuracy VF15%	
Fold 1	92.30
Fold 2	92.59
Fold 3	85.19
Avrg	90.03

5.2. Experiment to find the optimum feature VFI5 weight with PSO algorithm

The representation of particles used is real number encoding and decoding. Real Number Encoding is used to represent the weight of each feature to be optimized. Decoder process is done by multiplying the weight of each feature with the feature vote of each class in each fold to find the vote value of each class. After searching the vote value of each class counts the percentage of each vote for each class, the process is a decode process presented in Appendix 1, Table 3 for weighting VFI5 and Table 5 for weighting with PSO Algorithm.

5.2.1. Representation of Particles

This experiment aims to determine the parameters of the most optimum PSO algorithm, which consists of testing particle size parameters, learning rate, inertia to obtain the best global best match until the calculation process stalled.

5.2.2. Determination of the optimum parameters In this experiment,

The relationship of PSO algorithm parameters such as population, inertia, personal learning rate with global learning rate is very influential.

5.2.3. . First Experiment

In the first experiment using parameter W = 1, Max Iteration = 1000, C1 = 1.0, C2 = 1.0. Global Best earned 2.69. Predictive errors are in the class, so in Table 4 the DM ID ID values on features with greater values in dyspepsia ID are nausea, fever, heartburn, chest pain, diarrhea, and cough. The second error in VFI5 is in the 80 dyspepsia ID data predicted into the DM class on features with a larger vote value on ID DM is dizziness, weakness, shortness of breath, diarrhea and cough. In Appendix 2, comparing data of normalization data after using PSO, it is found that the 80th data ID with dyspepsia class is correctly classified into the actual class ie dyspepsia. The result of the calculation on the fold to-one is 92.31% with 1 prediction error. 5.2.4. The Second Experiment In the second experiment using parameters W = 1, Max Iteration = 1000, C1 = 1.5, C2 = 1.5. Global Best obtained by 0. Data prediction errors are in the 21st ID and 67th ID and are presented in Appendix 3 of Table 5. The first incorrect data vote prediction value in the 21st ID of DM, where ID DM is predicted to ID Dyspepsia because of this DM ID that features dizziness, nausea, vomiting and weakness is actually the main characteristic of DM disease, but ID DM also features with a higher vote value on features of fever, heartburn, diarrhea and cough. Error in the ID of the 21st DM can be said that this ID has DM disease that has been complicated with dyspepsia. Accuracy in the second experiment was 92.59% with 2 data prediction errors.

5.2.5. The Third Experiment In the third experiment use parameter W = 1, Max Iteration = 1000, C1 = 2.0, C2 = 2.0. Global Best earned 3,704. Error Dyspesia vote ID value on feature with bigger value in ID DM. The data are presented in Appendix 4 of Table 6. Prediction errors on classes in the VFI5 algorithm are not always the same as PSO algorithm results. The accuracy of the third experiment was 96.29% with one prediction error. From the above experiments, each of the accuracies presented in

Table 3.Accuracy of all experiments

Accuracy VF15 + PSO %	
Fold 1	96.15
Fold 2	92.59
Fold 3	96.29
Avrg	95.01

5.3. Optimization Results

In each experiment the parameter size of the PSO algorithm produces better weight to produce better accuracy. In this research we get the pattern of weight of each feature after optimized with PSO algorithm. This study shows that if the population size is greater than the size of the generation the weights are more diverse, and the determination of the parameters greatly affects the search on global best.

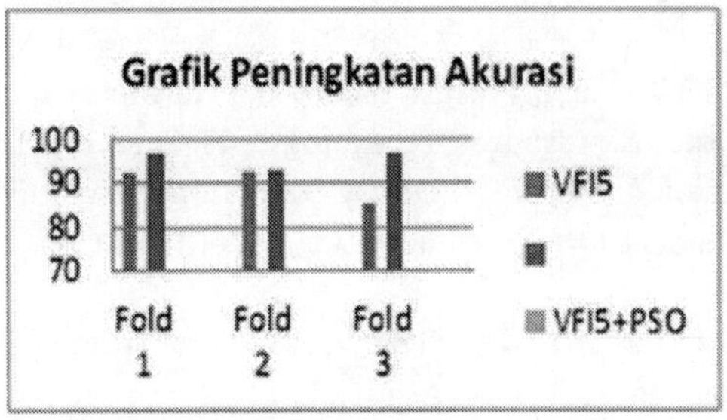

Figure 1. Graph of Increased accuracy

From all experiment in this research, it is found that the value of hose result from training generally have the same pattern between the first iteration to the third iteration that is for the features of dizziness, nausea, vomiting and weakness are common symptoms of ID DM, while for fever feature, Heartburn, shortness of breath, chest pain, diarrhea and cough are symptoms of dyspepsia ID. But in the second iteration, the results of the training value are quite similar to the chest pain feature so that the value of the interval is not much different between DM class and dyspepsia class which means that for chest pain feature can be owned by both ID. The results of the training hose can be seen in Appendix 5.In Figure 3 shows an improved graph of experimental accuracy using the VFI5 algorithm and the algorithm to optimize the PSO. The accuracy of the previous VFI5 algorithm in fold 1 is 96.15% with 1 prediction error of 3.84% accuracy, in the second fold the accuracy of the VFI5 algorithm is 92.59%, but no increase in the PSO calculation. The calculation on the third-fold accuracy generated by the VFI5 algorithm of 85.19% increased to 96.29% and an increase of 11%.

6. Conclusions

This research has implemented the Optimization of Voting Feature Algorithm Interval5 (VFI5) using PSO Algorithm.In general, the Partical Swarm Optimation algorithm has improved accuracy for several folds, therefore it can be concluded that the PSO algorithm is well used in optimizing the VFI5 Classification Algorithm.Increased accuracy by using PSO Algorithm ie fold 1 from 92.31% to 96.15% increase accuracy of 3.8%, accuracy of fold 2 on Algorithm VFI5 of 92.52% as well as generated on PSO Algorithm means accuracy fixed, then in fold 3 increase accuracy of 85.19% Increased to 96.29% Accuracy increased by 11%. The total accuracy of all three trials is 14%.

7. References

[1] Rosyid, Abdul 2008. Diagnosis of Pulmonary Tuberculosis Using VFI5, Bogor: Department of Computer Science, Bogor Agricultural University.
[2] Guvenir HA, Emeksiz N, 2000. An expert system for the differential diagnosis of Erythemato-squamous diseases. Www.elsevier.com/locate/eswa
[3] Hayana, Eka 2010. Diabetes Mellitus Disease Prediction Using Algorithm Classification Voting Feature Intervals 5, Bogor: Department of Computer Science, Bogor Agricultural University.
[4] Kristin, Ervina. 2010. Optimization of Voting Feature Intervals Algorithm 5 Using Genetic Algorithm On Pulmonary Tuberculosis Data, Bogor Department of Computer Science, Bogor Agricultural University.
[5] D. Dutta, R. Roy and K. Choudhury, Training artificial neural network using Particle swarm.International Journal of Advanced Research in Computer Science and Software Engineering, vol. 3, no 3, pp. 430- 434, 2013.
[6] M. Nayak, GPAC-APSO Clustering using modified s-transform for data mining. International Journal Of Engineering Science & Advanced Technology, vol. 2, no.1, pp. 38 - 48,2012
[7] H. Yang, C-H. Hsiao and L-Y Chuang, Linearly Decreasing Weight Particle Swarm Optimization with AcceleratedStrategy for ata Clustering. International Journal of Computer Science, vol. 37, no. 3, pp. 3-9, 2010.
[8] Lane, M.C., Xue, B. & Liu, I. 2013. PSO based selection of spectral features for remotely sensed image classification. International Journal on Artificial Intelligence Tools 22 (4): 214218.
[9] Wang, Yuhuai, Wang Qihui, Zhang Huixi, An Kang, Ye Xia and Sun Yaping 2016, Improved Particle Swarm Optimization Algorithm for Optimization of Power Communication Network. International Journal of Grid and Distributed Computing Vol. 9, No. 1 (2016), pp.225-236 http://dx.doi.org/10.14257/ijgdc.2016.9.1.23
[10] Demiroz G, Guvenir HA, 1997. Classification by Voting Feature Intervals. http://www.cs.ucf.edu/~ecl/papers/miros97classification.pdf.
[11] Y., Wang, G., Chen, H., Dong, H., Zhu, X., & Wang, S. Liu, "An Improved Particle Swarm Optimization for Feature Selection," Journal of Bionic Engineering, pp. 8(2), 191–200. doi:10.1016/S1672-6529(11)60020-6, 2011.
[12] R. L., &Haupt, S. E. Haupt, Practical Genetic Algorithms. United State of America: A John Wiley & Sons Inc Publication, 2004.
[13] Z.-L. Gaing, "A particle swarm optimization approach for optimum design of PID controller in AVR system," IEEE Trans. Energy Conversion, vol.19, pp. 384-391, June 2004.
[14] Kennedy, J. and Eberhart, C., "Particle Swarm Optimization," Proceedings of the 1995 IEEEInternational Conference on Neural Networks, Australia, pp. 1942 -1948, 1995.

Improved Fuzzy K-Nearest Neighbor Using Modified Particle Swarm Optimization

Jamaluddin[1], **Rimbun Siringoringo**[2]

[1, 2]Universitas Methodist Indonesia Jl. Hang Tuah No 8 Medan-Indonesia

[1]jac.satuno@gmail.com

Abstract- Fuzzy k-Nearest Neighbor (FkNN) is one of the most powerful classification methods. The presence of fuzzy concepts in this method successfully improves its performance on almost all classification issues. The main drawbackof FKNN is that it is difficult to determine the parameters. These parameters are the number of neighbors (k) and fuzzy strength (m). Both parameters are very sensitive. This makes it difficult to determine the values of 'm' and 'k', thus making FKNN difficult to control because no theories or guides can deduce how proper 'm' and 'k' should be. This study uses Modified Particle Swarm Optimization (MPSO) to determine the best value of 'k' and 'm'. MPSO is focused on the Constriction Factor Method. Constriction Factor Method is an improvement of PSO in order to avoid local circumstances optima. The model proposed in this study was tested on the German Credit Dataset. The test of the data/The data test has been standardized by UCI Machine Learning Repository which is widely applied to classification problems. The application of MPSO to the determination of FKNN parameters is expected to increase the value of classification performance. Based on the experiments that have been done indicating that the model offered in this research results in a better classification performance compared to the Fk-NN model only. The model offered in this study has an accuracy rate of 81%, while. With using Fk-NN model, it has the accuracy of 70%. At the end is done comparison of research model superiority with 2 other classification models;such as Naive Bayes and Decision Tree. This research model has a better performance level, where Naive Bayes has accuracy 75%, and the decision tree model has 70%.

Keywords: fuzzy k-nearest neighbor, modified particle swarm optimization, german credit data

1. Introduction

Non-performing loans is one of the most important issues on industrial and financialservices [1]. At a certain level, accumulation of loan defaults can trigger bankruptcy of a bank and other financial institutions.Learn about the background and characteristic of financial customers a factor that significant before deciding to give credit.To answer these problems, one concern of funding or banks is how to build an assessment technique to ensure eligibility of a customer on the filing of credit [2] Machine Learning is a field of information technology that has been widely applied to build decision support system, especially in the field of economics and finance. Machine learning plays a very important role and has resulted in various research studies that have been applied to related to credit risk assessment. The following is a variety of machine learning-based research that has succeeded in building the instruments and models used in the effective credit risk predictors and estimators. The researches are: K-Nearest Neighbor Method [3], [4], Fuzzy K-Nearest Neighbor [5], Bayesian

Network [6], Support Vector Machine [7], Artificial Neural Network [8], Fuzzy Immune Learning [9] and Logistic Regression [10].

The k-Nearest Neighbor (k-NN) method is the most popular machine-learning method, simple and easy to implement (Wang & Li, 2010), k-NN has two weaknesses. Firstly, the success of this method depends on the number of neighbors applied, so in order to produce a high degree of accuracy,researchers should try different values of k with varying amounts, of course this is not effective because it is done manually. This may be reflected in the research conducted [3], by applying the various k, the best accuracy is obtained at k = 3, while [5] obtains the best accuracy at k = 13. Second, in addition to the reliance on the k value, the relation between data with classes is rigid (crisp), where each instance/event/incident has only a relationship with one class exclusively, whereas with the other classes, it has no relationship at all.

Many attempts have been made to avoid the k-NN stiffness properties. One attempt is to combine fuzzy theory into k-NN. The application of the fuzzy theory produced a new method known as Fuzzy k-Nearest Neighbor (Fk-NN) [11]. In the Fk-NN model, the relationship between data and classes is not crisp, each data and class has membership relationship with a certain level. The strength of the relationship requires a fuzzy strengh (m) parameter. Compared to k-NN, Fk-NN always results in a higher degree of accuracy in all classification problems [12]. It is also the main reason why the Fk-NN method becomes the preferred method of interest in many researches, especially related to the economic issues. The fuzzy strength parameters (m) and the neighboring number parameters (k) are fundamental determinants of the Fk-NN model. It means that the values of m and k have a direct impact on the model accuracy. Determining the values of 'm' and 'k' is often not easy and difficult to control because there is no theory or guide that concludes how appropriate 'm' and 'k' should be [12]. To answer the problem, it is necessary to have another method that can help the Fk-NN model to find the value of m and k.

In this study,we offer an approach to assess the optimization parameter solution in order to determine the best value of m and k. In this study, weapplied Modified Particle Swarm Optimization (MPSO). The application of the method is based on some considerations. The first consideration, compared with similar algorithms, such as Genetic Algorithm (GA), PSO is relatively simple because it does not have many procedures such as selection, mutation and crossover procedures in GA. The second one, the PSO method has been proven to optimize the parameters of other machine learning methods. This is shown through the following studies. PSO is well-suited in combination with SVM [13], PSO and Artificial Neural Network [14], PSO with Self Organizing Map or SOM [15]. The results obtained in these models indicate improved accuracy through the application of the PSO in optimizing parameters. In this study, MPSO (Modified Particle Swarm Optimization) which is another variant of PSO was used to optimize Fk-NN parameters. This research built a model to evaluate the granting of credit based on FK-NN and MPSO classification.In other words, optimization of Fk-NN parameters by MPSO is expected to increase classification accuracy.

2. Research model

The FKNN-MPSO Classification Model This research was conducted in several stages, including pre-process data stage, selection of parameters k, m optimal using MPSO and FKNN. The procedures can be explained in the following steps:

Step 1 : Initialize the values of'k' and 'm'. In the FK-NN model, the values of 'k' and 'm' must be firstlyinitialized.

Step 2 : Generate the initial particles randomly. The values of 'k' and 'm' are generated in parallel as many as 8 particles.
Step 3 : The k and m values above are used to train the data.
Step 4 : Each particle is evaluated based on the fitness value. The fitness function is used to calculate fitness or the level of goodness of an individual to survive. This function takes individual parameters and produces the fitness value of the individual. For each problem to be solved with the PSO algorithm must be defined by a fitness function. In this study, the fitness function is presented in the equation (8) = (+) (+ + +) (8)
Step 5 : Update the position and the velocity of the particles. The PSO procedure requires that the position and the velocity of each particle be evaluated by equations (2.11) and (2.12).
Step 6 : At this stage, again testing FK-NN model to find the best particle. The best particles evaluated by the fitness value in step 4 were applied to train the training data and to calculate its fitness value
Step 7 : Update personal optimal fitness (pfit) and personal optimal position (pbest). So far all the particles in the first iteration have been evaluated, the first iteration yields the best particle (pbest).
Step 8 : At this stage it is evaluated whether all iterations have been completed. If it is not completed yet, resumed in the step 3, if it is completed then proceed to step 10.
Step 9 : Each iteration has the best particles. At this stage the best particles of all iterations will be evaluated to determine the best global particles optimal position (gbest).
Step 10 : If the gbest value in step 10 is the expected value then at that moment has obtained the optimum k and m value and continued to step 12. If the gbest value has not matched the expected criterion yet, then the new resurrected population is back to the step 4.
Step 11 : Perform data classification testing by entering the data test/the tested data.
Step 13 : Cross validation to see if the classification result in the step 12 has the best accuracy. If yes then the whole process is completed, if not then do the evaluation the value of 'k' ntil the smallest accuracy value is obtained

3. Experimental Design

A. Dataset

The dataset used in this study is German Credit Data, from UCI Machine Learning Repository. This dataset was chosen for several considerations. Firstly, it does not contain missing values [22]; secondly, this dataset also belongs to a high-dimensional dataset with 21 attributes (Al through A21) and 1000 data. The dataset consists of 1000 data, 21 attributes including class attributes. Datasets are grouped into 2 classes, ie 700 instances for the good classes and 300 instances for the bad ones. This dataset does not contain missing value or an empty value [23]. Table 2 shows the description of the dataset and the description of its attributes is shown. The dataset are available in the .arff format (attribute relation file format attribute).

Table 1.German credit data description

Number	Dataset	German Credit Data
1	Attribute type	Category
2	Number of attribute	20
3	Number of instance	100
4	Missing value	No
5	Number of class	2

Table II german credit data description Dataset German Credit Data Atributte type Categoryand Number of atributte 20 Number of instance 1000 Missing value No Number of class 2

B. Modified Particle Swarm Optimization

1) Parameter Setting: The MPSO (Modified Particle Swarm Optimization) algorithm is a parametric method, where its application requires parameter setting. In table III below is given a list of parameters of MPSO applied.

2) Particle Encoding: A particle encoding scheme is performed by generating random numbers ranging from 0 to 10 with a length of 20 numbers. Of the 20 numbers, the numbers 1 through 15 are used for the generation of parameters m, numbers 16 to 20 are applied for the generating parameters k. Figure 2 shows a particle coding scheme.

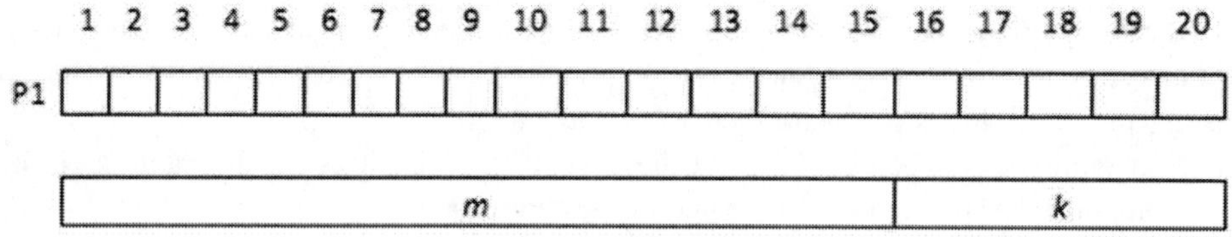

Figure 1. Particle coding scheme

4. Research Results

MPSO algorithm begins with swarm inizilization process. In the table IV, the swarm initialization scheme is displayed. The table segmentation consists of 3 parts: particle composition, parameters (k, m) and fitness value. Swarm consists of 10 particles (P1-P10), each particle contains of 20 random numbers, k and m values and the fitness.For example, particle-1 consists of 20 random numbers (7 1 2 1 3 5 3 9 0 4 4 6 6 9 2 4 0 9 7 0). The value of k, m which is raised on the particle is 13, 2.31,and the value of k, m obtained the fitness value 0.76 or 76%. Update the particles: In the next swarm, each particle will be updated until it reaches the target fitness value of 1.00 (100%). If this value can not be achieved then the highest value will be taken as a solution. Table V displays the particle to swarm-100, and resulted in the improvement of k, m and fitness values . Renewal of k and m value and fitness value improvement Based on the Table V below, , the best fitness value is 0.81 (81%) and the value of k, m (31, 13.23).

Table 2. Global Best

S	*G_{BEST}*	*k, m*	*fitness*
100	91 85 86 85 87 89 87 93 84 88 88 90 90 93 86 88 84 93 91 84	31, 13.23	0,81

Global Best (GBest): Global Best or GBest isthe particles with the greatest fitness value of all swarm available. Tableable VI shows GBest. The table contains information about swarm, particle and fitness values (Global Best fitness). The GBest position is achieved in the 100th swarm (S100),k and mvalues (k, m) are31, 13.23 and the fitness is 0.81 (81%). Table III Global Best S GBest k,m fitness. 100 91 85 86 85 87 89 87 93 84 88 88 90 90 93 86 88 84 93 91 84 31, 13.23 0,81

Influence of the Particle Number: To examine whether there is an effect of applying the number of particles to the increase of fitness value (in this case is the accuracy), this study applied the number of particle varies i.e 10 particles, 20 particles and 40 particles. The results are shown in the table VII, below. Based on the three tests, it can be indicated that the increase in the number of particles has no significant effect on fitness improvement.

Table 3. Test Result 10 Particle 20 Particle 40 Partikel No. k,m fitness No. k,m fitness No. k,m fitness

10 Particle			*20 Particle*			*40 Particle*		
No	***K, m***	***Fitness***	**No**	***K, m***	***Fitness***	**No**	***K, m***	***Fitness***
1	15, 3.09	0,76	1	9, 3.87	0,77	1	13, 3.87	0,78
2	15, 3.09	0,76	2	17, 4.65	0,77	2	13, 3.87	0,78
3	15, 5.43	0,79	3	17, 6.21	0,77	3	19, 3.87	0, 78
4	31, 12.45	0,8	4	23, 9.33	0,77	4	25, 6.21	0,8
5	15, 7.77	0,8	5	25, 9.33	0,77	5	29, 10.89	0,8
6	17, 7.77	0,8	6	25, 10.11	0,8	6	29, 10.89	0,8
7	23, 8.55	0,8	7	31, 10.11	0,8	7	31, 12.45	0,8
8	27, 9.33	0,8	8	31, 10.11	0,8	8	31, 12.45	0,8
9	31, 11.67	0,8	9	31, 10.11	0,8	9	31, 12.45	0,8
10	31, 11.67	0,8	10	31, 10.11	0,8	10	31, 12.45	0,8
11	31, 12.45	0,8	11	31, 10.11	0,8	11	31, 12.45	0,81
12	31, 13.23	0,8	12	31, 10.11	0,8	12	31, 13.23	0,81
13	31, 13.23	0,81	13	31, 13.23	0,81	13	31, 13.23	0,81
14	31, 13.23	0,81	14	31, 13.23	0,81	14	31, 13.23	0,81
…	…	…	…	…	…	…	…	…
100	31, 13.23	0,81	100	31, 13.23	0,81	100	31, 13.23	0,81

6) Global Seeking: Global Seeking is the GBest tracing process starting from the first swarm to the last swarm. In this penance GBest is the fitness with the greatest value, so the search process will form the ascending graph as shown in figure 3.

C. Testing Performance model classification Fk-NN + MPSO

The results showed that the Fk-NN + MPSO model resulted in better classification performance than the Fk-NN model. The comparison of both performance model is presented in the table X and Figureure 4.

Table 4. Comparison of Fk-NN model performanced with Fk-NN + MPSO model

Performance (0 to 1)	**Model**	
	Fk-NN	**Fk-NN + MPSO (This Experiment)**
Accurancy	0,71	0,81
Precision	0,78	0,93
AUC	0,77	0,84

Fk-NN + MPSO (This Experiment) Accuracy 0,71 0,81 Precision 0,78 0,93 AUC 0,77 0,84

D. The Comparison with2 SuperiorityModels

To validate the superiority of the FK-NN + MPSO model in predicting creditworthiness, the performance of FK-NN + MPSO method iscompared twoother models; the Naive Bayes method and the Decision Tree method. Theperformance results using the Naive Bayes method, the Decision Tree method is presented in table XI. Figure 5 shows the comparison in a graphical form.

Table 5. Comparison of model superiority

Performance (0 to 1)	**Model**			
	Fk-NN	**Naïve Bayes**	**Decission tree**	**Fk-NN + MPSO (This Experiment)**
Accurancy	0,71	0,75	0,70	0,81
Precision	0,78	0,74	0,68	0,93
AUC	0,77	0,78	0,64	0,84

5. CONCLUSION

This research offers a model of Fuzzy k-Nearest Neighbor (Fk-NN) classification based on the Modified Particle Swarm Optimization (MPSO) algorithm. The application of MPSO to Fk-NN aims to improve classification performance through the optimization of adjacent (k) and fuzzy (m) parameters. MPSO application can eliminate the aspect of sujektifitas in determining parameters k and m. The result of the research shows that the MPSO application in the classification process has improved the accuracy of Fk-NN, thereby affecting the performance improvement of creditworthiness classification. The Fk-NN and MPSO models can work on high-dimensional data such as germancredit datasets. This study also compared the superiority of the FK-NN + MPSO model with other classification models namely Naive Bayes and Decision Tree. The result shows that FK-NN + MPSO model provides superior performance compared to the two models.

6. BIBLIOGRAPHY

[1] Lee,. M-C.Enterprise Credit Risk Evaluation models: A Review of Current Research Trend. International Journal of Computer Applications, 44(11) : 0975 – 8887.2012

[2] Ghatasheh, A. Business Analytics using Random Forest Trees for Credit Risk Prediction: A Comparison Study. International Journal of Advanced Science and Technology, 72 (2014) : 1930.2014.

[3] Abdelmoula, A. K.Bank credit risk analysis with k-nearestneighbor classifier: Case of Tunisian banks. Accounting and Management Information Systems, 14 (1) : 79-106.2015.

[4] Rahman, M.M., Ahmed, S. & Shuvo, M.H. Nearest Neighbor Classifier Method for Making Loan Decision in Commercial Bank. I.J. Intelligent Systems and Applications, 4 (8) : 60-68.2014

[5] Kurama, O., Luukka, P. & Collan, K. Credit Analysis Using a Combination of Fuzzy Robust PCA and a Classification Algorithm. Advances in Intelligent Systems and Computing-Springer, 3(15) : 19-29. 2015.

[6] Mortezapour, R. & Afzali, M. Assessment of Customer Credit throughCombined Clustering of Artificial Neural Networks, Genetics Algorithm and Bayesian Probabilities. International Journal of Computer Science and Information Security, 11(12) : 1-5. 2013.

[7] Chen, Q., Xue, H.F. & Yan, L. Credit risk assessment based on potential support vector machine. International Conference on Natural Computation (ICNC) : pp. 1-25. 2011.

[8] Karimi, A. Credit Risk Modeling for Commercial Banks. International Journal of Academic Research in Accounting, Finance and Management Sciences, 4 (3) : 1-6. 2014.

[9] Kamalloo, E. & Abadeh, M.S. Credit Risk Prediction Using Fuzzy Immune Learning.Advances in Fuzzy Systems-Hindawi, 3 (2014) : 1-12. 2014

[10] Takyar, S.M.T., Nashtaei, R.A. & Chirani, E. The comparison of credit Risk between Artificial Neural Network and Logistic regression Models in Tose-Taavon Bank in Guilan. International Journal of Applied Operational Research, 5(1) : 63-72. 2014.

[11] Rosyid, H., Prasetyo, E. & Agustin, S. Perbaikan Akurasi Fuzzy K-Nearest Neighbor In Every Class Menggunakan Fungsi Kernel. Prosiding Seminar Nasional Teknologi Informasi dan Multimedia 2013 : pp. 13-18. 2013.

[12] Derrac, J., Chiclana, F., Garcia, F. & Hererra, F. Evolutionary Fuzzy K-Nearest Neighb ors Algorithm usingInterval-Valued Fuzzy Sets. Centre for computation intelligent : 1-28. 2014.

[13] Danenas, P. & Garsva, G. Credit risk evaluation modeling using evolutionary linear SVM classifiers and sliding window approach. Proceeding of International Conference on Computational Science-ICCS : pp. 1324 – 1333. 2012.

[14] Li, S., Zhu, Y., Xu, C. & Zhou, Z. Study of Personal Credit Evaluation Method Based on PSORBF Neural Network Mode. American Journal of Industrial and Business Management, 3(2013) : 429-434. 2013.

[15] O'Neill, M. & Brabazon, A. Self-Organizing Swarm (SOSwarm) for Financial Credit-Risk Assessment. Proceeding on 2008 IEEE Congress on Evolutionary Computation : pp. 3087 – 3093. 2008.

[16] Keller, M. James, Michael R Gray, James A. Givens. A Fuzzy K-Nearest Neighbor. IEEE Transactions On Sistem and Cybernetics, 15(4) : 1-8. 1985.

[17] Jacubcoca, M., Maca, P. & Pech, P. A comparison of selected modifications of the particle swarm optimization algorithm. Journal of Applied Mathematics., 14(2014) : 10-15. 2014.

[18] Guo, H. & He, J. Amodified particle swarm optimization algorithm. Journal of Computer Science10(2) : 341-346.2013.

[19] Yang, C.H., Hsiao, C-H. & Chuang, L-Y. Linearly decreasing weight particle swarm optimization with accelerated strategy for data clustering.International Journal of Computer Science, 37(3) : 3-9. 2010

[20] Clerc, M. The swarm and the queen: towards a deterministic and adaptive particle swarmoptimization. Proceeding of Congress on Evolutionary Computation : pp. 19511957.1999.

[21] Han, J., Kamber, M. & Pei, J. Data mining tecniques and concepts. Morgan Kaufman publisher. Watham : USA. 2012 [22] Ramya, R., S. Analysis of feature selection techniques in credit risk assessmen. Proceedings of International Conference on Advanced Computing and Communication Systems (ICACCS -2015) : pp. 1-6. 2015

IOP Conf. Series: Journal of Physics: Conf. Series **930** (2017) 012025 doi:10.1088/1742-6596/930/1/012025

Modification Of Learning Rate With Lvq Model Improvement In Learning Backpropagation

Jaya Tata Hardinata[1], [2]Muhammad Zarlis, Erna Budhiarti Nababan[1], Dedy Hartama[2] and Rahmat W Sembiring[3]

[1,2,3]Department Of Computer Science, Faculty of Computer Science and Information Technology, Universitas Sumatera Utara, Medan,Indonesia, 20155
[2]STIKOM Tunas Bangsa, Jl. Sudirman Pematangsiantar Indonesia
[3]AMIK Tunas Bangsa, Jl. Sudirman Pematangsiantar Indonesia

Email : jayatatahardinata@gmail.com[1], m.zarlis@yahoo.com[2], ernastp@yahoo.com[3]

Abstract. One type of artificial neural network is a backpropagation, This algorithm trained with the network architecture used during the training as well as providing the correct output to insert a similar but not the same with the architecture in use at training.The selection of appropriate parameters also affects the outcome, value of learning rate is one of the parameters which influence the process of training, Learning rate affects the speed of learning process on the network architecture.If the learning rate is set too large, then the algorithm will become unstable and otherwise the algorithm will converge in a very long period of time.So this study was made to determine the value of learning rate on the backpropagation algorithm. LVQ models of learning rate is one of the models used in the determination of the value of the learning rate of the algorithm LVQ.By modifying this LVQ model to be applied to the backpropagation algorithm. From the experimental results known to modify the learning rate LVQ models were applied to the backpropagation algorithm learning process becomes faster (epoch less).

1. Introduction

Backpropagation is a systematic method on neural network using a supervised learning algorithm and is typically used by perceptron with many layers of the screen to change the existing weights in the hidden layer. The standard backpropagation method (Gradient Descent) has its limitations of fairly slow convergence rates, This poor convergent rate is due to the parameters in determining the backpropagation algorithm depending on the selection of the network architecture, the initial weights, the bias, learning rate, cooefisien momentumAnd activation functions [2,3]. Backpropagation algorithm is an algorithm that has a network, Learning rate and momentum as the main components that will make the learning process, the length of the training process (the formation of a network model), therefore the selection of the network configuration (number of hidden layer neurons, the value of momentum, learning rate, activation function) the exact required for the training process.Learning Vector Quantization (LVQ) is known for his ability in the classification that have a high degree of accuracy and speed in terms of process [7]. The LVQ method tries to eliminate data that has a lot of noise that can accelerate the convergence rate in forecasting systems or predictive backpropagation predictions[13].

In this study,the author uses the data input in the process of back propagation neural network. The data used as training data and test data in this study isdata benchmark UCI Mechine Learning Repository Wine recognition data.

2. Backpropagation

In designing the neural network simulations covering several levels is divided into two parts, namely: data for training data for testing. In this study, the network architecture used consists of three layers, namely : 13inputneuron, 10hidden neuronand 3 outputneuron. In this simulation the limit valueerror

Used for 0,01and maximum epoch 10000. The selected error limit value of little value to maintain the accuracy of the results obtained.

Table 1. Network Architecture

No	Characteristics	Description
1	Architecture	3 Layer
2	Neuron input	13
3	Neuron hidden	10
4	Neuron output	3
5	Initialization weights	Random
6	Target error	0.01
7	Maximum Epoch	10000
8	Learning Rate	0.01, 0.03, 0.06, 0.1, 0.13, 0.16, 0.2, 0.23, 0.26, 0.3

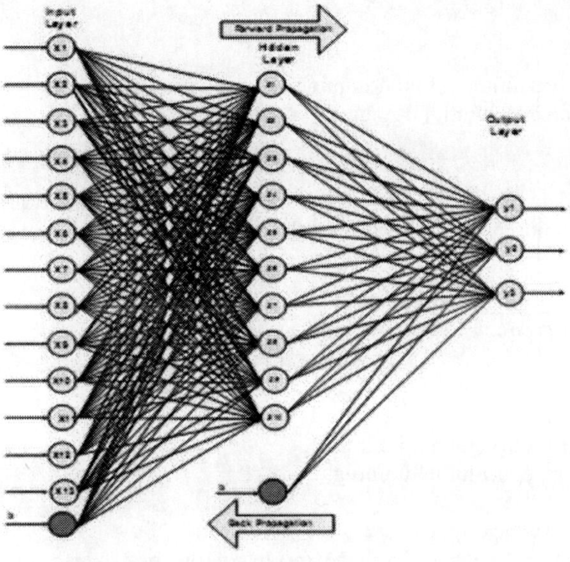

Figure 1. Network Architecture Used13 – 10 – 3

3. Learning Rate

Early learning rate selected from 0.01 - 0.3 with value 0.01, 0.03, 0.06, 0.1, 0.13, 0.16, 0.2, 0.23, 0.26, 0.3. The selection of this learning rate is randomly selected to be used for the research process. Learning rate is selected from 0.01 due to checking local minimum occurrence in network training process, while the highest at 0.3 learning rate is too high because if the training process more unstable and more the case converges on the training process. So that this process can be set to be faster is to determine the value of learning rate is based on the manufacture of tissue parameters previously.

1. Learning Rate LVQ Max
   Learning rate is determined based on the results of the initial learning rateOn standard backpropagation training process. Then the calculation of the learning rate can be calculated by (1.1.)

$$\delta\ new = \alpha\left(1 - \frac{epoch}{maxepoch}\right)$$

Where :
δ : learning rate new
α : initial learning rate
epoch : Value iteration resulting in initial training
maxepoch : Maximum epochinitial training

2. Learning Rate LVQ Min
   Learning rate is determined based on the results of the initial learning rateOn standard backpropagation training process. Then the calculation of the learning rate can be calculated by (1.2).

$$\delta\ baru = \alpha\left(1 - \frac{epoch}{minepoch}\right)$$

Where :
δ : learning rate new
α : initial learning rate
epoch : Value epoch resulting in initial training
minepoch : Minimum epochinitial training

3. Learning Rate LVQ
   Learning rate is determined based on the results of the initial learning rateOn standard backpropagation training process. Then the calculation of the learning rate can be calculated by (1.3).

$$\delta\ baru = \alpha\left(1 + \frac{epoch}{maxepoch}\right)$$

Where :
δ : learning rate new
α : initial learning rate
epoch : Value epoch resulting in initial training
maxepoch : Maximum epochinitial training

4. Learning Rate LVQ AverageEpoch
   Learning rate is determined based on the results of the initial learning rateOn standard backpropagation training process. Then the calculation of the learning rate can be calculated by (1.4).

$$\delta\ baru = \alpha\left(1 + \frac{epoch}{averangeepoch}\right)$$

Where :
δ : Learning Rate Baru
α : initial learning rate
epoch : Value epoch resulting in initial training
avaregeepoch : Average overall training

4. Experiments & Results
For the results of research using standard backpropagation where learning 0.01, 0.03, 0.06, 0.1, 0.13, 0.16, 0.2, 0.23, 0.26, 0.3, And subsequent research using initial learning rate and outcomes from initial learning rate training with modified learning rate of LVQ model that has been set. The results of this study based on standard backpropagation testing compared with backpropagation learning rate test of

LVQ model of learning acceleration seen from the number of iterations (epoch) produced with 10 experiments can be seen in the table 2.

Table 2. Learning Outcomes

No	Epoch (iteration)				
	BP Standart	LVQ Max	LVQ Min	LVQ Average Plus	LVQ Plus
1	10000	10000	352	1756	5268
2	3512	5268	364	1318	2634
3	1756	2108	406	958	1506
4	1054	1171	480	703	958
5	811	879	555	586	753
6	659	703	659	502	620
7	528	555	879	422	502
8	459	480	1171	377	440
9	406	422	1756	341	391
10	352	364	5268	302	341

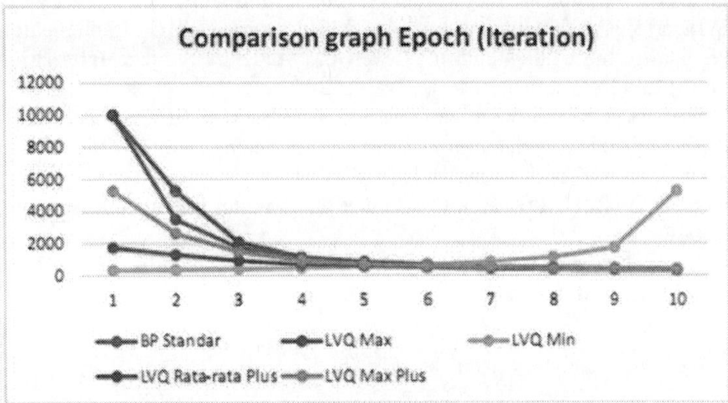

Figure 2. Comparison graph Epoch (Iteration)

5. Conclusion

Learning rate selection factor is very influential on the accuracy of better iteration results with the same level of accuracy. By looking at the test results, It can be concluded that there is an increase in speed and result of the same accuracy on the average LVQ Plus and LVQ Max Plus. Decreased iteration speed and decreased accuracy results in LVQ Max. Further research is needed on backpropagation methods in faster classification and better accuracy.This needs to be done because of the other factors which affect results Accuracy pace and extent of the resulting. Trying to convert existing data into complete so as to improve the accuracy of the results. The more data used for the learning process the more accurate the level of accuracy that artificial neural networks will produce.

References

[1] Andrian, Y. & Putra, H.P. 2014. Analysis of Addition of Momentum to Prediction of Rainfall in Medan Using Backpropagation Neural Network.Seminar Nasional Informatika 2014: 165 – 172.
[2] Brian, T. 2016.Analysis Learning Rates On Backpropagation Algorithm For Classification of Diabetes.Jurnal Ilmiah Edutic3(1):21 – 27.

[3] Brownlee, J. 2016. Learning Vector Quantization For Mechine Learning. (Online) http://machinelearningmastery.com/learning-vector-quantization-for-machine-learning/ (12 April 2017).
[4] Dhanewara, G. & Moertini, V. 2004. Artificial Neural Networks Back Propagation For Integral Data Classification9(3):117 – 131.
[5] Hamid N.A. & Nawi, N.M(2011).The Effect of Adative Gain And Adative Momentum in Improving Training Time Of Gradient Descent Backpropagation Algorith on Classification Problems. Proceedings of the Internasional Conference on Advanced Sciense, Engineering And Information Technology 2011.ISASEIT:pp. 178 – 184
[6] Hamid N.A, Nawi, N.M, Ghazali, R. & Saleh, M N.M. (2011) Accelerating Learning Performance of Backpropagation Algorithm by Using Adaptive Gain Together with Adaptive 0 Momentum and Adaptive Learning Rate on Classification Problems. International Journal of Software Engineering and Application.8(4) : 31 – 43.
[7] Huang, D., Wu, Z. 2017. Forecasting Out Patient Visits Using Empirical Mode Decomposition Coupled With Backpropagation Artificial Neural Networks Optimized By Particle Swarm Optimization. Journal Plos One 12(2): 1-18.
[8] Shanmuganathan, S. & Samarasinghe, S. (Editor). 2016. Studies in Computational Intelligence: Artificial Neural Network Modelling. Springer: Switzerland
[9] Sitanggang, I.S., Hermadi,I., Edward. 2007. Implementation of Neural Networks in Predicting the Understanding Level of Students Subject. Jurnal Ilmiah Ilmu komputer. 5(2): 124-143
[10] Sumijan, Windarto, P.A., Muhammad, A. & Budiharjo. 2016. Implementation of Neural Networks in Predicting the Understanding Level of Students Subject. International Journal of Software Engineering and Its Applications10(10): 189-204.
[11] Sumijan, Windarto, P.A., Muhammad, A. & Budiharjo. 2016. Implementation of Neural Networks in Predicting the Understanding Level of Students Subject. International Journal of Software Engineering and Its Applications10(10): 189-204.
[12] Wang, J. & Zhang, Y. 2011. Research On Prediction of Water Resource Based on LVQ Network, IEEE, pp. 4047- 4049
[13] UCI Machine Learning Repository. 2017. Glass Identification Dataset. (Online) https://archive.ics.uci.edu/ml/datasets/Wine (1 Juni 2017)

Optimizing Support Vector Machine Parameters with Genetic Algorithm for Credit Risk Assessment

Jonson Manurung[1], Herman Mawengkang[2], Elviawaty Zamzami[3]

[1,2,3] Department of Computer Science,
Faculty of Computer Science and Information Technology,
Universitas Sumatera Utara,
Medan, Indonesia, 20155
Email: Jhonson.geo@gmail.com[1], hmawengkang@yahoo.com[2], elvi_zamzami@usu.ac.id[3]

Abstract. Support vector machine (SVM) is a popular classification method known to have strong generalization capabilities. SVM can solve the problem of classification and linear regression or nonlinear kernel which can be a learning algorithm for the ability of classification and regression. However, SVM also has a weakness that is difficult to determine the optimal parameter value. SVM calculates the best linear separator on the input feature space according to the training data. To classify data which are non-linearly separable, SVM uses kernel tricks to transform the data into a linearly separable data on a higher dimension feature space. The kernel trick using various kinds of kernel functions, such as : linear kernel, polynomial, radial base function (RBF) and sigmoid. Each function has parameters which affect the accuracy of SVM classification. To solve the problem genetic algorithms are proposed to be applied as the optimal parameter value search algorithm thus increasing the best classification accuracy on SVM. Data taken from UCI repository of machine learning database : Australian Credit Approval. The results show that the combination of SVM and genetic algorithms is effective in improving classification accuracy. Genetic algorithms has been shown to be effective in systematically finding optimal kernel parameters for SVM, instead of randomly selected kernel parameters. The best accuracy for data has been upgraded from kernel Linear: 85.12%, polynomial: 81.76%, RBF: 77.22% Sigmoid: 78.70%.
Keyword: *support vector machine (SVM), genetic algorithm, parameter optimization, credit risk.*

1. Introduction

Classification is the process of finding models or functions that explain and distinguish classes or concepts, with the aim that the obtained model can be used to identify classes or objects that have unknown class labels. The derived model is based on the analysis of the training data. Classification falls into the category of predictive data mining [1].

In the classification process is divided into two phases namely, learning and testing. In the learning phase, some data that has been known to the data class (training set) is used to form the model. Furthermore, in the testing phase, the established model is tested with some other data (test set) to determine the accuracy of the model. If the accuracy is sufficient then the model can be used for prediction of unknown data class [2].

1.1. Support Vector Machine

Support Vector Machine (SVM) is a learning system for classifying data into two groups of data using hypothetical space in the form of linear functions in a feature space (feature space) high dimension. SVM has properties that are not owned by the learning machine in general that is in the process of finding the best dividing line (hyperplane) so as to obtain the maximum margin size between the non-linear input space with the feature space using the kernel rules [3]. Margin is twice the distance between hyperplane and support vector. The point closest to the hyperplane is called the support vector. Suppose the data is denoted as $x_i \in \Re^n$, for the class label of data xi denoted y ε {+1,1} with i = 1,2, ..., l where l is a lot of data. Separation of data linearly on the SVM method can be seen in Figure 1.

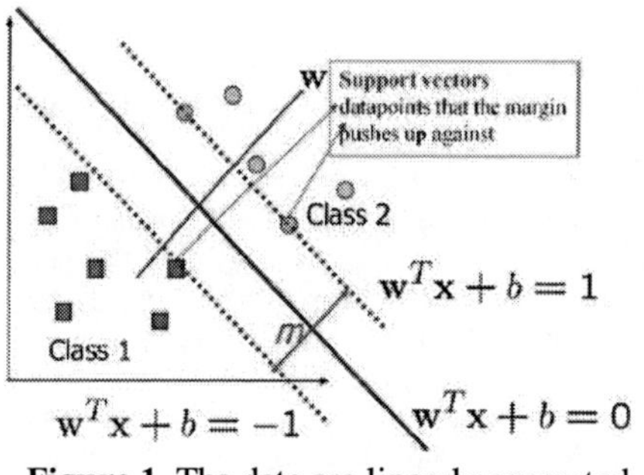

Figure 1. The data are linearly separated

The example above shows that the two classes can be separated by a pair of parallel bounding plane. The first delimiter field limits the first class while the second delimiter field limits the second class, so that it is obtained:

$$x_i.w + b \geq +1 \; for \; y_i = +1 \quad (1)$$

$$x_i.w + b \leq -1 \; for \; y_i = -1 \quad (2)$$

w = Vector weights that are perpendicular to the hyperplane (normal plane)

b = Position of the field relative to the coordinate center

The margin value between the two classes is

$m = \frac{2}{\|w\|}$. *Margin* Can be maximized using the optimization function

Lagrangian:

$$min_{w,b} \, L\,(w, b, \propto) = \frac{1}{2}\|w\|^2 - \sum_{i=1}^{l} \propto_i \left(y_i((x.w + b) - 1)\right) \quad (3)$$

By minimizing L against w and b, then it is obtained:

$$\frac{\partial L}{\partial w} = w - \sum_{i=1}^{l} \propto_i y_i x_i = 0 \quad (4)$$

$$\frac{\partial L}{\partial w} = w - \sum_{i=1}^{l} \propto_i y_i = 0 \quad (5)$$

The above equation can be modified as L maximization containing only αi as the equation below.

$$max_{\propto} \, L = \sum_{i=1}^{l} \propto_i - \frac{1}{2}\sum_{i=1}^{l}\sum_{j=1}^{l} \propto_i \propto_j \, y_i y_j x_i x_j \quad (6)$$

$$s.t.\sum_{i=1}^{l} \alpha_i \, y_i = 0, \alpha_i \geq 0 \quad (7)$$

The resulting α is used to find w. Data that have αi value ≥ 0 is support vector while the rest have value αi = 0.
After the αi value is found, then the class of the test data x can be determined based on the value of the decision function:

$$f(x_d) = \sum_{i=1}^{NS} \alpha_i \, y_i x_i . x_d + b \quad (8)$$

dengan
x_i = *support vector*
n_s = amount *support vector*
xd = Data to be classified.

If data can not be perfectly separated by linear separation, SVM is modified by adding variables ξ_i ($\xi_i \geq 0$, $\forall_I$;$\xi_i = 0$ if x_i Classified correctly) so that the best divider form search formula becomes:

$$min\frac{1}{2}\|w\|^2 + C(\textstyle\sum_{i=1}^{n} \xi_i) \quad (9)$$

$$s.t. y_i(w.x_i + b) \geq 1 - \xi_i, \xi_i \geq 0 \quad (10)$$

The search for the best splitter field with the addition of ξi variable is called soft margin hyperplane. C is the parameter that determines the magnitude of the penalty due to errors in the classification. Thus, the dual problem generated in a non linear problem equals the dual problem generated by linear problem only the αi range between

$$0 \geq \alpha_i \geq C.$$

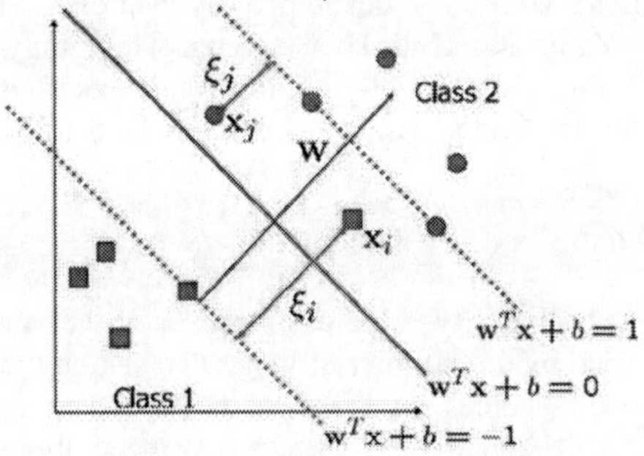

Figure 2. Soft margin hyperplane

Some commonly used Kernel functions are:

1. Linear kernel
$$K(x_i, x) = x_i^T . x \quad (11)$$
2. Polynomial kernel
$$K(x_i, x) = (\gamma . x_i^T . x + r)^d, \gamma > 0 \quad (12)$$
3. Radial Basic Function
$$K(x_i, x) = exp(-\gamma|x_i - x|^2), \gamma > 0 \quad (13)$$
4. Sigmoid kernel
$$K(x_i, x) = tanh(\gamma . x_i^T . x + r)^d, \gamma > 0 \quad (14)$$

In this case γ, r, and d are kernel parameters, as well as parameter C as a penalty due to errors in the classification for each kernel.

1.2. Algoritma Genetika

Defines a genetic algorithm as a search algorithm based on the mechanism of natural selection and genetics [4].Genetic algorithms are widely used on practical problems that focus on the search for optimum parameters. The emergence of this algorithm is inspired by theories in the biological sciences because initially genetic algorithms are used by biologists to simulate biological systems, so many terms and concepts of biology are used in genetic algorithms [5].

2. Computational Analysis

Data consists of 14 attributes with Categorical, Integer, and Real type. The amount of data is 690 lines. The data are classified into 2 classes, namely Positive (+) and Negative (-) classes. Class data + consists of 307 data or 44.5% of the total data, while the class data - consists of 383 data or as much as 55.5% of the overall data. Australian Credit Approval data does not have the value of a missing value attribute or a missing value label. So the missing missing value technique in pre-processing data is not done.

2.1. Clasification

In SVM training process for Linear Kernel function, Polynomial Kernel, Radial Basis Function (RBF) Kernel and Sigmoid Kernel, parameters C, γ, r, and d. Parameter C for Linear Kernel function, parameters C, γ, r, and d for Kernel Polynomial function, and parameters C and γ for RBF kernel function. As for the function of Sigmoid kernel required parameter values C, γ, and r. The SVM training process uses the LIBSVM library developed by National Taiwan University Computer Science students, Chih-Yuan Yang and Chih-Huai Cheng. This study classifies Australian Credit Approval Data with all four Kernel functions by inputting the specified parameter values. From the classification of data with SVM yields an accuracy level using the 5-fold cross validation method.

2.2. Parameter SVM Optimation

For this study we determine the probability value of crossover of 0.8, and the probability of mutation is also very small, about 1 divided by the number of genes. To search the parameter values C, γ, and r are limited in range 2-10 to 20, whereas parameter d is limited in range 2 to 5. The process of calculating the accuracy level will stop if on 10 successive generations have a small difference of accuracy value equal to 0.1. This means that an increase in the accuracy level of 0.1 in a generation will be considered the same as the preceding generation. Here is the best fit for SVM classification, obtained using the genetic algorithm along with the values of the kernel parameters.

From program execution using Australian Credit Approval data, Linear yield accuracy equal to 85.12% with parameter C worth 0.1503. Program execution with Polynomial kernel yields the highest accuracy value reaching 81.76% with parameter values C, γ, r and d respectively 0.0302, 0.0146, 0.5024, and 2. With RBF kernel yield accuracy of 77.22% and parameters C and γ obtained 0.7473 and 0.9883, while the execution result with optimum Sigmoid at 78.70% accuracy with parameter values C, γ, and r respectively 0.8781, 0.8215, and 0.0556. The increase in fitness values can be seen in Figure 2 and the best fitness for each kernel can be seen in 1.

Table 1. Best fitness value of Australian Credit Approval data in Linear, Polynomial, RBF and Sigmoid

Kernel	Best Fitnes
Linear	85.12%
Polynomial	81.76%
RBF	77.22%
Sigmoid	78.70%

Based on the search results of SVM parameter values using genetic algorithm, there is an increase of accuracy value in the data used. Table 2 shows the magnitude of the increase in accuracy value.

Table 2. The magnitude of the increase in fitness value on SVM parameter optimization using genetic algorithm

Kernel	Kenaikan
Linier	5.29%
Polynomial	8.68%
RBF	5.89%
Sigmoid	10.56%

When compared to the usual SVM classification by inputting parameter values arbitrarily, the results of the SVM parameter value search that optimizes the fitness value by using the genetic algorithm show an increase in accuracy. This means, increased accuracy in data classification. shows the comparison of accuracy values that perform parameter optimization with the accuracy of the inputs of predefined parameter values. The result of comparison of accurate value obtained shows that the search for SVM parameter values using genetic algorithm yields greater accuracy value. This, proving that the optimization of SVM parameters using genetic algorithms succeeds in achieving optimal accuracy.

3. Conclution

The result of the research shows that accurate value by searching the SVM parameter value systematically using genetic algorithm yields a better accuracy value compared to the accuracy value of SVM execution result without optimization with the highest accuracy increase that is 10.56%. Thus, the genetic algorithm successfully searches for SVM parameter values that optimize the accuracy value in the SVM classification method. This proves that genetic algorithm is effective in increasing the accuracy value of a classification model in the SVM classification method. Genetic algorithms are very effective and efficient for small data. However, the timing of genetic algorithm execution will increase with increasing population size, generation, parameter value range as well as data size. This causes the length of the execution process, making it less efficient when compared to the increase in accuracy value.

4. References

[1] Larose, D. T. 2006. *Data Mining Methods and Models*. John Wiley & Sons, Inc. Hoboken.
[2] Han J, Kamber M. 2006. Data Mining : *Concepts and Techniques.* USA : Morgan Kaufman Publishers.
[3] Cortes, C. & Vapnik, V. 1995. Support-Vector Networks. *Machine Learning,* Vol. 2, pp 273-297.
[4] Goldberg, D. E. 1989. *Genetic Algorithms in Search, Optimization, and Machine Learning.* England : Addison-Wesley Publishing Company.
[5] Michalewicz Z. 1996. *Genetic Algorithms + Data Structures = Evolution Programs*. USA : Department of Computer Science, University of North Carolina.

IOP Conf. Series: Journal of Physics: Conf. Series **930** (2017) 012027 doi:10.1088/1742-6596/930/1/012027

The Expansion of Initial Point Algorithm for K-Modes Algorithm

Juliandri[1], Zarlis M[2], and Situmorang Z[3]

[1] Department of Computer Science and Information Technology, Universitas Sumatera Utara, Medan, Indonesia.
[2] Department of Computer Science and Information Technology, Universitas Sumatera Utara, Medan, Indonesia.
[3] Department of Computer Science, Universitas Katolik Santo Thomas, Medan, Indonesia.

me@juliandri.com[1]

Abstract. The determination of the starting point in the k-modes algorithm is taken by random. Of course, such a thing can lead to an iteration of unpredictable numbers and accuracy. Therefore, it is necessary to develop different algorithms that are used to determine the starting point with hierarchical agglomerative clustering approach instead of randomly selecting the starting point in the initial iteration. At the end of this research is expected clustering process can produce more efficient iteration. The result of determining the value generated in this algorithm is the incorporation of a number of cluster central points on the variables based on the calculation of the approach which has the average linkage algorithm. Followed by calculating the difference of objective function on each iteration, of course, finished clustering process on k-modes. Iteration will stop after the difference of objective function is smaller than the specified limit.

1. Introduction

Dividing a group of objects into several groups and inserting them according to their homogeneity level is the basic operation in data mining. The initial step is the exploration of data in terms of forming a model used to recognize data patterns. Instinctively, it becomes the daily habit of each person to sort by the same value and classify or classify the object into different groups. Objects can be anything people, products, and services, objects and territories.

In relation to the object clustering method, data clustering problems are more widely studied in data mining and machine learning literature, because so diverse applications can be implemented, among others, exploring data information, learning systems, segmentation and others [1]. In addition, this method is also widely used to solve research problems from various disciplines such as medicine, data security and so on [9].

The problems within each study are almost always different and the use of various techniques in the clustering method is done in the hope of finding the right technique for each data type and output to be generated. In the classification of a proper technique used, the clustering method is divided into two types of methods, namely hierarchical clustering and non-hierarchical clustering [11]. The best-known technique in non-hierarchical clustering is the k-means algorithm.

Characteristics of clustering method operation in the most prominent data mining is not a problem if the existing data set contains a numeric value or a category type. The capability of the algorithm in the clustering method is expected to solve the problem with the level of complexity that has various types of numerical attributes or category-shaped [10]. To that end (Huang, 1997) in his research describes k-modes algorithm to handle boundaries on category-specific datasets and develop new algorithms adapted from k-means algorithms in the hope that the process has the same efficiency level with the algorithm.

But in the k-modes algorithm, the determination of the starting point is done randomly. This results in the coordinates of the center point not accurately representing every object contained in each cluster. The discussion of the percentage of the total improper object enters into each cluster is exposed by (Sun et al, 2002) and at the end of the calculation, iteratively k-modes substitutes the cluster center based on the number of frequencies of the object entering the cluster member [6]. In addition, this algorithm can also be trapped in local optima [3].

The problem that arises is how to determine the starting point for making the cluster central coordinates on this algorithm not determined by random. This is done because the selection of the cluster center is very important. The cluster center will directly impact the cluster information at the end of the iteration [2]. In his research, Cao also discussed the use of the MaxMin algorithm by developing the use of the average number of frequencies to replace random initialization of the MaxMin algorithm used in generating the cluster center that would be used to replace the cluster center on k-modes algorithms.

The same is also done by using Centroid Linkage algorithm which is a variant of Average Linkage algorithm to form the center point and Density-Based Multiscale Data Condensation algorithm to calculate the density at One particular area. The use of centroid linkage modification is also done by (Hadinata et al, 2017) to help determine more precisely the focal point of the fuzzy c-means algorithm. (He et al, 2011) in his research developed k-modes by changing distance calculations and incorporating weighted value attributes so as to produce a higher degree of accuracy whereas (Sangam, 2015) describes a concept of equality calculation using the coefficient basis of information entropy and successfully adds Value accuracy.

Therefore, in this research is presented about the formation of starting point by using the average linkage algorithm which is part of the linkage method and give weighting to the distance calculation to obtain higher accuracy. From the previously described method, it is expected to determine the value of the centroid more precisely without having to do the random process at the initialization stage and produce a more accurate distance accuracy calculation.

2. Basic Concept of Clustering Method

2.1. K-Modes

The k-modes algorithm is a simpler form of k-prototype described in the paper (Huang, 1997). In this algorithm there are three important modifications to the k-modes algorithm, including the use of differences in non-measurement, substituting k-means with k-modes and using frequency methods to update the mode values (Huang, 1997).

The process that occurs in the k-modes algorithm is $\{S_1, S_2, ..., S_k\}$ is the partition of X, where Si ¾ for $lm1mk$, and $\{Q_1, Q_2, ..., Q_k\}$ are values that often arise from $\{S_1, S_2, ..., S_k\}$. Thus, the total can be defined as follows:

$$E = \sum_{i=1}^{k} \sum_{i=1}^{n} y_{i,l} d(X_i, Q_i) \quad (1)$$

Where y_i, l is the element of the partition matrix $Y_{n\,x\,l}$ and d is like the equation below:

$$d(X, Y) = \sum_{j=1}^{m} \delta(x_j, y_j) \quad (2)$$

Where:

$$\delta(x_j, y_j) = \begin{cases} 0 & (x_j = y_j) \\ 1 & (x_j \neq y_j) \end{cases} \tag{3}$$

The function d (X, Y) provides an equation that is important for each category of each attribute. If calculating the frequency of the categories in the data set can be defined the following inequalities:

$$d_{\chi^2}(X, Y) = \sum_{j=1}^{m} \frac{(n_{x_j} + n_{y_j})}{n_{x_j} n_{y_j}} \delta(x_j, y_j) \tag{4}$$

Where $n_{xj}\, n_{yj}$ is a numeric value that is an object within a dataset having categories x_j and y_j. Since, $d\chi^2$ (X, Y) is the equation for chi-square distance. Similar to the k-means algorithm, the purpose of clustering X is to find a set $\{Q_1, Q_2, ..., Q_k\}$ that can minimize the value of E.

The k-modes algorithm contains steps that can be followed as follows:

1. Select k initials modes, one for each cluster.
2. Allocate the object into the nearest cluster based on the value d. Update the mode values of each cluster once allocated by theory.
3. After each object is allocated into the cluster, re-test the inequality and compare it with the existing mode. If, an object is found to be closer to another cluster then move the object into the cluster. Update the mode values of each cluster.
4. Repeat step 3 until no object changes the cluster after the whole testing process is undertaken by the entire data set.

Just as with the k-means algorithm, the k-modes algorithm also produces a local optima solution tied to initialized initial mode values.

2.2. Clustering with the Hierarchy Approach

There are two types of hierarchical clustering, including agglomerative and divisive. Agglomerative determines the number of clusters based on the number of objects in the dataset and is followed by the search for the most similar pair. Then the similar pair is put together into a single cluster process continued until the desired number of clusters is obtained [1]. Therefore, the measure to determine the inequality of each cluster must be determined first [4]. This method has three types of variations including single linkage, complete linkage, and average linkage.

The divisive hierarchical clustering method has a process opposite to agglomerative clustering, starting with a large cluster and dividing the cluster into sections as desired [4].

2.3. Average Linkage

In the average linkage, the distance between two clusters is defined as the overall mean distance of each member of the object, especially the average linkage distance between C_i and C_j.

$$D_{AV}(C_i, C_j) = avg\ x\ \in\ C_i, y\ \in\ C_j(dist(x, y)) \tag{5}$$

Where, the function of dist is the function of choosing distance calculation [1].

2.4. Criteria Reject Measure

To evaluate if in the cluster studied has good quality or not until now still an issue. In the final stages, the evaluation process is always done by every researcher. Therefore, several evaluation criteria have been developed. These criteria are usually divided into two categories, namely internal and external.

In this study, the use of internal criteria as a benchmark for comparison between clusters using some similarity rules. This process usually measures the level of homogeneity of objects between clusters and measures the level of homogeneity of objects within a cluster. This process does not use external information other than the data contained in the dataset itself [9].

2.5. Data Set
The dataset in this study is Internet advertisements obtained from the UCI Learning Machine Repository or UCI Dataset. This dataset contains online advertisements contained on internet pages. This data set was donated by Nicholas Kushmerick in 1998 with the Learning to remove Internet advertisements published in 1999.

This study uses a C4.5 algorithm with an accuracy level of 97% in predicting the state of the image in an advertisement on the website page. The number of data sets consists of 2,821 nonads and 458 ads.

3. Methodology

3.1. Introduction
At this time practitioners and academics have been doing a lot of testing and research on new methodologies as well as mixing methodologies that exist in the soft computing and machine learning sections including combining both fields.

For that reason, the process in this methodology incorporates a detailed draft plan that is required to complete the research and also includes the basics of the method used. In the process of describing any limitations in research methods used then included in the research stage and work diagram.

3.2. Research Framework
There are some stages carried out in this Study as followed:

1. *Initial Research.* At this stage collected research materials from various sources of literature, such as books, journals (both print and online), proceedings, magazines, articles and other relevant sources. In addition, it also conducts consultations with supervisors in the preparation and thesis writing.
2. *Data Collection.* The dataset used in this study is Internet Advertisements obtained from the UCI Learning Machine Repository or UCI Data set.
3. *Early Point Determination.* The initial initialization process of the k-modes algorithm will be replaced by the determination of the starting point by using the average linkage approach and to evaluate the distance will be the addition of weighted values to the distance measurements to produce a more appropriate cluster center.
4. *Formation of New Algorithms.* The starting point has been obtained based on the algorithm adapted from the average linkage and its output is the coordinate center or centroid point which is more representative of the cluster center. Then, the cluster center is used as a cluster center at the initial initialization stage.
5. *Testing and Evaluation.* In this section will be discussed the value of homogeneity and heterogeneity of each cluster, by looking at and analyzing the cluster whether the cluster is ideal or not. These results are compared with k-modes that are not modified (native) methods. The results are the number of iterations, the average objective function and standard deviation and the homogeneity and heterogeneity of each cluster.
6. *Algorithm Creation.* After producing a new method that matches an ideal cluster value next is to write the work stages in the form of pseudocode.
7. *Conclusion.* From the output that has been produced, it can be concluded by writing it into report form, starting from the theoretical study, methodology procedure, testing, and suggestion that can be given for further research process.

4. Experimental Result and Discussion

4.1. Introduction
In this stage the authors will describe the results and discussion of the development of the starting point in the clustering method of the k-modes algorithm, the starting point which was originally formed by random, is now changed by using agglomerative clustering average linkage method.

4.2. Performance of K-Modes Algorithm

In the k-modes algorithm clustering method, the initial membership function is first initialized by means of a random way. The result is that all of the *V* vectors that become cluster centers become random at the beginning of the iteration. The size of the distance used for the data is categorical as in k-modes is by replacing the average with clusters with the modes or values that often appear.

This modification removes restrictions that only exist in k-means when using category values in the clustering process. For the measurement of the previously mentioned distance is simple mismatching or it can be defined as the number of unequal attributes between two objects as in equation 3.

4.3. K-Modes Pre-processing Data Set and Largest Frequency Strategy

Before performing the process of data classification, it is necessary to analyze the data to be used as test data. The dataset is obtained from a trusted source in data collection. In this research, data retrieval through UCI Machine Learning Data set site has been known as the curator of research data. Data used as the dataset is the data "Learning to Remove Internet Advertisements" which has been studied previously. Consisting of 3,279 rows and 1,558 columns.

In the dataset there are thousands of rows and columns which are divided according to the type of content: *url*image+button*, *origurl*labyrinth*, *ancurl*search+direct*, *alt*your*, and *caption*and*. Based on the classification of types of ads on the data then it can be done the category division, namely "ad" and "nonad". And as an additional display that affects the media appear, whether enlarged or not. It is characterized by a ratio of magnitude or aratio whose value is more than 1.

The ratio of 1 and 0 in k-modes is calculated based on the frequency of the number "1". Therefore, if all the data that have the type of url * image + button, origurl * labyrinth, ancurl * search + direct, alt * your and caption * and, but is the type of ad "ad" which has the most frequencies summed and so With "nonad" and which has a "ratio> 1" summed in cross.

After the data containing the 1 is summed then proceeded to form a percentage value based on the sum of data 1, divided by the total of all data in rows and columns counted cross-tabulated. Then, calculate the entire existing data to be the percentage value of the comparison between the amount of data that is worth 1 and the total data. Change of category data into percentage form is done, because the procedure of starting point determination is done by using agglomerative clustering average linkage algorithm, where the processing contained in the algorithms is not categorical data.

4.4. Testing K-Modes Algorithm

The testing stage is done by using the dataset which has been done percentage calculation. Testing process and calculation with percentage data, in this case, a k-modes algorithm using distance calculation Euclidean Distance as a benchmark of capture value similarity.

The testing process to be performed is to compare the quality of the cluster and the number of iterations between the two types of a k-modes algorithm that have changed the starting point using the average linkage algorithm with the k-modes algorithm whose starting point is still using the random way. The test also uses the same parameter, where the maximum iteration is 200 with the allowable error value is 0.05 and the cluster number = 3. Next, do the random value generation process to generate the random point of the center. Then, calculate the distance of each row of data to the cluster center by using the Euclidean Distance calculation.

Then, calculate the distance on the next lines until the overall distance of each cluster is calculated. After calculating the entire distance is done then the next step is to calculate the minimum distance to each cluster. The minimum distance or the smallest in each cluster is expressed as the cluster center on the data. If the minimum is in the first cluster the data is declared a member of that cluster. Furthermore, the results can be presented in a graphical form shown in the figure below.

IOP Conf. Series: Journal of Physics: Conf. Series **930** (2017) 012027 doi:10.1088/1742-6596/930/1/012027

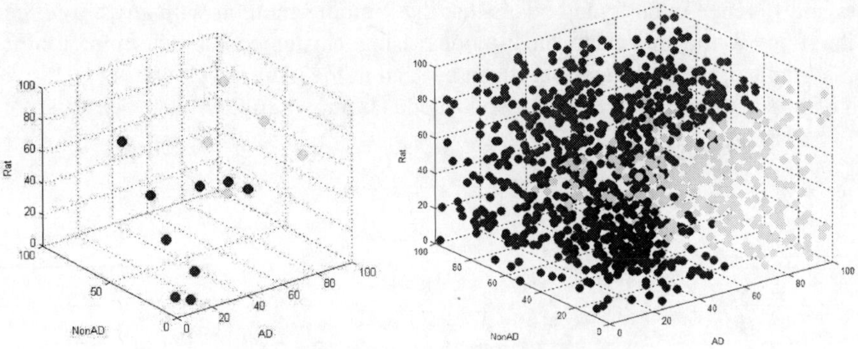

Fig. 1 Scattering 3D 15 lines of data.

4.5. Testing K-Modes Algorithm with Early Point Determination

The k-modes testing stage using the starting point determination begins by searching for variables that are always changed to be fixed in the next iteration.

In k-modes that have been shown previously iterative changes are done in the centroid. Where the section determines the position of the member, whether it is part of one cluster or another part. For that in this study, the centroid is searched by using agglomerative clustering average linkage algorithm.

4.6. Average Linkage Algorithm In Search Centroid Value

The calculation phase for the initial point search on this algorithm using agglomerative average linkage algorithm approach. The agglomerative average linkage algorithm in the cluster center search to determine the starting point begins by determining the number of clusters = number of rows or $c = n$. The agglomerative average linkage algorithm is expected to reduce the number of iterations contained in the k-modes algorithm by substituting an arbitrarily selected cluster center in the initial iteration.

In the average linkage selection algorithm done column by column and cluster center search begins by taking the first column as observation data, that is ad column. The value of the index for each cluster is used to unify the cluster that has the smallest overall value. Combine the C value with index i, j the smallest. Calculate the value of combinations i and j with an average of two values to fill in the new value. Next, calculate the smallest value between two-row groups whose numbers are not equal to the value itself. Having obtained the smallest value as a whole then combine the column that has the smallest value. In this explanation combine columns C and J.

Then, to fill the confused value with that column, calculate the value by using the mean between two values. Calculate the average value tangent to the combined column into one.

After the entire row is filled with an average of two values for the first column then the next step of the contents of the column is not tangent to the initial value or the contents back with the previous value. When this process is done then the first iteration of this algorithm is done. The next step repeats the process as in the previous process, which is to find the smallest value of the line and the smallest overall value to unify the columns.

4.7. Centroid Initialization On Calculation K-Modes Algorithm

The testing stage of k-modes calculation is done by counting all data in the same way and parameters. The parameters used are standard error value 0.05 with truth value 95%, the maximum iteration count is 200 iteration besides the same amount of data with previous and cluster number as much 3.

Next, calculate the distance of the center point to the observed data in the same way.

4.8. Comparison Between the Two Methods

For comparison between k-modes using the determination of the center point of the average linkage and using only any number in determining the center of the cluster is shown in the graph.

The average difference in both methods is that the k-modes method with any cluster center changes to the 4th iteration. While the mean on k-modes using cluster center determination at the start of initialization is stable in the 2nd iteration. It can be seen in the picture below.

The ratio of fitness values between the two k-modes is shown in the figure below.

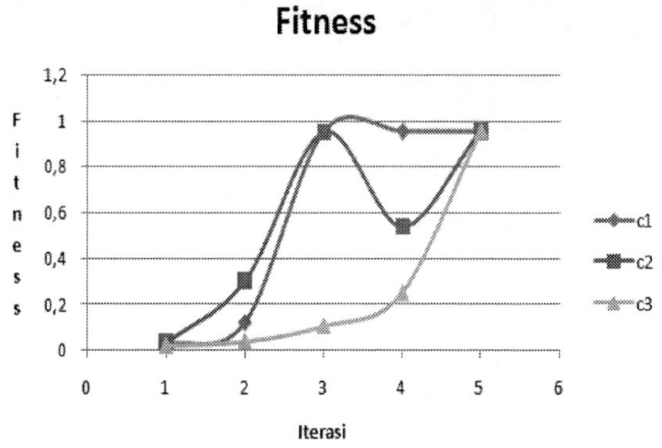

Fitness k-modes with any cluster center.

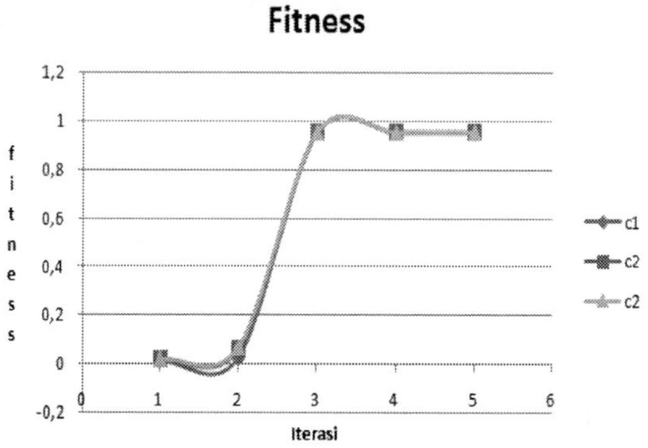

Fitness k-modes by using cluster center determination at the beginning of the iteration.

Fig. 2 Fitness value.

5. Conclusion

Dividing a group of objects into groups and inserting them according to their homogeneity level is the basis of operations in data mining. The initial step is the exploration of data in terms of forming a model used to recognize data patterns.

The use of starting point determination in the classification process is the focus of this research. Especially in the k-modes algorithm, the use of any value in determining the center point produces a number of fluctuating iterations from several experiments in this study. On the other hand, the use of procedurally determined procedural center values in k-modes is a starting point in obtaining better cluster quality without generating a number of fluctuating iterations.

From the result of the research, the starting point determined by using agglomerative clustering average linkage is more appropriate to fill the center point of k-modes algorithm compared with any value. In addition, after passing the Sum of Square Error calculation, the determination of the center point using any value is much greater than the central point determined through the agglomerative clustering average linkage.

The use of more diverse data sets would be better so that it is more representative of different types of datasets. In addition, cluster quality analysis process will be better if reviewed not only using one point of view.

References

[1] Aggarwal, C. C. 2014. Data Clustering Algorithms and Applications, Data Mining and Knowledge Discovery Series. Chapman& Hall/CRC Book., CRC Press, New York.

[2] Cao, Fuyuan., Liang, Jiye & Bai, Liang. 2009. A New Initialization Method for Categorical data Clustering. An International Journal : Expert System with Application. Elsevier. Vol. 36, DOI:10.1016/j.eswa.2009.01.060, pp. 10223-10228.

[3] Chaturvedi, Anil., Foods, Kraft., Green, Paul.E & Carrol, J. Douglass. 2001. K-modes Clustering. Journal of Classification. Vol. 18, DOI: 10.1007/s00357-001-0004-3. Pp. 35-55.

[4] Everitt, B.S., Landau,S., Leese,M. & Stahl, D. 2011. Cluster Analysis 5th Edition. King's College, London. Willey Series in Probability and Statistics. Jhon Willey & Sons, Ltd. UK.

[5] Hadinata, Edrian., Sembiring, Rahmat.W., Kusumasari, Tien Fabrianti & Herawan, Tutut. 2017. The Algoritm Expantion for Starting Point Determination Using Clustering Algorithm Method with Fuzzy C-Means. Recent Advances on Soft Computing and Data Mining. Springer.
[6] He, Zengyou., Xu, Xiaofei & Deng, Shengchun. 2011. Attribute Value Weighting in k-modes Clustering. Elsevier, Expert System With Applications. Vol. 38, DOI:10.1016/j.eswa.2011.06.027, pp. 15365-15369.
[7] Huang, Zhexue. 1997. A Fast Clustering Algorithm to Cluster Very Large Categorical Data Sets In Data Mining. Cooperative Research Centre for Advanced Computational System(ACSys). Autralia.
[8] Huang, Zhexue. 1997. Clustering Large Datasets With Mixed Numeric and Categorical Values. Cooperative Research Centre for Advanced Computational System(ACSys). Autralia.
[9] Maimon, O., & Rokach, L(Editor). 2010. Data Mining and Knowledge Discovery Handbook Second Edition. Springer. DOI. 10.1007/978-0-387-09823-4.
[10] Michalski, R., Bratko, I., Kubat, M. 1998. Machine Learning and Data mining:Methods and Applications. Wiley, New York.
[11] Yim, O., & Ramdeen, Kayle T. 2015. Hierarchical Cluster Analysis : Comparison of Three Linkage Measures and Application to Psychological Data. The Quantitative Method for Psychology. TQMP. Vol.11, No.1, DOI: 10.20982/tqmp.11.1.p008, pp.8-22.
[12] Sun, Ying., Zhu, Quiming. & Chen, Zhenxin. 2002. An Iterative Initial Points Refinement Algorithm for Categorical Data Clustering. Pattern Recognition Letters. Elsevier Science.Vol. 23, pp. 875-884.

PERFORMANCE ANALYSIS OF ENTROPY METHODS ON K MEANS IN CLUSTERING PROCESS

Mhd. Dicky Syahputra Lubis[1], Herman Mawengkang[2], Saib Suwilo[3]

[1]Departement of Computer Science,
[2,3]Department of Mathematics Science,
University Of Sumatera Utara
Medan, Indonesia, 20155

Email: [1]dickylubis91@gmail.com

Abstract. K Means is a non-hierarchical data clustering method that attempts to partition existing data into one or more clusters / groups. This method partitions the data into clusters / groups so that data that have the same characteristics are grouped into the same cluster and data that have different characteristics are grouped into other groups.The purpose of this data clustering is to minimize the objective function set in the clustering process, which generally attempts to minimize variation within a cluster and maximize the variation between clusters. However, the main disadvantage of this method is that the number k is often not known before. Furthermore, a randomly chosen starting point may cause two points to approach the distance to be determined as two centroids. Therefore, for the determination of the starting point in K Means used entropy method where this method is a method that can be used to determine a weight and take a decision from a set of alternatives. Entropy is able to investigate the harmony in discrimination among a multitude of data sets. Using Entropy criteria with the highest value variations will get the highest weight. Given this entropy method can help K Means work process in determining the starting point which is usually determined at random. Thus the process of clustering on K Means can be more quickly known by helping the entropy method where the iteration process is faster than the K Means Standard process. Where the postoperative patient dataset of the UCI Repository Machine Learning used and using only 12 data as an example of its calculations is obtained by entropy method only with 2 times iteration can get the desired end result.

Keywords: K Means, Entropy, Clustering, Data Mining, Weight

1. Introduction

The purpose of this data clustering is to minimize the objective function set in the clustering process, which generally attempts to minimize variation within a cluster and maximize the variation between clusters. (YudiAgusta 2007). Where, the difference between the nearest point and the farthest point in the large amount of data becomes rather weak. In, the method based on the vector angle is suggested to know the angle between the two vectors of the data object. This step seems much more sensitive to high dimension data sets based on distance. (Shu-yin Xia 2015). In two dimensions, the length of the straight line connection becomes two points x and y. In n dimensions, one problem with this method is that the Euclidean distance is sensitive to large values; In other words, sensitive to outliers. In addition, it will lose negative correlation because they will provide a great distance. (Bernard Chen et al 2005). Entropy method is a method that can be used to determine a weight and make a decision from a set of alternatives. Entropy is able to investigate the harmony in discrimination among a multitude of data sets. By using Entropy criteria with the highest value variations will get the highest weight (Abbas, 2004). Another study mentions that Entropy can take decisions in the selection of subcontracted gloves production (Jamila, 2012). In that study, weighting each criteria and determining which subcontracts are the best choice. Where the selection of the entropy method is to determine the center of the initial centroid to obtain the weight of the value in making Clustering on the K-Means algorithm.

1.2 K means

K-Means is a non-hierarchical data clustering method that attempts to partition existing data into one or more clusters. This method partitions the data into clusters / groups so that data that have the same characteristics are grouped into the same cluster and data that have different characteristics are grouped into other groups. The purpose of this data clustering is to minimize the objective function set in the clustering process, which generally attempts to minimize variation within a cluster and maximize the variation between clusters. (YudiAgusta 2007)

Data clustering using K-Means method is generally done with the following basic algorithm:

1. Determine the number of clusters
2. Allocate data into the cluster randomly
3. Calculate the centroid / average of the data in each cluster
4. Allocate each data to the nearest centroid / average
5. Return to Step 3, if there is still data that moves the cluster or if the centroid value changes, some are above the specified threshold value or if the value changes on the objective function used above the specified threshold value

1.2.1 Distance Space To Calculate Distance Between Data and Centroid

Some distance space has been implemented in calculating the distance between data and centroid. For Euclidean distance space, the distance between two points is calculated using the following formula:

$$D_{L2}(x_2, x_1) = \|x_2 - x_1\|_2 = \sqrt{\sum_{j=1}^{p} (x_{2j} - x_{ij})^2}$$

Where : P: Dimensions of data

Euclidean is often used because the distance calculation in distance space is the shortest distance that can be obtained between two points taken into account, whereas Manhattan is often used because of its ability to detect special circumstances such as the existence of outliers better. (YudiAgusta 2007)

1.3 Entropy

Entropy method is a method that can be used to determine a weight. Entropy is able to investigate harmony in discrimination between data sets. A set of alternative value data on certain criteria is described in the Decision Matrix (DM). Using the Entropy method, the criteria with the highest value variations will get the highest weight. Thus, the Entropy method can calculate the maximum likelihood (maximum Entropy) for each single data in a set of entities that have different possibilities. Specifically, Entropy is also able to adapt to a set of multiple plural data that have variations varying from one criterion to another. The weighting steps using the Entropy method are as follows (Tiyaswiyoso, 2012)

1.3.1 Normalization of Data Criteria

The normalization formula is as follows:

$$d_i^k = \frac{x_i^k}{x_{i\,max}^k} \quad d_i = d_i^1, \dots\dots, d_i^m \qquad\qquad D_i = \sum_{k=1}^{m} d_i^k \text{ , i =1,2,....., n}$$

Where :

d_i^k = Normalized data values $x_{i\,max}^k$ = Unranked value that has the highest value
x_i^k = Unnormalized data values D_i = The amount of data values that have been digitized
m = Number of alternatives

2. Entropy Calculation

The next step is the measurement of Entropy for each of the i-th criteria.

The formula is:

$$e_{max} = \ln m, \text{ then calculate}$$

$$K = \frac{1}{e_{max}}$$

$$e\,(d_i) = -K \sum_{k=1}^{m} \frac{d_i^k}{D_i} \ln\left(\frac{d_i^k}{D_i}\right),\ K>0$$

Where :
e_{max}= Maximum entropy K = Entropy constant
e (d_i)= Entropy for every attribute / criterion i
After obtaining e (d_i) for each criterion, then can be determined total Entropy for each criterion, the formula is:

$$E = \sum_{i=1}^{n} e(d_i) \quad \text{, n is the number of criteria}$$

3. Calculation of Entropy Weight

The next step is to calculate the weights by using the following formula:

$$\overline{\lambda_1} = \frac{1}{n-E}[1 - e\,(d_i)],\ 0 \le \overline{\lambda_1} \le 1$$

$$\sum_{i=1}^{n} \overline{\lambda_1} = \pm 1$$

Where : $\overline{\lambda_1}$ = Temporary Entropy weight n = number of attributes / criteria

2 Research methodology

In the process of determining the initial weight value for centroid center with Entropy method is done in two ways:
The direct weighting resulted from the calculation using Entropy of the dataset in each class.
a. Entropy calculation for each of the I-criteria

$$e_{max} = \ln m$$
$$= \ln 12 = -0.402429$$
$$K = \frac{1}{e_{max}} = \frac{1}{0.4024} = 2{,}48242$$

$e(d_i) = -K \sum_{k=1}^{m} \frac{d^k{}_i}{D_i} \, ln\left(\frac{d^k{}_i}{D_i}\right),\quad K > 0$ $\qquad e(d_1) = 0{,}39516$ $\qquad e(d_2) = 0{,}98521$

with :

$E = \sum_{i=1}^{n} e(d_i)$, then : $\quad E = 1{,}38037$

b. Calculation of Entropy Weight for each of the i-th criteria

$\overline{\lambda_\iota} = \frac{1}{n-E}[1 - e(d_i)], \quad 0 \leq \overline{\lambda_\iota} \leq 1$

Then the result:

$\overline{\lambda_1} = 0.97613$ $\quad\quad \overline{\lambda_2} = 0{,}02386$

The value of the data is multiplied by the final entropy weight. Then the value is searched for the highest, middle and lowest values to serve as the center of the initial cluster.

Table 1. first centre centroid

	Patient temperature	Patient comfort
C1 (Max)	43.92588614	0.358037953
C2 (Median)	36.11683972	0.238691969
C3 (Min)	35.14070891	0.119345984

5. Result and Discussion

Following calculations on its iteration:

Where on iteration 1

Input set ke-1 : (36.116, 0.358)

Centre cluster 1: (43.925, 0.358) Centre cluster 2: (36.116, 0.238)

Centre Cluster 3: (35.140, 0.119)

The iteration:

C1=$\sqrt{(36{,}116 - 43{,}925)^2 + (0{,}358 - 0{,}358)^2}$

=$\sqrt{(-7{,}809)^2 + (0)^2}$

= 7,809

C2= $\sqrt{(36{,}116 - 36.116)^2 + (0{,}358 - 0.238)^2}$

=$\sqrt{(0)^2 + (0.12)^2}$

= 0.12

C3=$\sqrt{(36{,}116 - 35.140)^2 + (0{,}358 - 0.119)^2}$

=$\sqrt{(0{,}976)^2 + (0.239)^2}$

= 1,005

And so on so that the results obtained are shown in the table 2

Table 2 result of first iteration

Patient Number	Temp of Patient	Comfortable of Patient	C1	C2	C3	Shortest Distance
1	36.11683972	0.358037953	7.809046425	0.119346	1.004891	0.119345984
2	35.14070891	0.238691969	8.785987844	0.976131	0.119346	0.119345984
3	37.09297052	0.238691969	6.833957811	0.976131	1.955906	0.976130803
4	36.11683972	0.358037953	7.809046425	0.119346	1.004891	0.119345984
5	36.11683972	0.238691969	7.809958357	0	0.9834	0
6	43.92588614	0.358037953	0	7.809958	8.788419	0
7	35.14070891	0.119345984	8.788419243	0.9834	0	0
8	41.97362453	0.238691969	1.955906144	5.856785	6.833958	1.955906144
9	35.14070891	0.238691969	8.785987844	0.976131	0.119346	0.119345984
10	35.14070891	0.238691969	8.785987844	0.976131	0.119346	0.119345984
11	36.11683972	0.358037953	7.809046425	0.119346	1.004891	0.119345984
12	36.11683972	0.238691969	7.809958357	0	0.9834	0

Based on the above clustering obtained 2 data on cluster 1, 6 data on cluster 2 and 4 data on cluster 3 then resumed in the iteration process-2 which its way almost the same as the first iteration that only distinguish it is on the determination of its center C Where in this 2nd iteration the center C is determined by searching for the average of each previous cluster. In the 2nd iteration the new cluster center is determined based on the average of each value on each new cluster where in C1 there is 1 member, C2 is 6 members and C3 there are 4 members, then obtained a new cluster center. Where in the 2nd iteration process gets the desired final result because the clusterization result on the iteration to 1 and the 2nd iteration remain or not change as the calculation process stops here.

Table 3 comparison methods K-Means

Patient	**K Means With Entropy**				**K Means Standart**			
Number	Iteration	C1	C2	C3	Iteration	C1	C2	C3
1.	2		1		3		1	
2.	2			1	3			1
3.	2		1		3			1
4.	2		1		3		1	
5.	2		1		3			1
6.	2	1			3	1		
7.	2			1	3			1
8.	2	1			3	1		
9.	2			1	3			1
10.	2			1	3			1
11.	2		1		3		1	
12.	2		1		3			1

From the results of research that has been done that Entropy method can be used as a determinant of initial centroid value in K Means. Centroid value obtained from Entropy method shows good performance in studying data during training process. Entropy method yields centroid value that can represent the whole data of rice productivity with a little amount of training data that is only 12 data although with number of iteration 2 times. But the learning process is getting better too. Thus the initial centroid value also yields a better level of effectiveness when compared to standard K Means which takes a random value to determine the initial centroid value. This is due to the nature of the Entropy method of calculation capable of investigating the degree of harmony in a set of data and being able to adapt to a set of plural-plural data that has variations in values that differ from one data to another. So the initial centroid value obtained can learn and represent data from each class quickly. Therefore, the initial centroid value is very influential from the learning process data and the level of effectiveness gained. But overall testing, the initial centroid value generated by the Entropy method has a success rate in clustering data compared to other ways. So it can be concluded that the value of early centroid is very influential from the process of learning data and the level of accuracy obtained. Then the determination of initial centroid value on K Means by using Entropy method can be said better when compared with the determination of the initial centroid value by K Means standard.

6. References

[1] Abbas, A.R. 2004. *Entropy*Methode for Adaptive Utility Elicitation. *IEEE Transaction on System, Man, and Cybernetics***34**(2): 169-178.

[2] Agusta, Yudi PhD, "*K-Means – Penerapan, PermasalahandanMetodeTerkait", JurnalSistemdanInformatika Vol. 3 (Pebruari 2007), 47-60*

[3] Chen , Bernard, Phang C. Tai, R. Harrison and Yi Pan, "Novel Hybrid Hierarchical-K-means Clustering Method (H-K-means) for Microarray Analysis"

Department of Computer Science, Department of Biology, Georgia State University, Atlanta, USA 2005

[4] Chen , Bernard, Phang C. Tai, R. Harrison and Yi Pan, "Novel Hybrid Hierarchical-K-means Clustering Method (H-K-means) for Microarray Analysis" Department of Computer Science, Department of Biology, Georgia State University, Atlanta, USA 2005

[5] Jamila. 2012. SistemPendukungKeputusanPemilihanSubkontrakProduksiSarungTanganMenggunakanMetode*Entropy*dan TOPSIS. *Seminar NasionalInformatika UPN Yogyakarta.* **1**(2).

[6] Nurul C., Wiharto, &Umi S. (2012). PengaruhNormalisasi Data padaJaringanSyarafTiruanBackpropation Gradient Descent Adaptive Gain (BPGDAG) untukKlasifikasi. *JURNAL ITSMART*. 2301-7201.

[7] Xia, Shu-yin,Zhong-yang Xiong, Yue-guo Luo, Wei-Xu, "Effectiveness of the Euclidean distance in high dimensional spaces", *College of Computer Science,* 2015

IOP Conf. Series: Journal of Physics: Conf. Series **930** (2017) 012029 doi:10.1088/1742-6596/930/1/012029

Comparative Analysis of Membership Function on Mamdani Fuzzy Inference System for Decision Making

Putri harliana1*, Robbi Rahim[2]

[1]Department of Computer Engineering, Sekolah Tinggi Teknik Harapan, Jln. H.M. Joni No. 70 C Medan, Medan, Indonesia
[2]School of Computer and Communication Engineering, Universiti Malaysia Perlis, Malaysia

*putri.harliana.stth@gmail.com

Abstract. Membership function is a curve that shows mapping the input data points into the value or degree of membership which has an interval between 0 and 1. One way to get membership value is through a function approach. There are some membership functions can be used on mamdani fuzzy inference system. They are triangular, trapezoid, singleton, sigmoid, Gaussian, etc. In this paper only discuss three membership functions, are triangular, trapezoid and Gaussian. These three membership functions will be compared to see the difference in parameter values and results obtained. For case study in this paper is admission of students at popular school. There are three variable can be used, they are students' report, IQ score and parents' income. Which will then be created if-then rules.

1. Introduction

Fuzzy logic is a methodology of problem-solving control systems, suitable for implementation on systems, from simple systems, small systems, embedded systems, PC networks, multi-channel or workstations based data acquisition, and control systems. This methodology can be applied to hardware, software, or a combination of both. In a more specific sense, Fuzzy logic is an extension of multivalued logic whose objective is approximate reasoning rather than exact solution [1].

Unlike traditional Crisp Logic, such as Binary Logic where variables may only take on truth values true and false represented by 1 and 0 respectively, the variables in fuzzy logic may have a truth value that ranges in degree between 0 and 1. Instead of describing absolute yes or no, the truth value, or membership in fuzzy logic explains a matter of degree. 0 shows completely false, while 1 expresses completely true, and any value within the range indicates the degree of true. Furthermore, the concept of membership in fuzzy logic is close to human words and intuition, so the number and variety of applications of fuzzy logic have increased significantly in recent years [4].

Membership function is a graph which represents the magnitude of the degree of membership of each input variable that is in the interval between 0 and 1. The degree of membership of a variable x is denoted by the symbol (x). The rules use the value of membership as a weighting factor to determine its effect when making inferences to draw conclusions [1]

In this paper will only discuss the comparison of three membership functions so that it can determine decisions in the determination of students accepted or rejected at favorite schools. To get the output value is done the main approach of fuzzy by taking the input variable consisting of the if-then rule. The If-Then rules are expressed directly by human words, and each of the word is regarded as a fuzzy set.

2. Theory

A. Logical Operation

Because the standard binary logic is a special case of fuzzy logic where the membership values are always 1 (completely true) or 0 (completely false), fuzzy logic must hold the consistent logical operations as the standard logical operations. The most foundational logical operations are AND, OR and NOT. Unlike standard logical operation, the operands A and B are membership values within the interval [0, 1]. In fuzzy logical operations, logical AND is expressed by function min, so the statement A AND B is equal to min (A, B). Logical OR is defined by function max, thus A OR B becomes equivalent to max (A, B). And logical NOT makes operation NOT A become the operation 1 – A [2][3].

B. Fuzzy Inference and Mamdani-Type Fuzzy Inference

Fuzzy Inferece System has 3 basic structure that is Rule Base which is used to do selection to fuzzy rule. The database, this component is used to define the membership value of the fuzzy set and the reasoning mechanism used to generate the output of the operations performed on the fuzzy set. Basically the input given on the fuzzy inference system is in the form of a firm set and will produce the output of a fuzzy set depending on the situation where the fuzzy inference system is used [1].

Generally, there are three types of fuzzy inference system, they are Mamdani, Sugeno, and Tsukamoto. All of these three methods can be divided into two processes. The first process is fuzzifying the crisp values of input variables into membership values according to appropriate fuzzy sets, and these three methods are exactly the same in this process[5]. While the differences occur in the second process when the results of all rules are integrated into a single precise value for output. Mamdani is often known as the Max-min method. This method was introduced by Ebrahim Mamdani in 1975. To get the output required 4 stages are the establishment of the fuzzy set: definition of the fuzzy set and the determination of the degree of membership of the input crisp on a fuzzy set, application function implication: evaluation of rules fuzzy to generate output from each rule, composition rule: aggregation or combination of outputs of all rules, defuzzification: calculation of output crisp. The reasoning with the Sugeno method is almost the same as the Mamdani method, only the output system is not a fuzzy set, but rather a constant or a linear equation[6]. Then a weighting mechanism is implemented to work out the final crisp output [2].

3. Result and Discussion

In this section will be explained result of comparison three membership functions are triangular, trapezoid, and Gaussian. Samples of data taken from a school by input student's report score, IQ score and parents' income in order to get a decision whether the student is accepted or rejected at the popular school. Only three data samples of students are used to see the comparison parameters of each membership functions. The following is result of three membership functions use matlab:

A. Triangular Variable of Students' Report

Triangular variable of students's report has 3 parameters, are low, middle and high.

- low parameter : 55, 55, 70
- middle parameter : 60, 77, 95
- high parameter : 85, 100,100

B. Triangular Variable of IQ

Triangular variable of IQ has 3 parameters, are low, middle and high.

- low parameter : 90, 90, 100
- middle parameter : 90, 110, 130
- high parameter : 120, 130,130

C. Triangular Variable of Income

Triangular variable of Income has 3 parameters, are low, middle and high.

- low parameter : 0, 1000000, 4000000
- middle parameter : 2000000, 5500000, 9000000
- high parameter : 7000000, 10000000, 10000000

The result use matlab as follows:

Figure 1. Report

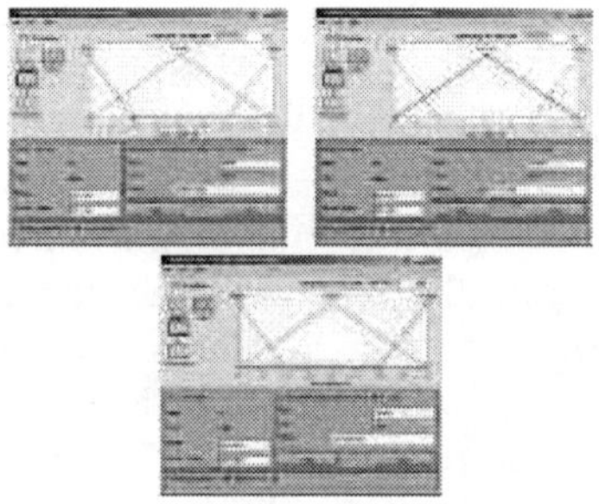

Figure 2. IQ

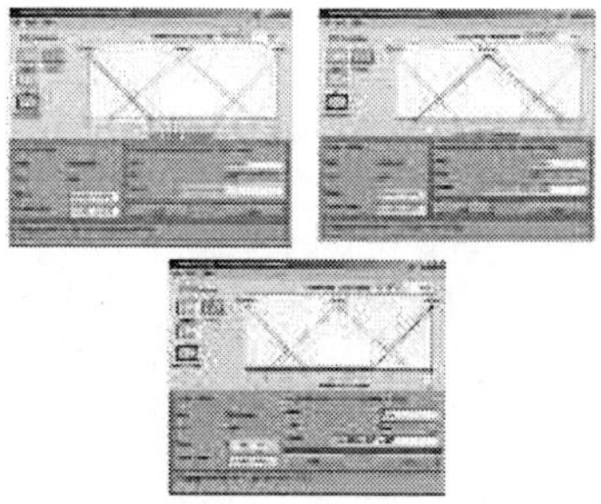

Figure 3. Income

Output Triangular Curve

The output of the triangle curve in the form of rejection and acceptance of prospective students in popular schools by including data that become the reference assessment such as report score, IQ and income parents. Data input is as follows :

Report	IQ	Income
81.65	90	5000000
90.45	120	2000000
74.91	90	10000000

Result of data input use matlab as follows:

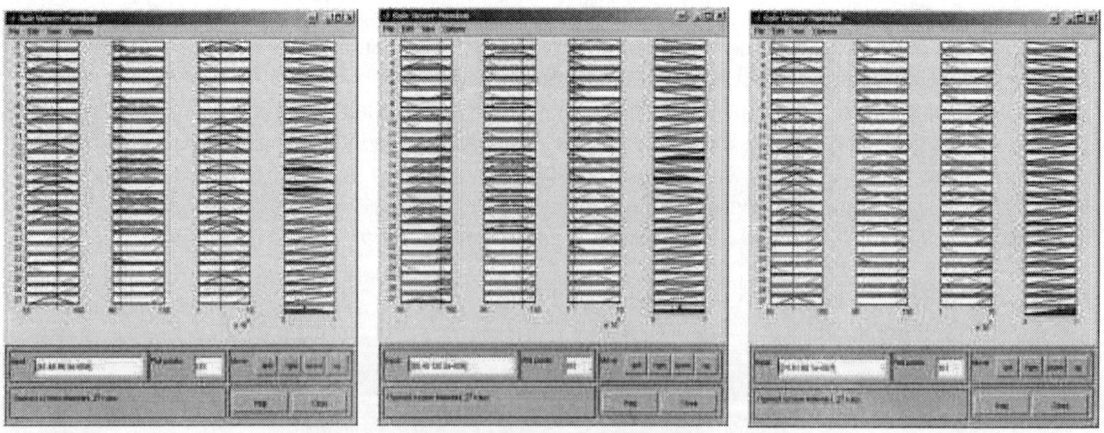

Figure 4. Rejected and Accepted Output of Triangular

D. Trapezoid Variable of Students' Report

Trapezoid variable of students's report has 3 parameters, are low, middle and high.

- low parameter : 55, 55, 56.5, 68.5
- middle parameter : 61.7, 75.3, 78.8, 93.2
- high parameter : 86.5, 98.5, 100, 100

E. Trapezoid Variable of IQ

Trapezoid variable of IQ has 3 parameters, are low, middle and high.

- low parameter : 90, 90, 91, 99
- middle parameter : 92.11, 108.1, 112.1, 128.1
- high parameter : 121, 129, 130, 130

F. Trapezoid Variable of Income

Trapezoid variable of Income has 3 parameters, are low, middle and high.

- low parameter : 100000, 900000, 1300000, 3700000
- middle parameter : 2350000, 5150000, 5850000, 8650000
- high parameter : 7300000, 9700000, 10000000, 10000000

The result use matlab as follows:

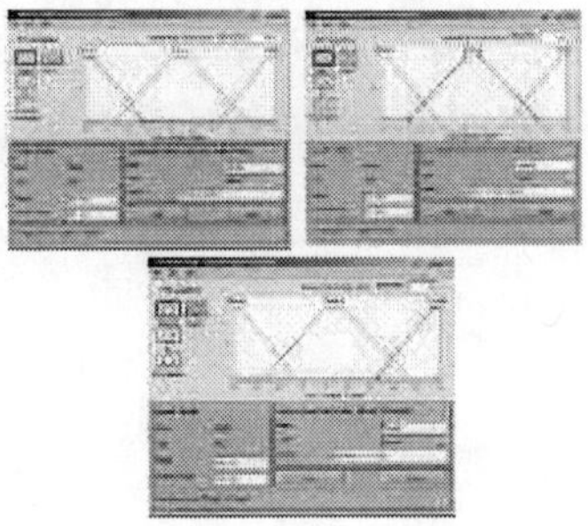

Figure 5. Report

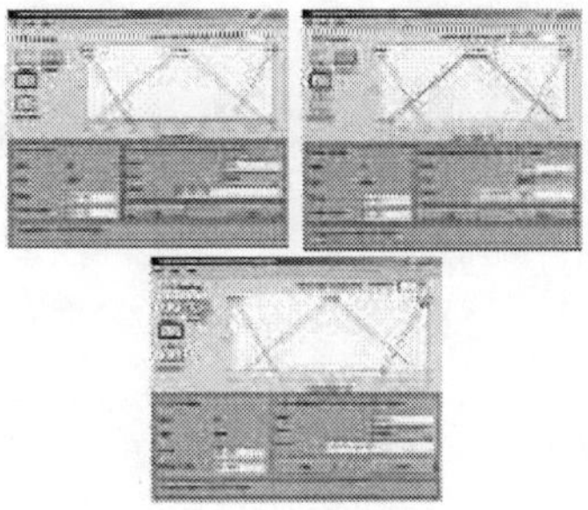

Figure 6. IQ

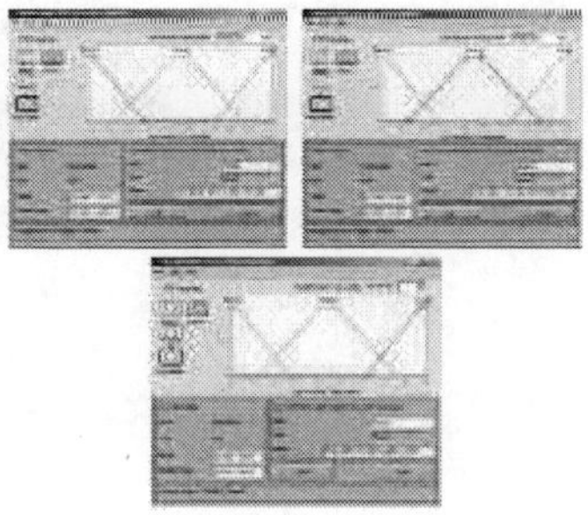

Figure 7. Income

Output Trapezoid Curve

The output of the trapezoid curve in the form of rejection and acceptance of prospective students in popular schools by including data that become the reference assessment such as report score, IQ and income parents. Data input is as follows :

Report	IQ	Income
81.65	90	5000000
90.45	120	2000000
74.91	90	10000000

Result of data input use matlab as follows:

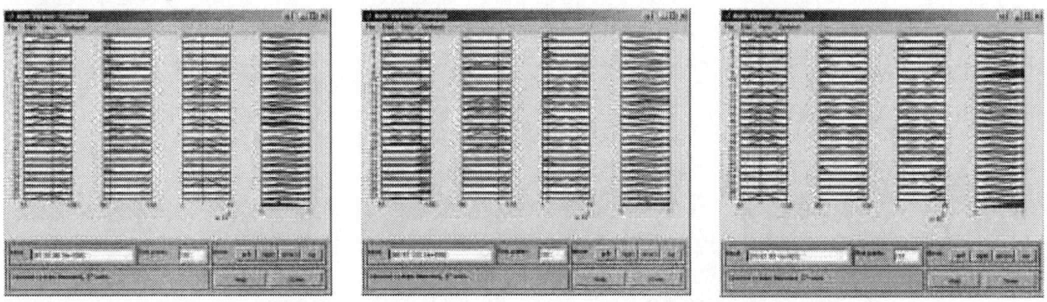

Figure 8. Rejected and Accepted Output of Trapezoid

G. Gaussian Variable of Students' Report

Gaussian variable of students's report has 3 parameters, are low, middle and high.

- low parameter : 7.644, 55
- middle parameter : 7.644, 77
- high parameter : 7.644, 100

H. Gaussian Variable of IQ

Gaussian variable of IQ has 3 parameters, are low, middle and high.

- low parameter : 7.644, 90
- middle parameter : 7.644, 110
- high parameter : 7.644, 130

I. Gaussian Variable of Income

Gaussian variable of Income has 3 parameters, are low, middle and high.

- low parameter : 1.147000, 1000000
- middle parameter : 1.147000, 5000000
- high parameter : 1.147000, 10000000

The result use matlab as follows:

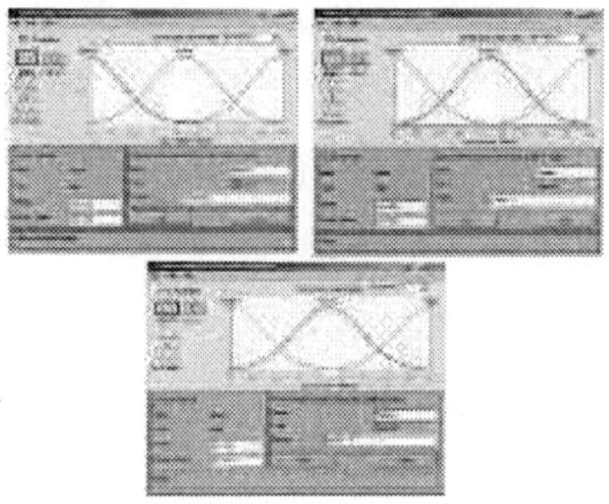

Figure 9. Report

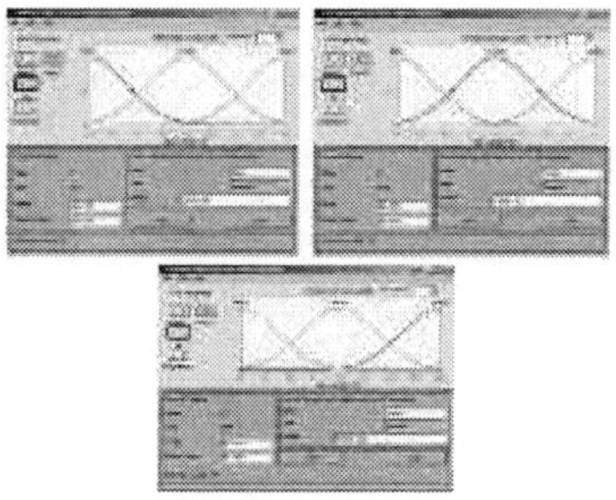

Figure 10. IQ

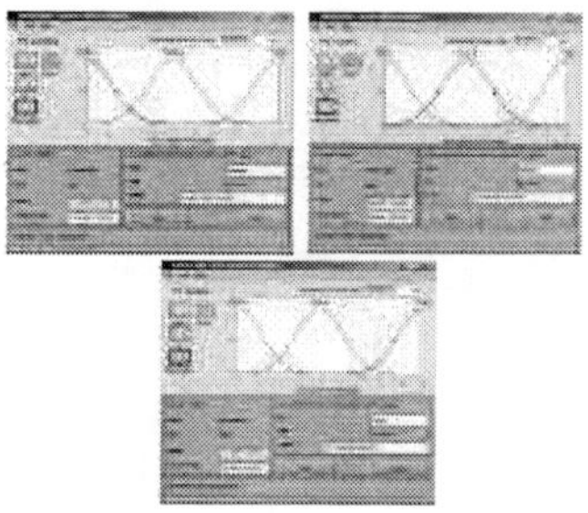

Figure 11. Income

Output Gaussian Curve

The output of the gaussian curve in the form of rejection and acceptance of prospective students in popular schools by including data that become the reference assessment such as report score, IQ and income parents. Data input is as follows :

Report	IQ	Income
81.65	90	5000000
90.45	120	2000000
74.91	90	10000000

Result of data input use matlab as follows:

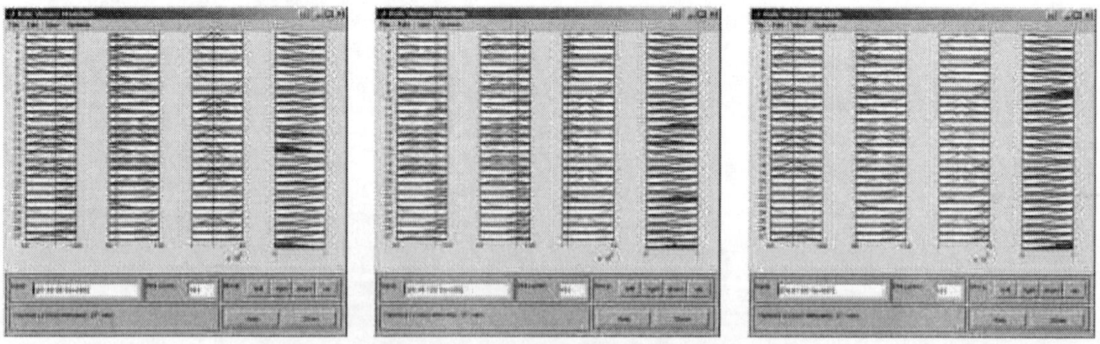

Figure 12. Rejected and Accepted Output of Gaussian

IOP Conf. Series: Journal of Physics: Conf. Series **930** (2017) 012029 doi:10.1088/1742-6596/930/1/012029

4. Conclusion

Based on experiments conducted in this paper it can be concluded that three of membership functions are triangular, trapezoid, and Gaussian have the same result. But the parameters of them are different. For decision making case in popular school, three of membership functions are equally good. But for taking parameter values, triangular membership function is easier than trapezoid and Gaussian. So the three of membership functions have the disadvantages and advantages depending on the case study conducted.

References

[1] Wang, Chonghua. 2015. A Study of Membership Functions on Mamdani-Type Fuzzy Inference System for Industrial Decision Making. Theses. Lehigh University

[2] Saepullah, A., Wahono, R.S. 2015. Comparative Analysis of Mamdani, Sugeno and Tsukamoto Method of Fuzzy Inference System for Air Conditioner Energy Saving. *Journal of Intelligent Systems* 1(2) : 143-147.

[3] Sivarao, Brevern, P., El-Thayeb, N.S.M., & Vengkatesh, V.C. 2009. GUI Based Mamdani Fuzzy Inference System Modelling to Predict Surface Roughness in Laser Machining. *International Journal of Electrical & Computer Science IJECS-IJENS* 9(9) : 37-43.

[4] Sumiati, Sigit, H.T., & Kapuji, A. 2014. Mamdani Fuzzy Inference System Application Setting for Traffic Light. *International Journal of Application or Innovation in Engineering & Management (IJAIEM)* 3(10) : 56-62

[5] Torrubia, G.S. & Blanc, C.T. 2010. A Mamdani-Type Fuzzy Inference System to Automatically Asses Djikstra's Algorithm Simulation. *International Journal Information Theories and Application* 17(1): 35-99.

[6] Simon, D. 2002. Sum Normal Optimization of Fuzzy Membership Functions. *International Journal of Uncertainty, Fuzziness, and Knowledge Base System* 10(4): 363-368

[7] Poonam, Tripathi, P.S., Shukla, K.P. 2012. Uncertainty Handling using Fuzzy Logic in Rule Based System. *International Journal of Advanced Science and Technology* 45: 31-46

Mechanism of Food Ordering in A Restaurant Using AndroidTechnology

Rachmat Auli[1], Ahmad Zakir[2], Haida Dafitri[3], Dodi Siregar[4], and Hasdiana[5]

[134]Departement of Informatics, Sekolah Tinggi Teknik Harapan Medan, Jl. H.M. JoniNo. 70, Kota Medan, Sumatera Utara, Indonesia2,5
[2 4]Departement of Information System, Sekolah Tinggi Teknik Harapan Medan, Jl.H.M. Joni No. 70, Kota Medan, Sumatera Utara, Indonesia

Abstract. A Restaurant is a gathering place for many people to taste the favorite foods are inthere.The restaurant which visited many people sure will increase the attraction of them tovisit it. Of course, the owner will get more benefit. However, what happens when a restaurantis famous still uses a service without technology, such as making orders using pens and paper,inspects the food stocks manually, and delivering orders to the kitchen using manpower, andmore. Therefore, it designed a system that can accelerate the ordering and processing food inthe restaurant. This system replaces the use of pen and paper with digital devices such astablets/smartphones based on Android. Not only that, order data can be sent through a wirelessnetwork which connects tablets/smartphones with the kitchen's computer. It can be read bykitcheners and showed directly on the LCD screen. By the application is expected to reduce thelevel of error in the processing of the consumer's order.
Keywords: Food Order, Wireless Technology, ICT Restaurant

1. Introduction

The evolution of technology is increasing rapidly. This is evidenced by many things can do usingtechnology such a mobile device. The mobile devices such as smartphones, tablets, and PDAs, can beintegrated with other devices like Laptop or LCD screen, may use to establish a network ofinformation and communication technology .

The steps of menu order, in the restaurant, generally will gain some problems, for examples: take along time because the food order is written use a pen and paper, there is a possibility that it exchangedwith the others consumers have, and no information about the stock of food whether available or not,that causing the waiter return to consumers and ask it again the replacement.

This research aims to explain how to implement an application that can be used to make the foodorder in the restaurant digitally. The procedure starts when the consumers are served by the waiterusing a tablet/smartphone. Then, the consumer's order is inserted through that device. After theordering finished, it sent directly to the kitchen through Wifi Hotspot which connected to a kitchen'scomputer server and also the tablets/smartphones held by the waiters. In order to make easykitcheners, it is displayed via LCD screen, so that the possibility of incorrect the order process willminimize. After it is over, kitcheners will post a status for the number of table's ordering that ready toserve to the system. Finally, the waiter will pick up the ordering and bring it according to the list of thenumber of the order that is in the system.

2. Objectives

The purpose of this research is to make simple the processing of the food order as well as speed upthe ordering meals that generally use a paper and pen changed to be Android's application.As for the goals to be achieved in this study, are:

a. Making easy and saving the time of the waiter to process the food order
b. Avoiding ozone depletion
c. Helping the chefs in knowing the consumers' order fastly

3. Literature Review

The restaurant is a favorite place to hang out with family and girlfriends. But, in the restaurantssometimes have a major problem such take a long time to serve the food order because getting a lot oforders. It can be confusing chefs in sorting which orders have to be processed first. Delays in servingfood which has requested by the consumers can disappoint and not satisfied them. That thing canmake a bad image of the restaurants so that they' appeal is reduced. This obstacle can be minimized byestablishing a system which speeds up the process of it using a mobile technology. The system namedis DORSUA [2]. It may help the management of restaurants by enabling they to make themselvesorder via tablets which have given. The order's data sent directly to the kitchen area that viewed on aserver computer through an internet network. After that, the order can be processed immediately bychefs.

Many of technologies can be applied in restaurants. One of them is a combination of technologyinternet network and mobile. By those technologies, the consumers may order a dining table andmenus before visiting the restaurants. Of course, it can help and easy to process the management inthere because the order can be done in long ways. Ease of this technology sure may make toconvenient them and speed up the serving of food. There are three points in this study among are [4]:

a. Able to make the order from long ways
b. The order's data get in immediately to the kitchen area
c. The manager of a restaurant can track the whole transaction

The growth of restaurants lately more rapidly. The use of information and communicationtechnology sue the owner restaurant for installing the technology. The application of ICT is intendedfor increasing the consumers' appeal so resulting in the increased revenue. Such as research whichbuilds a smart restaurant. It may exchange the use of pens and papers which used to write the foodorder.Besides, time of service applied is no wasting because they may reserve directly by accessingthe system through smartphones based on Android. In other hands, there are facilities in the systemamong a list of an empty dining table, instant e-billing, and the service of fast parking [5].

4. Research Method

In designing of a system needs a methodology. It has a role for analyzing and detecting the way of system processes. The methodology which used in this system is SDLC (software development life cycle). SDLC is the process stages used to develop an information system has a goal to explain theactivities related to project management [1]. The steps will give information to analyst and programmer how the system's works. The SDLC has four phases are: planning, analysis, design, and implementation.

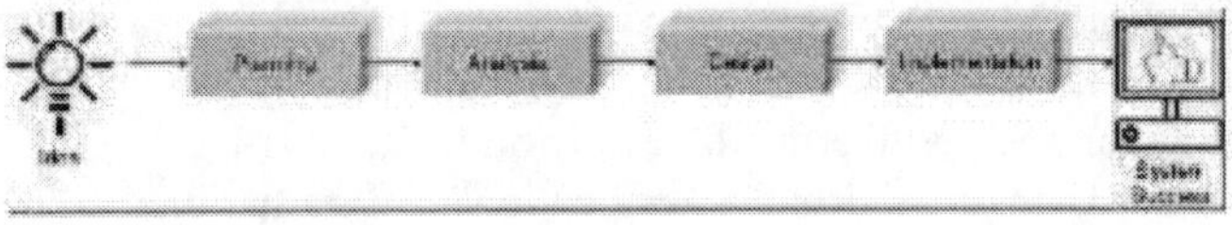

Figure 1. Four Steps of SDLC [3]

The elucidation of the figure 1, above as follows:A planning phase is used to know the reasons why has to build a system and also to understand astructure shaped in a project. While, for the analysis phase has some of the purposes among who willuse the proposed system, what the system does, where it is going to be applied, and when it begins todesign. Then, for design phase is purposed to conceive the way of how the system is going to work toface the problems. Last, the implementation phase gets a role for delivery the system. that means howto make end users become understand in operating it. Besides, this phase must focus to support thingsin order the system can run completely.

5. Proposed System

IOP Conf. Series: Journal of Physics: Conf. Series **930** (2017) 012030 doi:10.1088/1742-6596/930/1/012030

This study only designed a system that can facilitate the waiters in making food orders in arestaurant. Orders inputted to the system by using tablets/smartphones, then submit it to the kitchenwith a wireless network. It connects the tablets /smartphones device to the server computer located inthe kitchen. The incoming order data is displayed on the LCD screen to make it easier the chefs todetect it. After the order is processed, then the kitchen informs them by updating its status through thesystem. The status indicates that it is ready to be served to the consumers and they immediately picksup and delivers it. The previous elucidation is the mechanism proposed in this study. Many things thesystem can do in the restaurant business using ICT. However, this paper is limited to designingsystems for ordering food and order processing in the kitchen.
The system starts the kitchen's area has to log on into it. If successful then displayed the main menu.But, if not then the system will re-request the user id and password until correct. After getting anaccess, it can update the food stock and saved directly to the system. This is necessary in order thestock of food is not empty so the consumers can order it. While waiters must be connected to thesystem before they are in the restaurant. Here, the waiter needs to enter the server IP and user id andpassword. The main page for the waiter can be displayed if their verification is correct. It is intendedto speed up making the consumer's order. When they come, they look the restaurant menu. Waiterscan insert the customer's order. But if the food they want is not available, the system immediatelyinforms it and the waiter can immediately notify it to them. Orders did, be submitted directly to thekitchen. Furthermore, Kitcheners can directly display the orders on the LCD. The list of incoming itcan be directly processed. Completed orders, can be informed by Kitcheners to the system by updatingthe status. The waiter can view order status to the system using tablets/smartphones. Orders that areready to be served can immediately pick up and sent by the waiters to the consumers

6. Conclusion
This study only analyzes how food ordering mechanisms in restaurants use mobile-basedtechnology. The ease produced is to replace the use of paper and pen to digital devices when makingorders. Inaddition, this system can synchronize the server computer in the kitchen with digital devicessuch as tablets/smartphones that order data can be directly displayed on the LCD screen. Sure, this canreduce the level of errors that occur in the kitchen at the time of processing the consumer's order.

7. References

[1] Aulia, R. 2012. “Using An Online Auction System in Open Source”. Conference: SeminarNasional Teknologi Informasi & Komunikasi.
[2] Bhargave, A, Jadhav, N., Joshi, A., Oke, P. & Lahane, R. S. 2013. “Digital Ordering System forRestaurant Using Android”. International Jurnal of Scientific and Research Publications.
[3] Dennis, A., Wixom, H. B. & Roth, M. R. 2012. “System Analysis and Design”, Fifth Edition.
[4] Dhore, B. V., Thakar, S., Kulkarni, P. & Thorat, R. 2014. “Digital Table Booking and FoodOrdering System Using Android Application”. International Journal of Emerging EngineeringResearch and Technology
[5] Gundle, P. K., Harshe, A. A., Kinage, B. K. & Ghanawat, L. N. 2016. “Digital Smart System forRestaurants Using Wireless Technology”. International Research Journal of Engineering andTechnology.
[6] Pujari, S., Rawool, S., Malsawmkima, R. & Bodade, V. 2017. “Restaurant Booking and OrderingSystem”. Imperial Journal of Interdisciplinary Research.
[7] Raut, S. Ketaki & Dhonde, B. S. Mr. 2016. “Android Based Intelligent E-Restaurant OrderingSystem”. International Education & Research Journal

THE ANALYSIS PERFORMANCE METHOD NAIVE BAYESANDSSVM DETERMINE PATTERN GROUPS OF DISEASE

Rianto Sitanggang[1], Tulus[2], Zakarias Situmorang[3]

[123]Department of Computer Science and Information Technology University of NorthSumatra Medan, Indonesia

[1]rianto_sitanggang@students.usu.ac.id

Abstrack. Information is a very important element and into the daily needs of the moment, to get aprecise and accurate information is not easy, this research can help decision makers and make acomparison. Researchers perform data mining techniques to analyze the performance of methodsand algorithms naïve Bayes methods Smooth Support Vector Machine (ssvm) in the grouping ofthe disease.The pattern of disease that is often suffered by people in the group can be in thedetection area of the collection of information contained in the medical record. Medical recordshave infromasi disease by patients in coded according to standard WHO. Processing of medicalrecord data to find patterns of this group of diseases that often occur in this community take theattribute address, sex, type of disease, and age. Determining the next analysis is grouping of fourersebut attribute. From the results of research conducted on the dataset fever diabete mellitus,naïveBayes method produces an average value of 99% and an accuracy and SSVM method produces anaverage value of 93% accuracy
Keywords :Metod naïve Bayes, SSVM method, Diabetes Mellitus

1. Background

Data mining is a series of processes for adding additional value of a set of data in the form of knowledge that had been unknown to them manually. Keep in mind that the word mining itself meanslittle effort to get valuable data from a large number of basic data. Because the data mining actually has along root of disciplines such as artificial intelligence (artificial intelligent), machine learning, statistics anddatabases. Some techniques that are often mentioned in the literature among others, the data miningassociation rule mining, clustering, classification, neural networks, and others. Piles of data that exist bothin the health service and the hospital currently only limited to providing graphs or statistics on the numberof patients seeking treatment with disease in misery along with the report of the patient's departure.Reports from this data that is currently used by health authorities to carry out the policies when willprovide counseling to the public.

1.2. Formulation of the problem

With reference to the background of the above problems, the issues discussed in this study is to analyzethe naïve Bayes methods in getting polakelopok disease classification techniques datamining smoothsupport vector machine (ssvm).

1.3. Limitations of the study

The boundary problem in this research are:

a. Data on the disease General Hospital Sari Mutiara with interfal time data collection in themedicalrecord by 2016.

b. Disease information being sought is a grouping of diseases by age grouping and the grouping area.

1.4. Research purposes

This study has the objective to apply naïve Bayes methods in searching for the information containedin astack of medical records

I.5 Benefits Research

IOP Conf. Series: Journal of Physics: Conf. Series **930** (2017) 012031 doi:10.1088/1742-6596/930/1/012031

The benefits derived from this study are as follows:

a. Assist medical personnel in analyzing the pattern of the disease to the medical record data quicklyand accurately.
b. Become a comparison to researchers who focus on Datamining particular field of medicine.

2. LITERATURE REVIEW

2.1. Understanding Data Mining

In a simple data mining is mining or the discovery of new information by looking for patterns orparticularrules of a number of very large data (Davies, 2004). Data mining is also referred to as a series of processesfor adding additional value in the form of knowledge that had been unknown to them manually from adata set (Pramudiono, 2007). Data mining, often referred to as knowledge discovery in databases (KDD).KDD is an activity that includes the collection, use of data, historical to find regularities, patterns orrelationships in large data sets (Santoso, 2007).

2.2 Naïve Bayes

Naive Bayes is a method that has no rules, Naive Bayes using a branch of mathematics known as thetheory of probability to find the greatest opportunities of the possibilities of classification, by looking atthe frequency of each classification on the training data.NaiveBayes classification is a statistical classification that can be used to predict the probability ofmembership of a class. Classification bayesiandidasarkan on Bayes' theorem, named for themathematician who is also minister Prebysterian England, Thomas Bayes (1702-1761) [8].

Bayesianclassification has the capability similar to that classification decision tree and neural network [17].Bayes rule is used to calculate the probability of a class. Naive algorithm Bayesmemberikan away to combine the opportunities foregone on condition likely to be a formula that can be used tocalculate the chance of each possibility occurring. The general form of the Bayes theorem as follows:

$$P(H \mid X) = \frac{P(X \mid H)P(H)}{P(X)}$$

Where:
X : Data with unknown class
H : Hypothesis X data is a specific class.
P (H | X): The probability of the hypothesis H based on the conditions X (a posteriori probability)
P (H) : The probability of the hypothesis H (prior probability)
P (X | H): Probability X based on the conditions on the hypothesis H
P (X): The probability of XNaïve Bayes is a simplified method of Bayes.
Bayes theorem simplifies to:
P (H | X) = P (X | H) P (X)

2.3 Definition of SSVM (Smooth Support Vector Machine).

SSVMis a new development of SVM with kernel and non-linear functions for classification analysis usingsmoothing methods. SVM was first introduced by Boser, Guyon and Vapnik in 1992 as a series ofharmonic concepts featured in the statistical learning theory. The basic principle of SVM is a linearclassifier and further developed in order to work on a non-linear problem by incorporating the concept ofkernel trick on high-dimensional workspace (Vapnik, 1995).

Standard SVM to this problem is given by the following quadratic program:

$$min_{(\omega,\gamma,y)\epsilon R^{n+1+m}},\ D(A\omega - e\gamma) + y \geq e$$

Where :
v = Sebuahberatyangpositif
y = variableSlack
e = Kolomvectorsatudimensi
In SSVM approach (Lee et al., 2001), modified SVM problem is generated as follows:

$$min_{(\omega,\gamma,y)\in R^{n+1+m}} \frac{v}{2} y'y + \frac{1}{2}(w'w + \gamma^2)$$

Thus, we can replace y in constraints with and convert into SVM SVM similar problem that anoptimization problem without constraints as follows:

$$min_{(\omega,\gamma)} \frac{v}{2} \|(e - D(Aw - e\gamma))_+\|_2^2 + \frac{1}{2}(w'w + \gamma^2)$$

This function with smoothing parameter used in the replace function hereto plus to get Smooth VectorMachine (SSVM)

$$min_{(\omega,\gamma)\in R^{n+1}} \frac{v}{2} \|p(e - D(Aw - e\gamma)), \alpha\|_2^2 + \frac{1}{2}(u'u + \gamma^2)$$

Just as before, it is obtained SSVM for an integral problem:

$$min_{(\omega,\gamma)} \frac{v}{2} \|p(e - D(Aw - e\gamma)), \alpha\|_2^2 - \frac{1}{2}(uu + \gamma^2)$$

2.4.Klasifikasi

Classification is the process of finding a set of models / functions that define and distinguish the data intocertain classes, with the aim of using the model to determine the class of an unknown object class (Han, etal2000). There are two processes in classification, namely:

a. The process of learning or training do the construction model using training data. In this study,using the model of Smooth Support Vector Machine. At Smooth Support Vector Machine(SSVM), this is done by labeling each feature value
b. The process of testing conducted tests on the data of testing using a model that has been obtainedfrom the training process. The use of the model to classify new data. Here, a records are fed intothe model, and the model will provide answers grade calculation results.Accuracy (%) = x 100% (2.7)

3. RESEARCH METHODOLOGY

3.1 Diagram Research

The diagram of the overall study can be seen in Figure 1.

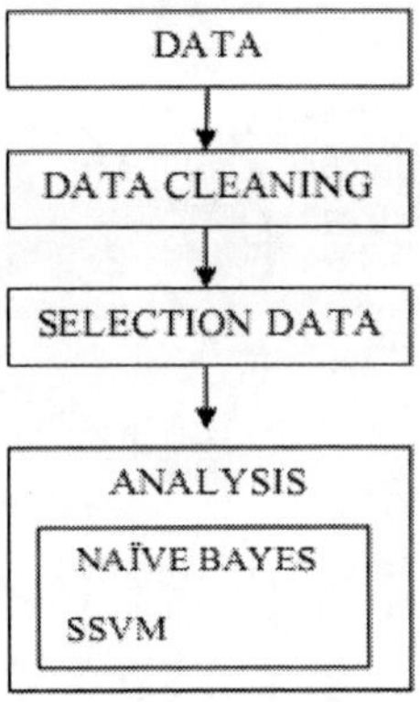

Figure 1. Diagram Research

3.2 Data Collection

Medical record data to be processed is RSU Sari Mutiara Medan patient medical records. period in 2016.Before the process of data mining, medical records obtained from the data base hospital performed thepreprocessing stage. Preprocessing mechanism at this stage of integration and transformation has beencollecting data from the tables contained in the formal application in hospitals into the parent table.

3.3. Data cleaning

The initial stage of data mining preposessing are cleaning the data, this is done to discard data thathaveincomplete information. It is possible that a lot of data have incomplete information, such as the exclusionof the information age, information about the area, information about sex, disease grouping information,asshown in

3.4 Selection of Data

Data that already have complete information in any subsequent attribute such data be selected. Selection isdone to classify the attribute in accordance with the information needed. From attribute in the selection isdone at the address attribute, attribute age. While the gender attribute is the target attribute. In this study,the data selection is done on the data that has information area in the city field. Because hospitals areconcerned also receive patients from outside the field, the data containing information about the areaoutside the city of Medan will be discarded.

4. RESULTS AND DISCUSSION

4.1. Application of Naive Bayes Method

Naive Bayes is based on the simplifying assumption that the attribute values are conditionallyindependentif given a value of output. In other words, given the value of output, the probability of observing togetheris the product of the individual probabilities. The advantage of using Naive Bayes is that this method onlyrequires the amount of training data (training data) are small to determine the required parameterestimation in the classification process.

4.2 Read Data Training

To specify the data that will be analyzed by the method of Naive Bayes then the first step is to readtraining data. The training data used can be seen in Figure 1.

4.1.2.Testing Methods Naïve Bayes

From the above probability value will be tested as much data as 60 data and solved by using the tools sothat the resulting classification results weka frequent users such as in Figure 2.

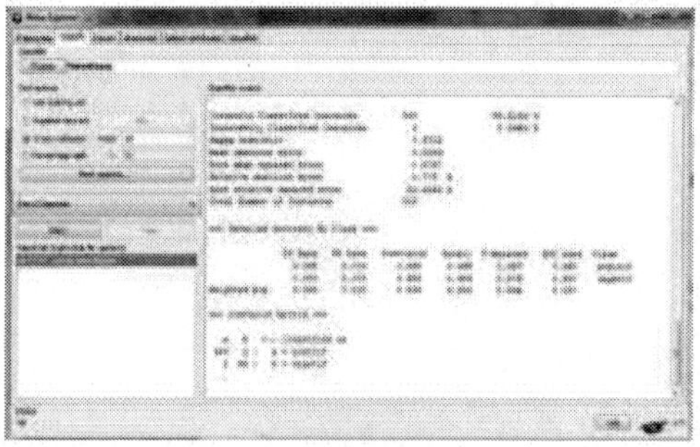

Figure2. Results of Treatment Naïve Bayes

4.3 Method of Testing Results SSVM

The test results of 10 experiments above can be displayed in the table below 4:21. The tables show thepercentage of accuracy values for each - each training and testing. Testing and training value obtainedfrom the calculation as follows:

For training:

$$\text{Accuracy rate } (\%) = \frac{\text{jumlah pola yang dikenali}}{\text{jumlah pola pelatihan}} \times 100\%$$

For testing:

$$\text{Accuracy rate } (\%) = \frac{\text{jumlah pola yang dikenali}}{\text{jumlah polapengujian}} \times 100\%$$

Table1. Percentage of Training and Testing Data In 10 Times Trial

Percobaan	X1	X2	X3	Akurasi Training (%)	Akurasi Testing (%)
Fold01	14	18	23	99.63	90.00
Fold02	21	20	10	99.63	96.67
Fold03	30	16	4	99.63	90.00
Fold04	26	6	32	100.00	96.67
Fold05	8	3	17	99.63	93.33
Fold06	1	7	29	100.00	93.33
Fold07	5	22	23	99.63	93.33
Fold08	24	27	19	99.63	96.67
Fold09	12	9	25	99.63	93.33
Fold10	15	11	13	99.63	93.33
Rata-rata				99.70	93,67

5. CONCLUSIONS

5.1. Conclusion

The conclusion that can be drawn from this study are as follows:

1. The process of grouping with SSVM method of obtaining the average - average accuracy of thetraining as much as 99.70% and the average - average test as much as 93.67%.
2. The process of grouping with Naive Bayes algorithm obtain the average - average accuracy asmuch as 99.40%.
3. SSVM method provides better accuracy than the Naïve Bayes method in the clustering of thedisease.

BIBLIOGRAPHY

[1] Andari, S., 2012, Smooth Support Vector Machine and Multivariate Adaptive Regression Splines ToDiagnose Breast Cancer, Thesis, Faculty of Mathematics and Statistics Department Student ITS,Surabaya.

[2] Erdogan, SZ and M. Timor. 2005. A Data Mining Application In A Student Database. Journal OfAeronautics And Space Technologies. Volume 2 Number 2: 53-57.

[3] Furqan, M., Embong, A. (2009). Smooth Support Vector Machine For FaceRecognition UsingPrincipal Component Analysis, International Journal of Science Engineering and Technology2 (3):1985-3785.

[4] Garcia, EPI and PM Mora. 2011. Prediction Model of Academic Performance for First YearStudents. IEEE Computer Society.

[5] Gunn, S. 1998.Support Vector Machines for Classification and Regression, Technical Report, ISIS4(3): 1985-3785 ..

[6] Habibi, R. 2011. Application of Data Mining Techniques with Smooth Method of Support VectorMachine (SSVM) to predict the chance of students who drop out (polytechnic student case studyfield). Thesis. University of Northern Sumatra.

[7] Han, Jiawei. &Kamber, Micheline. 2000 Data Mining: Concepts and Techniques, Morgan KaufmannPublishers, Urbana-Champaign

[8] Hien, NTN and P. Haddawy. 2007. A Decision Support System for Evaluating International StudentApplications. IEEE Frontiers in Education Conference.

Prime Number: an Experiment Rabin-Miller and Fast Exponentiation

Robbi Rahim[1*], Hendryan Winata[2], Iskandar Zulkarnain[2] and Hendra Jaya[3]

[1]Postgraduate Ph.D. Student Universiti Malaysia Perlis, Universiti Malaysia Perlis, Malaysia
[1]Department of Health Information, Akademi Perekam Medik dan Infokes Imelda, Jl. Bilal Ujung Medan 20116, Indonesia
[2]Department of Computer System, STMIK Triguna Dharma, Jl. Jenderal Abdul Haris Nasution No.73, Medan 20219, Indonesia
[3]Department of Computer Engineering, STMIK Triguna Dharma, Jl. Jenderal Abdul Haris Nasution No.73, Medan 20219, Indonesia

*usurobbi85@zoho.com

Abstract. Prime number is one of the most widely used names for various security purposes, especially in the process of encryption and decryption for asymmetric algorithms, the speed in determining prime number becomes one of the most important things, Rabin-Miller algorithm is used to test a number whether including prime or not, speed in determining prime number is very important and technique that could be using is fast exponentiation, this research get significant result in testing prime number with length of variation number.

1. Introduction

Public key cryptography has several issues such as factorization, Knapsack problems, elliptical curves, discrete algorithms, and so on. Factorization [1] [2] [3] [4], significant number modulo powers and prime numbers is a widely used mathematical function in public key cryptography [5] [6].

The process of factorization, large number modulo force and the formation and testing of prime numbers using conventional way often experience obstacles [7] [8], for example in determining the top value of 2^{2000} mod 5^{300}, to know the process of identifying prime numbers can be used algorithms such as Rabin-Miller and Fast Exponentiation, the use of algorithms using the Pseudo Code .Net Language is expected to allow users to design applications prime generator.

2. Theory

A. Basic Mathematical Concepts of Cryptography

The underlying mathematical basis of the process of encryption and decryption is the relation between two sets i.e. the set contains the plaintext element, and the set includes the ciphertext element [2]. Encryption and decryption is a transformation function between the two sets. If the plaintext set is indicated by P and the set of ciphertext is indicated by C, while E denotes the encryption function and

the decryption function with D [2] [8] [9] the encryption-decryption process can be expressed in mathematical notation by:
E (P) = C and
D (C) = P
Since the encryption-decryption process aims to recover the original data, then:
D (E (P)) = P
The relation between the plaintext set and the ciphertext set must be a one to one relation to preventing the occurrence of ambiguity in decryption i.e. one ciphertext element declares more than one plaintext element.

B. Factorization of Large Numbers

Number theory is the basis of the number itself, and the core blocks of numbers are primes so that primes are often said to be central to number theory [10]. A prime number is a number that only has two factors: the number one and the number itself [10] [11]. Numbers that have more than two elements called composite numbers. Numbers one has only one factor and is considered not prime and composite numbers [10] [12]. When a composite number is composed of a product of all prime factors, it is said to be the main factorization of the number. Factorization of a number into its primary constituent element also called prime decomposition. Typically, factors are listed from the smallest to the largest [9] [10].

C. Rabin-Miller

The simple algorithm used by everyone is designed by Michael Rabin based on some ideas from Gary Miller. The test algorithm is as follows:

1) Select a random number p to test.
2) Calculate b, where b is the number (p - 1) divided by 2 (i.e., b is the largest power of 2, such that 2b is a factor of p - 1).
3) Then calculate m, such that $p = 1 + 2^b.m$
4) Choose a random number *a* such that *a* is smaller than *p*.
5) Set j = 0 and set $z = a^m$ mod p.
6) If z = 1 or if z = p - 1, then p passes the test and may be a prime number.
7) If j> 0 and z = 1, then p is not a prime number.
8) Set j = j + 1. If j <b and $z \neq p - 1$, set $z = z^2$ mod p and return to stage 4. If z = p - 1, then p passes the test and may be prime.
9) If j = b and $z \neq p - 1$, then p is not a prime number

3. Result and Discussion

For testing module velocity with Fast Exponentiation can be taken case example 1123829 mod 95317 with the process as follows:

A1 = 11
B1 = 23829
Product = 1

While 23829 <> 0 → True

While 23829 mod 2 = 0 → False

B1 = 23829 - 1 = 23828
Product = (1 * 11) mod 95317 = 11

While 23828 <> 0 → True

While 23828 mod 2 = 0 → True

B1 = 23828 div 2 = 11914
A1 = (11 * 11) mod 2 = 121

While 11914 mod 2 = 0 → True
B1 = 11914 div 2 = 5957
A1 = (121 * 121) mod 2 = 14641

While 5957 mod 2 = 0 → False

B1 = 5957 - 1 = 5956

While 4 <> 0 → True

While 4 mod 2 = 0 → True
B1 = 4 div 2 = 2
A1 = (40123 * 40123) mod 2 = 46316

While 2 mod 2 = 0 → True
B1 = 2 div 2 = 1
A1 = (46316 * 46316) mod 2 = 62771

While 1 mod 2 = 0 → False

B1 = 1 - 1 = 0
Product = (11004 * 62771) mod 95317 = 65102

While 0 <> 0 → False

FastExp(11, 23829, 95317) = 65102

The core of the prime random generation process is to test whether the random number generated is either prime or not. In order to better understand the process of generating prime random numbers, taken for example number 95317 with A = 11, then the process of completion with Rabin Miller algorithm is as follows:

C = 24889 - 1 = 24888
nTemp = 0
24888 mod (2^0) = 0 And ((2^0) < 24889) - True
nTemp = 0 + 1 = 1
24888 mod (2^1) = 0 And ((2^1) < 24889) - True
nTemp = 1 + 1 = 2
24888 mod (2^2) = 0 And ((2^2) < 24889) - True
nTemp = 2 + 1 = 3
24888 mod (2^3) = 0 And ((2^3) < 24889) - True
nTemp = 3 + 1 = 4
24888 mod (2^4) = 0 And ((2^4) < 24889) - False
B = 4 - 1 = 3
M = 24888 / (2^3) = 3111
J = 0

Table 1. Running Time Execution

No	Length Number in digit (Random)	With Fast Exponentiation (second)	Without Fast Exponentiation (second)
1	2	0.09	0.08
2	3	0.247	0.161
3	5	1.45	3.14
4	7	11.39	19.43
5	9	22.51	35.19
6	13	73.02	108.49
7	17	165.58	301.11

Testing performed on a different number to determine prime numbers using fast exponentiation technique known faster if for large numbers compared without using fast exponentiation, for more details could see in the graph below following the resulting time difference.

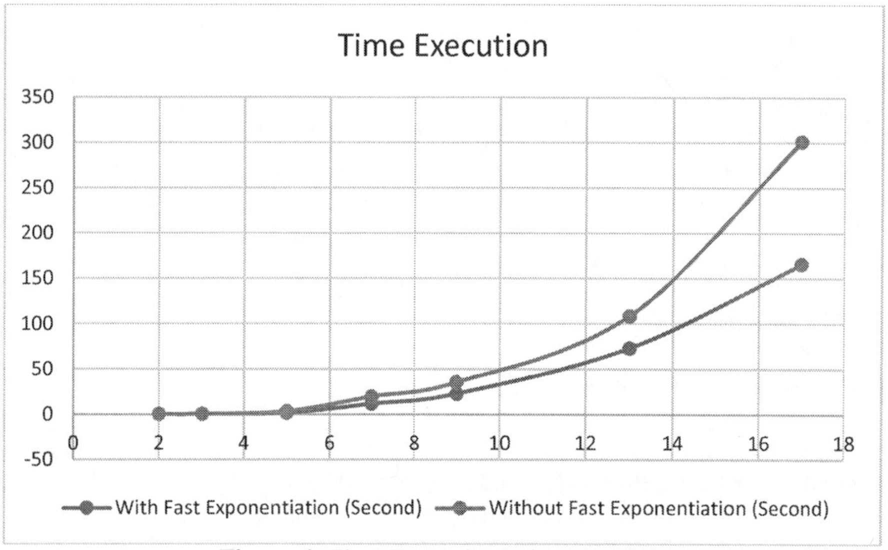

Figure 1. Time Execution Prime Number

The above graph shows the difference in execution time by using Rabin-Miller with fast exponentiation faster to determine prime numbers.

4. Conclusion

The prime Rabin-Miller and Fast Exponentiation test can be using for excellent test of randomly generated random numbers, and by using Rabin-Miller and Fast Exponentiation, algorithms can perform a large test process of up to 15 digits

References

[1] R. Rahim, "128 Bit Hash of Variable Length in Short Message Service Security," *International Journal of Security and Its Applications,* vol. 11, no. 1, pp. 45-58, 2017.

[2] R. Rahim and A. Ikhwan, "Cryptography Technique with Modular Multiplication Block Cipher and Playfair Cipher," *IJSRST,* vol. II, no. 6, pp. 71-78, 2016.

IOP Conf. Series: Journal of Physics: Conf. Series **930** (2017) 012032 doi:10.1088/1742-6596/930/1/012032

[3] R. Rahim and A. Ikhwan, "Study of Three-Pass Protocol on Data Security," *International Journal of Science and Research (IJSR),* vol. 5, no. 11, pp. 102-104, 2016.

[4] A. P. U. Siahaan and R. Rahim, "Dynamic Key Matrix of Hill Cipher Using Genetic Algorithm," *International Journal of Security and its Applications,* vol. 10, no. 8, pp. 173-180, 2016.

[5] D. Cries and J. Misra, "A Linear Sieve Algorithm for Finding Prime Numbers," *Communication of the ACM,* vol. 21, no. 12, pp. 999-1003, 1978.

[6] Y. Fu and S. Xu, "Generalized Fast Coding Steganographic Algorithm Based on Prime Number," in *IEEE International Conference on Information Theory and Information Security (ICITIS)*, Beijing, China, 2010.

[7] G. A. Jullien, "Implementation of Multiplication, Modulo a Prime Number, with Applications to Number Theoretic Transforms," *IEEE Transactions on Computers,* vol. 29, no. 10, pp. 899 - 905, 1980.

[8] Y.-S. Kim, S. R. Shrestha, and J.-W. Jang, "Refined Algorithm for Prime Number Generation in Embedded Security Systems," in *IEEE Asia-Pacific Services Computing Conference (APSCC)*, Jeju Island, South Korea, 2011.

[9] V. S. Igumnov, "Generation of the Large Random Prime Numbers," in *5th International Siberian Workshop on Electron Devices and Materials, 2004. Proceedings. 5th Annual. 2004 International Siberian Workshop on*, Erlagol, Altai, Russia, Russia, 2004.

[10] D. Liu, "Iteration Distribution of Prime Numbers," *Journal of Mathematics Research,* vol. 3, no. 4, pp. 112-117, 2011.

[11] Z. Qing and H. Zhihua, "The large prime numbers generation of RSA algorithm based on genetic algorithm," in *International Conference on Intelligence Science and Information Engineering*, Wuhan, China, 2011.

[12] R. R. Sarukkai, "Prime Numbers and Output Codes," in *IEEE International Conference on Neural Networks*, Perth, WA, Australia, 1995.

Use of One Time Pad Algorithm for Bit Plane Security Improvement

Suhardi[1], Saib Suwilo[2], Erna Budhiarti Nababan[3]

[123]Department of Computer Science and Information Technology University of North Sumatera Medan, Indonesia

[1]suhardi147038064@students.usu.ac.id

Abstract. BPCS (Bit-Plane Complexity Segmentation) which is one of the steganography techniques that utilizes the human vision characteristics that cannot see the change in binary patterns that occur in the image. This technique performs message insertion by making a switch to a high-complexity bit-plane or noise-like regions with bits of secret messages. Bit messages that were previously stored precisely result the message extraction process to be done easily by rearranging a set of previously stored characters in noise-like region in the image. Therefore the secret message becomes easily known by others. In this research, the process of replacing bit plane with message bits is modified by utilizing One Time Pad cryptography technique which aims to increase security in bit plane. In the tests performed, the combination of One Time Pad cryptographic algorithm to the steganography technique of BPCS works well in the insertion of messages into the vessel image, although in insertion into low-dimensional images is poor. The comparison of the original image with the stegoimage looks identical and produces a good quality image with a mean value of PSNR above 30db when using a large-dimensional image as the cover messages.

1. Introduction

Steganography is a technique that the secret data is hidden into vessel image without any suspicion[4]. Steganography is the art of hiding information that cover the messages in a vessel image [1][8]. Information security is very important in steganography because the information can only be known by the recipient. Bit-Plane Complexity Segmentation (BPCS) is one of the steganography techniques that replace all of the noise-like regions in the bit-planes of the vessel image with secret information[6]. Storing secret information into bit planes directly results in the process of extracting information becomes easier so that confidential information can be known by others. Therefore bits plane need to be modified. Security techniques to improve the security system can be done by modifying structure noise like regions[5]. One way to modify the bit plane is by combining it with encryption. It can generate a new bit order or cipher block that make extracting information more difficult.

2. Theories

2.1. *Bit Plane Complexity Segmentation (BPCS)*

Bit-plane complexity segmentation (BPCS) is a steganography technique introduced by Eiji Kawaguchi and Richard O. Eason in 1998. This technique is a steganograpy technique that has a large capacity, because it can accommodate confidential data with a relatively large capacity when compared with the steganography method Such as LSB (Least Significant Bit). This BPCS technique is a steganography technique that is not based on programming techniques, but a technique that uses the nature of human vision. The nature of human vision that is utilized is the inability of humans to interpret a very complicated binary pattern. The inserted data capacity can reach 50% from the size of the cover- image [3]. Eiji Kawaguchi and R. O. Eason introduced this BPCS technique for use on uncompressed color image documents with BMP format. The image document is divided into segments of 8x8 pixels per segment [3]. In an 8-bit image document, each segment will have 8 plane bits representing the pixels of each bit. The process of dividing the 8x8 pixel segment into 8 pieces of bit plane is called bit slicing process. Representation of the eight bit plane is a PBC system (Pure Binary Code). In BPCS, the insertion process is performed on a bit plane with a CGC (Canonical Gray Code) system because bit slicing process on CGC tends to be better than PBC [3]. So in the insertion process, the bit plane with the PBC representation is converted into a plane bit with a CGC representation. The main principle of BPCS technique is the binary image is divided into informative region and noise-like region. The secret information is hidden in noise-like region of the vessel image[4].

2.2. *One Time Pad (OTP)*

One Time Pad algorithm (OTP) is a symmetric key type algorithm which means that the key used to perform the encryption and decryption is the same key. In order for OTP not to be solved, the private key must be as long as the message, very random and can only be used once therefore the name "One-Time" Pad[7]. In the encryption process, this algorithm uses a stream cipher derived from the XOR result between the plaintext bits and the key bits. In this method plain text is converted into ASCII code and then subjected to XOR operation against the changed key into ASCII code. Data encryption is the first part of the data security process. In this encryption process the original data will be process randomization with a predetermined algorithm, while the encryption process can be done by using the formula below:

$$Ciphertext(i) = Plaintext(i) \oplus Key(i) \ldots\ldots\ldots\ldots\ldots\ldots (1)$$

Decryption process is a process done to restore files from the form of symbols back to their original form. Decryption can be done using the formula below

$$Plaintext(i) = Ciphertext(i) \oplus Key(i) \ldots\ldots\ldots\ldots\ldots\ldots (2)$$

3. Bit Plane Security Improvement

For security enhancements to BPCS we utilize the One Time Pad cryptography algorithm to modify the original message bits before the message is inserted. Combination of Algorithm can be seen in fig.1

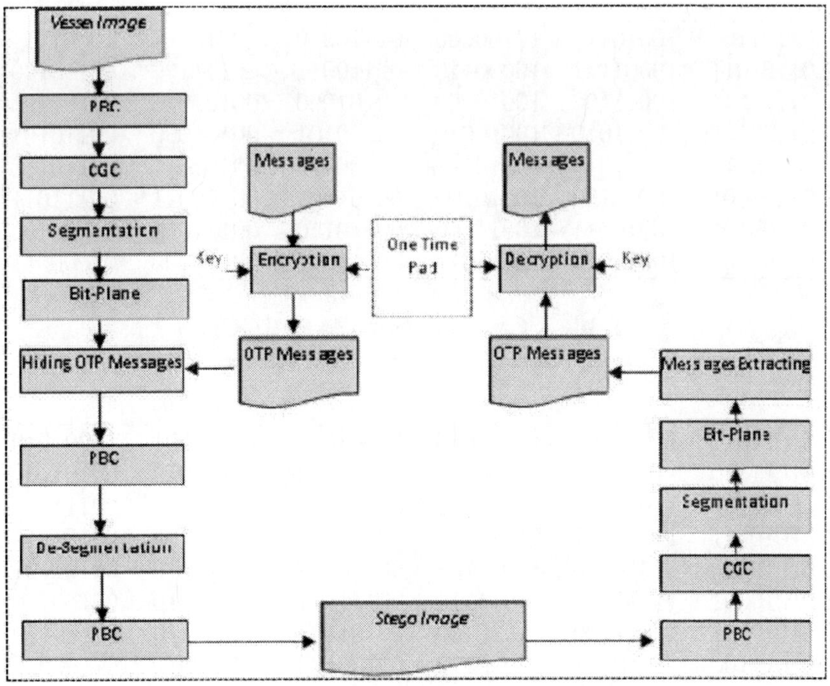

Figure 1. The technique of improved Bit Plane using OTP.

4. Result and Discussion

The image pixels are converted to numbers 0 through 255 that represent the intensity of the color. They are converted into pure Binary Code (PBC) systems. Each row and column represents the pixel position in the image. Image segmentation can be seen in Table 1. Each segment consists of 64 pixels.

Table 1.8x8 pixel of Image Light Intensity.

Pixel Value							
143	108	150	160	114	145	143	142
142	107	143	134	51	60	54	51
145	111	156	144	40	42	45	48
135	101	154	146	37	40	53	58
144	103	155	150	39	43	54	55
144	97	148	148	45	51	59	58
147	101	151	152	54	59	68	70
148	104	153	155	56	60	67	72

All pixels must be converted into binary numbers to be a Pure Binary Code (PBC). Once converted into a PBC form then converted again into Canonical Gray Code (CGC). Changes from PBC to CGC because in BPCS Steganography, image representation with CGC is better than using PBC system. Each pixel in the CGC segment will be split into bits to form several bit-planes. These changes will be shown in Table 2 and Table 3 and PBC Change Process to CGC example from row 0 and column 0 can be seen in fig 2

Table 2. Pure Binary Code (PBC).

PBC							
10001111	01101100	10010110	10100000	01110010	10010001	10001111	10001110
10001110	01101011	10001111	10000110	00110011	00111100	00110110	00110011
10010001	01101111	10011100	10010000	00101000	00101010	00101101	00110000
10000111	01100101	10011010	10010010	00100101	00101000	00110101	00111010
10010000	01100111	10011011	10010110	00100111	00101011	00110110	00110111
10010000	01100001	10010100	10010100	00101101	00110011	00111011	00111010
10010011	01100101	10010111	10011000	00110110	00111011	01000100	01000110
10010100	01101000	10011001	10011011	00111000	00111100	01000011	01001000

Table 3. Canonical Gray Code (CGC).

CGC							
11001000	01011010	11011101	11110000	01001011	11011001	11001000	11001001
11001001	01011110	11001000	11000101	00101010	00100010	00101101	00101010
11011001	01011000	11010010	11011000	00111100	00111111	00111011	00101000
11000100	01010111	11010111	11011011	00110111	00111100	00101111	00100111
11011000	01010100	11010110	11011101	00110100	00111110	00101101	00101100
11011000	01010001	11011110	11011110	00111011	00101010	00100110	00100111
11011010	01010111	11011100	11010100	00101101	00100110	01100110	01100101
11011110	01011100	11010101	11010110	00100100	00100010	01100010	01101100

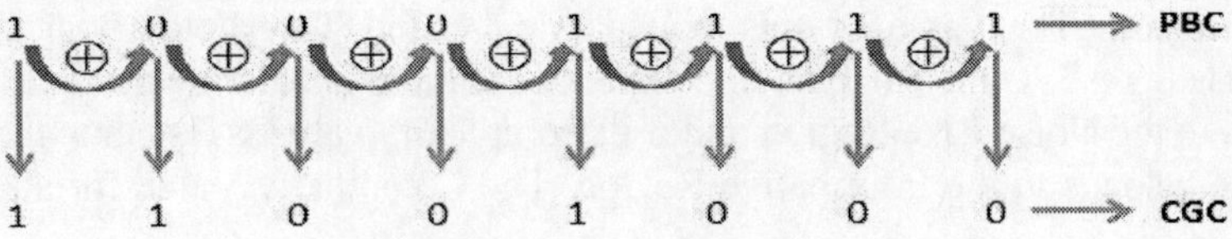

Figure 2. PBC to CGC process.

ᴐı

$\frac{k}{n}$

Wheı
∝= C
k= T
n= M

Bit Plane 1

1	0	1	1	0	1	1	1
1	0	1	1	0	0	0	0
1	0	1	1	0	0	0	0
1	0	1	1	0	0	0	0
1	0	1	1	0	0	0	0
1	0	1	1	0	0	0	0
1	0	1	1	0	0	0	0
1	0	1	1	0	0	0	0
						k	28
						∝	0,25

Bit Plane 2

1	1	1	1	1	1	1	1
1	1	1	1	0	0	0	0
1	1	1	1	0	0	0	0
1	1	1	1	0	0	0	0
1	1	1	1	0	0	0	0
1	1	1	1	0	0	0	0
1	1	1	1	0	0	1	1
1	1	1	1	0	0	1	1
						k	15
						∝	0,1339285

Bit Plane 3

0	0	0	1	0	0	0	0
0	0	0	0	1	1	1	1
0	0	0	0	1	1	1	1
0	0	0	0	1	1	1	1
0	0	0	0	1	1	1	1
0	0	0	0	1	1	1	1
0	0	0	0	1	1	1	1
0	0	0	0	1	1	1	1
						k	14
						∝	0,125

Bit Plane 4

0	1	1	1	0	1	0	0
0	1	0	0	0	0	0	0
1	1	1	1	1	1	1	0
0	1	1	1	1	1	0	0
1	1	1	1	1	1	0	0
1	1	1	1	1	0	0	0
1	1	1	1	0	0	0	0
1	1	1	1	0	0	0	0
						k	27
						∝	0,2410714

Bit Plane 5

1	1	1	0	1	1	1	1
1	1	1	0	1	0	1	1
1	1	0	1	1	1	1	1
0	0	0	1	0	1	1	0
1	0	0	1	0	1	1	1
1	0	1	1	1	1	0	0
1	0	1	0	1	0	0	0
1	1	0	0	0	0	0	1
						k	46
						∝	0,410714286

Bit Plane 6

0	0	1	0	0	0	0	0
0	1	0	1	0	0	1	0
0	0	0	0	1	1	0	0
1	1	1	0	1	1	1	1
0	1	1	1	1	1	1	1
0	0	1	1	0	0	1	1
0	1	1	1	1	1	1	1
1	1	1	1	1	0	0	1
						k	44
						∝	0,3928571

Bit Plane 7

0	1	0	0	1	0	0	0
0	1	0	0	1	1	0	1
0	0	1	0	0	1	1	0
0	1	1	1	1	0	1	1
0	0	1	0	0	1	0	0
0	0	1	1	1	1	1	1
1	1	0	0	0	1	1	0
1	0	0	1	0	1	1	0
						k	59
						∝	0,526785714

Bit Plane 8

0	0	1	0	1	1	0	1
1	0	0	1	0	0	1	0
1	0	0	0	0	1	1	0
0	1	1	1	1	0	1	1
0	0	0	1	0	0	1	0
0	1	0	0	1	0	0	1
0	1	0	0	1	0	0	1
0	0	1	0	0	0	0	0
						k	61
						∝	0,5446428

Figure 3. Bit Planes.

The value of complexity for all Bit Plane has a boundary value between the Informative Region and the Noise-Like Region. The value of this boundary is called Threshold. Basically, the standard threshold value used is 0.3. If the Bit Plane's complexity value is greater than the threshold value, the Bit Plane belongs to the Noise-Like Region and if its complexity value is less than the threshold value, then the Bit Plane belongs to the Informative Region. The Complexity Value for all Bit Plane can be seen in Table 3.

Table 4. Complexity Value of All Bit Planes.

Bit Plane	Total of Bit Change	Complexity Value
1	28	0,250000000
2	15	0,133928571
3	14	0,125000000
4	27	0,241071429
5	46	0,410714286
6	44	0,392857143
7	59	0,526785714
8	61	0,544642857

In Table 4 can be seen that Bit Plane 1, 2, 3 and 4 are Informative Region so that secret message cannot be inserted into this Bit Plane. The secret message will be inserted into the next Bit Plane which is a Noise-Like Region.

At the next stage the secret message to be inserted will be represented first in the binary value so it will be a Bit Plane. In this research we will utilize the One Time Pad algorithm to encrypt the Bit Plane first before it is inserted. In this case, the word to insert is "KOMC2015". The word is eight characters. This will be one message block. The plain message and the bit-plane can be seen in Table 4 danTable .

Table 5. Plain Message

Char	ASCII	Biner
K	75	01001011
O	79	01001111
M	77	01001101
C	67	01000011
2	50	00110010
0	48	00110000
1	49	00110001
5	53	00110101

Table 6. Bit Plane of Message

Bit Plane Message							
0	1	0	0	1	0	1	1
0	1	0	0	1	1	1	1
0	1	0	0	1	1	0	1
0	1	0	0	0	0	1	1
0	0	1	1	0	0	1	0
0	0	1	1	0	0	0	0
0	0	1	1	0	0	0	1
0	0	1	1	0	1	0	1

The next step is to set the key for the message to be encrypted. The key used in this algorithm is symmetric key. The keys used for encryption and decryption are the same. The key must also be converted into ASCII then converted to binary value.The key example used is "FASILKOM".

Table 7. Key.

Char	ASCII	Biner
F	70	01000110
A	65	01000001
S	83	01010011
I	73	01001001
L	76	01001100
K	75	01001011
O	79	01001111
M	77	01001101

Table 8. Bit Plane of Key

Bit Plane of Key							
0	1	0	0	0	1	1	0
0	1	0	0	0	0	0	1
0	1	0	1	0	0	1	1
0	1	0	0	1	0	0	1
0	1	0	0	1	1	0	0
0	1	0	0	1	0	1	1
0	1	0	0	1	1	1	1
0	1	0	0	1	1	0	1

Each character of the secret message will be XOR with the key. This will produce a bit-plane cipher

Table 9. Cipher Bit Plane

Cipher Bit Plane							
0	0	0	0	1	1	0	1
0	0	0	0	1	1	1	0
0	0	0	1	1	1	1	0
0	0	0	0	1	0	1	0
0	1	1	1	1	1	1	0
0	1	1	1	1	0	1	1
0	1	1	1	1	1	1	0
0	1	1	1	1	0	0	0

Bit-plane ciphers will be inserted in the Noise-Like Region. After that, the bit-plane is restored to its original position and converted back into pure binary code before it is finally rewritten to the new stego image.

5. Experiment

In this paper, we experiment on 4 different resolution images as their vessel image and with 4 text files of different file sizes as secret messages. Each of secret message will be inserted into each of image file. The result of stego image and vessel image will be compared by calculating its PSNR value.

Table 10. Value of PSNR Results Comparison of Vessel Image and Stego Image.

Vessel Image	Message File	Stego Image	PSNR Value
Hello Kitty.jpg(48x48)	Pesan1.txt(26 bytes)	Hello Kitty+Pesan 1.jpg	39,7177
Hello Kitty.jpg(48x48)	Pesan2.txt(433 bytes)	Hello Kitty+Pesan 2.jpg	19,1265
Hello Kitty.jpg(48x48)	Pesan3.txt(1,7 kb)	Hello Kitty+Pesan 3.jpg	12,3566
Hello Kitty.jpg(48x48)	Pesan4.txt(3,55 kb)	Hello Kitty+Pesan 4.jpg	12,4213
KFC.jpg(96x96)	Pesan1.txt(26 bytes)	KFC+Pesan 1.jpg	46,4751
KFC.jpg(96x96)	Pesan2.txt(433 bytes)	KFC+Pesan 2.jpg	29,2047
KFC.jpg(96x96)	Pesan3.txt(1,7 kb)	KFC+Pesan 3.jpg	22,4068
KFC.jpg(96x96)	Pesan4.txt(3,55 kb)	KFC+Pesan 4.jpg	19,5847
My Son.jpg(128x128)	Pesan1.txt(26 bytes)	My Son+Pesan 1.jpg	53,0892
My Son.jpg(128x128)	Pesan2.txt(433 bytes)	My Son+Pesan 2.jpg	43,1413
My Son.jpg(128x128)	Pesan3.txt(1,7 kb)	My Son+Pesan 3.jpg	32,1467
My Son.jpg(128x128)	Pesan4.txt(3,55 kb)	My Son+Pesan 4.jpg	29,2466
Singapore.jpg(256x256)	Pesan1.txt(26 bytes)	Singapore+Pesan 1.jpg	59,9260
Singapore.jpg(256x256)	Pesan2.txt(433 bytes)	Singapore+Pesan 2.jpg	41,0297
Singapore.jpg(256x256)	Pesan3.txt(1,7 kb)	Singapore+Pesan 3.jpg	34,6222
Singapore.jpg(256x256)	Pesan4.txt(3,55 kb)	Singapore+Pesan 4.jpg	27,5152
		Average of PSNR value	**32,6256**

From table 9 has been proved that the utilization of one time pad to bit-plane in BPCS can improve the security of the message because the average value of PSNR resulting from testing of the stego image shows the value of 32.6256, in other words the resulting image has a good quality.

6. Conclusion

The One Time Pad cryptography algorithm can be combined well in steganography techniques of BPCS in terms of modifying message bits. Utilization of One Time Pad cryptography algorithm on steganography technique of BPCS is proven to increase security to message bits. This can be seen from the test results showing that the original message bits cannot be extracted directly but must go through the decryption stage first and the stego image looks identical to the original image and the value of PSNR resulting from the stego image shows the value above 30db or with Other words produce a good quality image.

References

[1] Johson N, Jajodia S 1998 Exploring Steganography: Seeing The Unseen *IEEE Computer* 26

[2] Katti J, Pote S, L B. K 2015 Two Level Encryption based on One Time Pad and Koblitz Method of Encoding *International Journal of Computer Applications.* **122** 0975

[3] Kawaguchi E, Eason R O 1999 Principle and Applications of BPCS-Steganography *SPIE*

[4] Khaire S S, Nalbalwar S L 2010 Review: Steganography-Bit Plane Complexity Segmentation (BPCS) Technique *International Journal of Engineering Science and Technology* **2** 4860

[5] Lahane P, Kumbhar Y, Patil S, More S, Barse M 2015 Visual Cryptography and BPCS Steganography for Data Shielding *International Journal Of Engineering And Computer Science.* **4** 11997

[6] Mehta S, Dighe K, Jagtap M, Ekre A Web Based BPCS Steganography *International Journal of Computer Technology and Electronic Engineering* **2**

[7] Nagaraj N 2012 One-Time Pad as a nonlinear dynamical system *Commun Nonlinear Sci Numer Simul* **17** 4029

[8] Rudramath P R, Madki M R 2012 High Capacity Data Embedding Technique Using Improved BPCS Steganography *International Journal of Scientific and Research Publications.* **2**

Optimizing Robinson Operator with Ant Colony Optimization As a Digital Image Edge Detection Method

Tarida Yanti Nasution[1], Muhammad Zarlis[2], Mahyuddin K.M Nasution[3]

[1,2,3]Department of Computer Science Faculty of Computer Science and Information Technology University of Sumatera Utara Medan, Indonesia, 20155

tarida.yanti.nst@students.usu.ac.id[1], m.zarlis@usu.ac.id[2], mahyuddin@usu.ac.id[3]

Abstract. Edge detection serves to identify the boundaries of an object against a background of mutual overlap. One of the classic method for edge detection is operator Robinson. Operator Robinson produces a thin, not assertive and grey line edge. To overcome these deficiencies, the proposed improvements to edge detection method with the approach graph with Ant Colony Optimization algorithm. The repairs may be performed are thicken the edge and connect the edges cut off. Edge detection research aims to do optimization of operator Robinson with Ant Colony Optimization then compare the output and generated the inferred extent of Ant Colony Optimization can improve result of edge detection that has not been optimized and improve the accuracy of the results of Robinson edge detection. The parameters used in performance measurement of edge detection are morphology of the resulting edge line, MSE and PSNR. The result showed that Robinson and Ant Colony Optimization method produces images with a more assertive and thick edge. Ant Colony Optimization method is able to be used as a method for optimizing operator Robinson by improving the image result of Robinson detection average 16.77 % than classic Robinson result.

Keywords : Edge Detection, Robinson Operator, Ant Colony Optimization

1. Introduction

The edge line of image is a fundamental component in image processing, particularly in the areas of detection and feature extraction. Edge detection of an image is the process which generates the edge line of objects in image that aims to improve the appearance of the boundary lines of an area or object. Up to this point has been known some classic edge detection methods, one of them is the Robinson operator. However, for some conditions, classic edge detection method still has weaknesses, namely the possibility of error detection of edge or a broken edge detection.

Based on the morphology of the edge line result, Robinson operator produces the thin, not assertive and grey edge line[4]. To overcome these deficiencies, the proposed improvements to edge detection method with the approach graph with Ant Colony Optimization algorithm[1]. Through this approach, graph was expected to yield a not optimal edge detection can be improved. The repairs may be performed are thicken the edge and connect the edges cut off.

Ant Colony Optimization (ACO) is a heuristic method that mimics the behavior of ants in order to solve the problem of discrete optimization[2]. ACO also be used to detect the edge of image, that to extract edge information from images. The proposed approach to harnessing the movement number of ants on the image based on the local variations in the value of the intensity of image. This information is used to build a matrix of pheromones, which provides information the edges of image.

ACO method are able to be used as a method for fixing a Canny edge detection method with fix image edge detection of Canny results average 12.05% of the results of detection of Canny. Canny methods and ACO are able to detect the edges of image with accuracy up to 82.7%[5]. This research discusses to do optimizing Robinson operator use Ant Colony Optimization then compare the output that is generated for the inferred the extent of Ant Colony Optimization can improve the results of edge detection that has not been optimized and improve the accuracy of the result of Robinson edge detection.

2. Robinson Operator

Robinson is the operator Compass that the type of operators who are looking for the edge by using eight wind direction, i.e. East, South East, South, South West, West, Northwest, North and North East. Robinson operators introduced by Robinson in 1977. This operator is identical to the 3 x 3 matrix or form window size 3 x 3 pixels, with r1 to r8 with calculated use mask as shown in Figure 1[3].

-1	0	1
-2	0	2
-1	0	1

r_1 Timur

0	1	2
-1	0	1
-2	-1	0

r_2 Timur Laut

1	2	1
0	0	0
-1	-2	-1

r_3 Utara

2	1	0
1	0	-1
0	-1	-2

r_4 Barat Laut

1	0	-1
2	0	-2
1	0	-1

r_5 Barat

0	-1	-2
1	0	-1
2	1	0

r_6 Barat Daya

-1	-2	-1
0	0	0
1	2	1

r_7 Selatan

-2	-1	0
-1	0	1
0	1	2

r_8 Tenggara

Figure 1. *Mask* of operator *Robinson*

2.1. *Algoritm of Robinson operator edge detection*

In detecting the edge of digital imagery, the Robinson operator had the following stages:

1. Read the value of the pixels of RGB image converted into grayscale image
2. Doing convolution between every pixel in the image to the eight mask of Robinson operator that already determined as in Figure 1.
3. In image, not all pixels can do convolution, that rows and columns which is located on the edge of the image have a neighbor is not complete. In this case, create additional rows and columns at edge is filled with a value of 0 so that the process convolution can be implemented.
4. In convolution, there are two possibilities if found, solved in the following manner, namely,:
   a. If the value result of convolution is negative, the value is made 0.
   b. If the value result of convolution is bigger than maximum grayscale degree (>255), the value is converted to the maximum grayscale degree.
5. Take the maximum value from eight convolution result.
6. Output image is the result of the convolution value.

3. Ant Colony Optimization

The ACO may be used to detect the image edge, that to extract edge information from images. The proposed approach to harnessing the movement number of ants on the image based on the local variations in the value of the image intensity. This information is used to build a matrix of pheromones, which provides information the image edge. On the ACO, artificial ants behave like intelligent agent made by memory and the ability to see. Edge detection approach based ACO aims to utilize a number of ants moving on a 2-D image to build a matrix of pheromones, that every entry that represents the information on each location pixel image. In addition, the movement of the Ant driven by a local variation of the value of the intensity of the image[6].

4. Simulation Results and Discussion

The proposed method was tested on four different images of 100 × 100 pixels resolution. The values of α and β were varied: $\alpha = \{1, 2, 0.1\}$; $\beta = \{0.1, 0.7, 1\}$.

IOP Conf. Series: Journal of Physics: Conf. Series **930** (2017) 012034 doi:10.1088/1742-6596/930/1/012034

Table 1. Result of Image 1

Image 1	Robinson	Robinson + ACO		
		α =1 , β= 0.1	α = 2, β=0.7	α = 0.1, β=1
	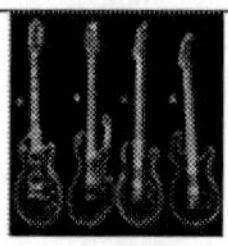			
MSE	129,89	96,25	94,86	96,03
PSNR	27,03	28,33	28,39	28,34

Table 2. Result Of Image 2

Image 2	Robinson	Robinson + ACO		
		α =1 , β= 0.1	α = 2, β=0.7	α = 0.1, β=1
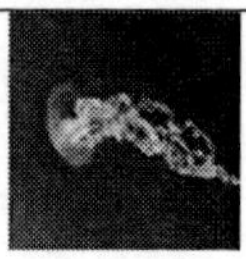	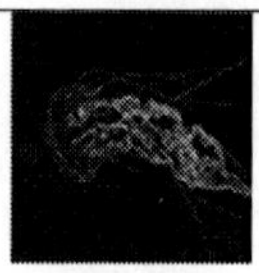			
MSE	134,65	96,98	95,29	100,35
PSNR	26,87	28,3	28,37	28,15

Table 3. Result of image 3

Image 3	Robinson	Robinson + ACO		
		α =1 , β= 0.1	α = 2, β = 0.7	α = 0.1, β=1

MSE	154,47	100,60	99,14	101,91
PSNR	26,28	28,14	28,20	28,08

Table 4. Result of image 4

Image 4	Robinson	Robinson + ACO		
		α =1 , β= 0.1	α = 2, β=0.7	α = 0.1, β=1

MSE	112,39	96,28	99,78	98,68
PSNR	27,66	28,33	28,17	28,22

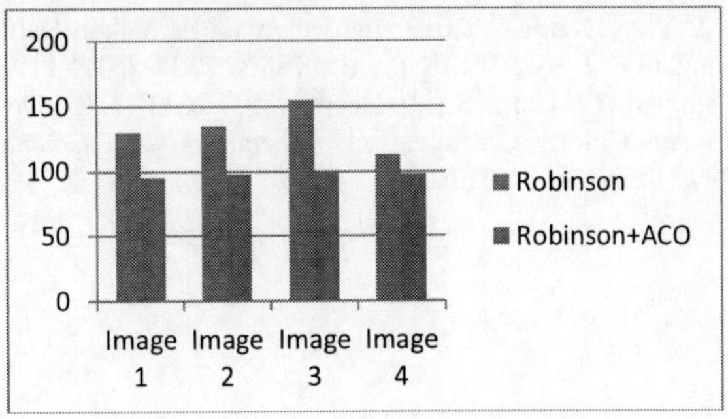

Figure 2. Comparison chart value MSE of Robinson and Robinson + ACO

From the graph in Figure 2 can be drawn the conclusion that the MSE value of the Robinson edge detection is higher than Robinson + ACO (132.85-> 98.0125). The higher MSE value, then the view of image result will be even worse. This proves that Robinson + ACO better than Robinson.

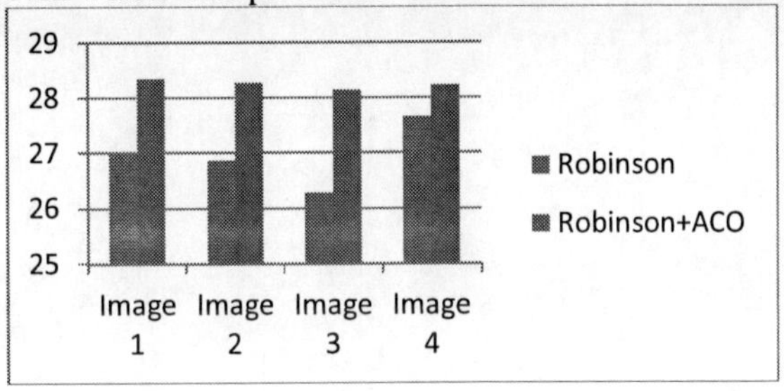

Figure 3. Comparison chart value PSNR of Robinson and Robinson + ACO

From the graph in Figure 3 can be drawn the conclusion that the PSNR value of the Robinson edge detection is lower than Robinson + ACO (26.96-> 28.25). The higher PSNR value, than the view of image result is better. This proves that Robinson + ACO better than Robinson.

5. Conclusions

The results of optimizing Robinson operator use Ant Colony Optimization produces images with a more assertive and thick edge. Ant Colony is able to be used as a method for optimizing Robinson operator by correcting the image of Robinson detection average 16.77% than the results of Robinson detection. Robinson and ACO methods are able to detect the image edge with accuracy up to 79.90%. On the method of Ant Colony, if seen from the parameters of testing that the best results if seen from the morphology of the resulting edge line, MSE and PSNR, yield the best edge detection at the time $\alpha = 2$, $\beta = 0.7$.

References

[1] Chen, C.-C. 2008. Edge Detection Improvement by Ant Colony Optimization. *Pattern Recognition Letters* **29**: 416-425.

[2] Dorigo, M., Birattari, M., & Stutzle, T. 2006. Ant Colony Optimization: Artificial Ants as a Computational Intelligence Technique. *IEEE Computational Intelligence Magazine*: 28–39.

[3] Kadir, A. dan Susanto, A., 2013. *Teori dan Aplikasi Pengolahan Citra.* Yogyakarta: Andi.

[4] Nasution, T.Y. 2014. Analisis Perbandingan Kinerja Deteksi Tepi Menggunakan Operator Kirsch dan Operator Robinson dengan Modifikasi Resolusi Citra Digital. *Seminar Nasional Literasi Informasi.* ISBN **979-458-762-1**: 135-140.

[5] Rohmad R., Umi L.Y., & Nanin S. 2013. Implementasi Perbaikan Deteksi Tepi Canny dengan Ant Colony Optimization. *Jurnal Teknik Pomits.* ISSN: **2337-35391** (1): 2301-9271. Vol 2.

[6] Verma, Om P., Singhal, P., Garg, S., & Deepti, S.C. 2011. Edge Detection Using Adaptive Thresholding and Ant Colony Optimization. *Information and Communication Technologies (WICT), 2011 World Congress on*. IEEE: 113–118.

Analysis Resilient Algorithm on Artificial Neural Network Backpropagation

Widodo Saputra[1, 3], Tulus[1], Muhammad Zarlis[1], Rahmat Widia Sembiring[2], Dedy Hartama[1, 4]

[1]Univesitas Sumatera Utara
[2]Politeknik Negeri Medan
[3]AMIK Tunas Bangsa
[4]STIKOM Tunas Bangsa

Abstract. Prediction required by decision makers to anticipate future planning. Artificial Neural Network (ANN) Backpropagation is one of method. This method however still has weakness, for long training time. This is a reason to improve a method to accelerate the training. One of Artificial Neural Network (ANN) Backpropagation method is a resilient method. Resilient method of changing weights and bias network with direct adaptation process of weighting based on local gradient information from every learning iteration. Predicting data result of Istanbul Stock Exchange training getting better. Mean Square Error (MSE) value is getting smaller and increasing accuracy.

1. Introduction

Previous Artificial Neural Network (ANN) backpropagation researches have some weaknesses, like as local minimum problem. Speed of convergence Backpropagation algorithm too slow, dependent on initial parameters, such as the number of hidden nodes, input, output, the learning rate and the weights of the connections in the network (Park, et.al, 2009). Another issue is the training time too long, especially if used parameters are less precise, for example that the selection of learning rate effect on changes in weights. Thus need algorithms to speed up the training. Many techniques applied to increase the speed of convergence in the process of training, such as optimization of gradient descent techniques (Kusumadewi & Sri, 2004).

Resilient method can be used to optimize accuracy. Resilient method is local adaptive learning algorithm, when compared with the backpropagation, its faster convergence and needs of training tends to be less. Compared with the backpropagation, resilient can provide faster of training and the rate of convergence and has the ability to stay away from the local minimum. Resilient is very strong with respect in internal parameters and its considered as one of the best learning method in ANN (Sheng, 2011).

2. Literature Review

2.1. Artificial Neural Network

ANN is a system of information processing that has certain performance characteristics as the neural network of the human brain, by doing the learning process in changes the weights. ANN is able to recognize activities with past data. The data of the past will be studied by ANN so it has the ability to give a decision of data that hasn't been studied (Hermawan, 2006).

2.2. Backpropagation

Backpropagation is a type of ANN used in solving problems of forecasting. This is possible because it is one type of training methods ANN with supervision. On the network was given a pair of patterns composed of input patterns and the desired patterns. When a pattern is given to networks, weights are modified to minimize the difference in patterns of output and the desired patterns. This exercise is done over and over again so that the entire patterns issued network can satisfy the desired pattern. This network architecture consists of the input layer, hidden layer, and output layer. Its architecture can be seen in Figure 2.1 (Fausett, 1994).

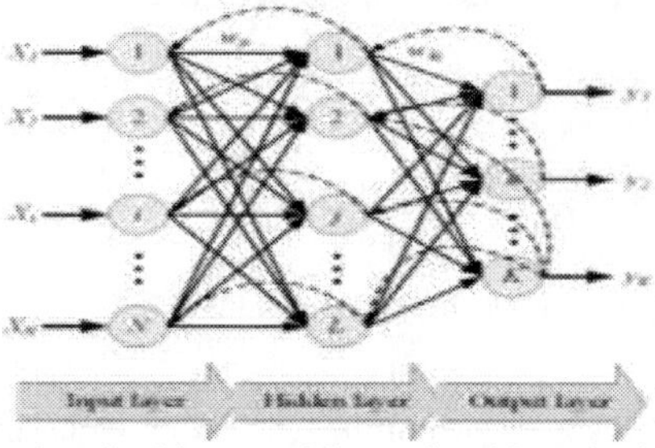

Figure 2.1. Backpropagation Architecture Network with a single Hidden Layer

Any some algorithms that can be used to make backpropagation, but was chosen because it may be the easiest to implement, while keeping the network efficiency. It's using more than one input layer (usually 3). Each of these layers:

a. Input Layer - This layer store the input to the network.
b. Output Layer - This layer store the output data, usually an input Identifier.
c. Hidden Layer - This layer is between the input and the output layer. They as a point of backpropagation to send data from the previous layer to the next layer

2.3. Resilient Backpropagation (Rprop) Algorithm

Rprop algorithm is the result of backpropagation. Change weights on backpropagation are affected of the learning rate and depending on the slope of the error curve. The smaller the learning rate, learning the longer. While the greater the learning rate, the weighting values will be far from the minimum weights. To overcome this, developed a new algorithm is Rprop. This algorithm uses the sign (positive or negative) of the gradient to show the direction of the adjustment weight. While size of the change the weights is determined by the value adjustment (0Δ) (Riedmiller, et.al, 1992).

The algorithm is Resilient to change weights and bias network with direct adaptation process of weighting based on local gradient information from iteration of learning, so that the number of iterations needed to reach the target (Sulistijanti, 2013).

For it, given the size of the change of weight is improvement value Δjk. The value of the repair adjusting the development during the learning process based on local vision towards the function of error with learning the rules as follows:

$$\Delta jk(m) = \Delta jk\ (m\text{-}1) * \eta^{+},\ \text{if}\ \frac{\partial E}{\partial wjk}(m) \times \frac{\partial E}{\partial wjk}(m-1) > 0$$

$$\Delta jk(m) = \Delta jk\ (m\text{-}1) * \eta^{-},\ \text{if}\ \frac{\partial E}{\partial wjk}(m) \times \frac{\partial E}{\partial wjk}(m-1) > 0$$

$$\Delta jk(m) = \Delta jk\ (m\text{-}1)\ ,\ \text{for more}$$

Where $0 < \eta^{-} < 1 < \eta^{+}$

The rules of adaptation works is every partial derivative of weights and bias on two different successive iterations in order to show that the improvements of the last value that is too large and the algorithm has to leap over a local minimum, the value of the repair Δ_{jk} lowered by a factor of η−. If in two iterations sequential derivative marks constant, the value of the repairs is raised with the factor η + to speed up convergence in the area of surface errors. And if its derivative is zero, then the value of the update stays the same. Each time the repair, oscillating weights reduced. If changing weights in the same direction for a few iterations, then the resulting increases in the magnitude of the change in weight (Sulistijanti, 2013).

After repair value for each of the weights are adjusted, improved weights follow the rules that simple if the derivative is positive (increasing errors) then the weighted down with the value of the value of the repair, if the derivative is negative then the value of the repairs is added as follows:

$\Delta w_{jk}(m) = -\Delta jk\ (m)$ jika $\frac{\partial E}{\partial wjk}(m) > 0$

$\Delta wjk(m) = +\Delta jk\ (m)$ jika $\frac{\partial E}{\partial wjk}(m) < 0$

$\Delta wjk(m) = 0$ For more

$Wjk(m+1) = wjk + \Delta wjk(m)$

W_{jk} (m) which is the weighting between j and k neurons in two successive layers on the iteration m, W_{jk} (m + 1) is the new weight.

3. Methods

3.1. Data

Required data from imkb.gov.tr and finance.yahoo.com. The data is compiled based on Labor Day in Istanbul's stock exchange. (http://archive.ics.uci.edu/ml/datasets/ISTANBUL+STOCK+EXCHANGE).

3.2. Data Retrieval

This study, created a pattern recognition and prediction of Istanbul Stock Exchange. This process has two stages where the first is to do pattern recognition by way of finding the best architecture of the ANN models. The process of training and testing data to get the best models from the Istanbul Stock Exchange data in working is 60 Patterns from January 5, 2009 – April 4, 2009. Criteria of pattern 1 to Pattern 30 is training data, while 31 Patterns up to 60 Patterns is test data. The second is to do predictions with the best architecture pattern obtained in the first. The process of Testing had done entering in Istanbul Stock Exchange data and to comparing the minimum error value from the best architectural patterns that are done on the first.

3.3. Training Network

Training in ANN needed input data and target data. Training covers the process of iterative input data are entered into the network so that the network can learn and adjust data that trained with the desired target data. Training is done to look for the values of the weights that connect all the neurons so that minimize the error output that is produced by the network.

Training using much as 50% are the amount of data that is comprised of data input and output targets that have been normalized. In this process will be initiated with the architecture of the ANN from a number of hidden neurons of different. Each of the tested architecture that will result in weight training that will be used as the initial weight in the process of testing. Then initialize the weights and biases to calculate the value of the output of each neuron will be multiplied by the function activation and learning rate.

After the value of the output of the network on the output layer is obtained, calculate the value of a network error, an error value is then compared to the value of the error the target that has been set. If a network error that is generated is not less than or equal to the value of the error that has been established, then the process will be done by modifying the backpropagation weighting network and a bias on a particular iteration until the minimum error value was obtained by approaching a predetermined target error before. When the error condition is less than the weighting then the target error will be stored as weights in training process.

3.4. Testing Network

Testing used as much as 50% ANN from amount of data that has been selected for each category. At this stage of the network will be in testing with the new data that hasn't been trained into a network to find out the capabilities of the network doing the generalizing the case at hand and then draw the inclination towards a specific output.

The results of the testing process in the form of a matrix which forms in compliance with output targets. The output of the network will then be determined on a particular pattern. If the output of the network is greater than or equal to 0.5 the network considered have of the output 1 and if the network has the output of less than 0.5 then it shall be deemed to have the output 0. The data is said to be known is when data normalization results are used as input values can produce the same output value chain with the desired target value.

4. Results and Discussion

At this point would do the testing system. The system created using MATLAB R2011b. System of Testing for the final review of the analysis and implementation.

Table 4.1 Arsitektur JST Backpropagation Standar

Karakteristik	**Spesifikasi**
Arsitektur	1 hidden layer
Neuron input	12
Neuron Hidden	10
Fungsi Aktivasi	Sigmoid Biner
Inisialisasi bobot	Random
Target *Error*	0.01
Maksimum *Epoch*	10000
Learning Rate	0.01

4.1. Results

Backpropagation algorithm is an algorithm that will produce the best value through potentially experiencing unplanned test trial and error. So this research, conducted a few experiments to increase the speed of learning (accelerated learning).

4.1.1. Training of Standard Backpropagation algorithm with Standard Backpropagation and Resilient Algorithm

In the process of training will be seen from Epoch and limits of error. In recent times an experiment with varying architecture that can be seen from table 4.2:

Table. 4.2. Training of Standard Backpropagation algorithm with Standard Backpropagation and Resilient Algorithm

Experiment		Architecture	Epoch	MSE	Accuracy (%)
Backpropagation Standard	1	12-5-1	37 Iteration	0.009926	6.7
	2	12-8-1	33 Iteration	0.009566	43.3
	3	12-10-1	32 Iteration	0.009923	30.0
	4	12-14-1	24 Iteration	0.009894	33.3
	5	12-20-2	36 Iteration	0.001356	56.7
Backpropagation Standard and Resilient	1	12-5-1	6 Iteration	0.009260	20.0
	2	12-8-1	3 Iteration	0.002731	56.7
	3	12-10-1	3 Iteration	0.002731	56.7
	4	12-14-1	6 Iteration	0.002223	63.3
	5	12-20-2	5 Iteration	0.004229	66.7

4.1.2. Testing of Standard Backpropagation algorithm with Standard Backpropagation and Resilient Algorithm

In the process of testing will be seen from Epoch and limits of error. In recent times an experiment with varying architecture that can be seen from table 4.3:

Table 4.3. Testing of Standard Backpropagation algorithm with Standard Backpropagation and Resilient Algorithm

Experiment		Architecture	Epoch	MSE	Accuracy (%)
Backpropagation Standard	1	12-5-1	37 Iteration	0.006238	10.0
	2	12-8-1	33 Iteration	0.000538	86.7
	3	12-10-1	32 Iteration	0.002255	60.0
	4	12-14-1	24 Iteration	0.031164	33.3
	5	12-20-2	36 Iteration	0.010406	56.7
Backpropagation Standard and Resilient	1	12-5-1	6 Iteration	0.008060	13.3
	2	12-8-1	3 Iteration	0.000771	90.0
	3	12-10-1	3 Iteration	0.000771	90.0
	4	12-14-1	6 Iteration	0.014335	63.3
	5	12-20-2	5 Iteration	0.000553	86.7

From the results of testing conducted as many as 5 experiments on backpropagation with varying architecture is 33 iteration. Whereas with same architecture of Backpropagation with Resilient Backpropagation (RPROP) is 5 iterations. Accuracy of training conducted the same experiment, an increase in the accuracy of training in experiment 1 with 12-5-1 network architecture. The same thing happens also on a second experiment with network architecture 12-10-1 to trial to 5 results accuracy Standard Backpropagation reaches 93% and results accuracy Standard Backpropagation Optimizations Resilient Backpropagation (RPROP) get the same results.

5. Conclusion

Neural Network algorithm gives satisfactory results in terms of predictions. That is as follows:

1. Changes to the parameters of the learning rate and the epoch affecting the changes up to the level of accuracy, so the selection of the parameters greatly influence on the results.
2. Artificial Neural networks are able to do predictions against data time series which is quite accurate. Looks at the cases taken, namely the Istanbul Stock Exchange.
3. By looking at the test results can be drawn a conclusion that is testing the resulting Backpropagation experience increased after optimized Resilient Backpropagation (RPROP).

Bibliography

Apriliyah, Mahmudy, WF and Widodo, AW 2008. Perkiraan penjualan beban listrik menggunakan jaringan syaraf tiruan resilent backpropagation (RProp), Kursor, vol. 4, no. 2, pp. 41-47.

Chen. C, and Lin. J, Applying Rprop Neural Network for the Prediction of the Mobile Station Location, www.mdpi.com/journal/sensors Communication 2011

Fausett, L. 1994. Fundamentals of neural networks : architectures, algorithms, and applications. Pretince Hall. New York.

Hermawan, A. 2006. Jaringan Saraf Tiruan, Teori, dan Aplikasi. Penerbit Andi. Yogyakarta.

Kim, J.H., Park, S.J., Kim, K.T., dan Hwang, S.C. 2003. Stock price prediction using Backpropagation neural network in KOSPI. Proceedings of the International Conference on Artificial Intelligence IC-AI 2003. Las Vegas, United Stated.

Kusumadewi, S. 2004. Membangun Jaringan Saraf Tiruan menggunakan Matlab & Excel Link, Graha Ilmu, Yogyakarta.

Muzakkir. I., Syukur. A., Dewi. I.K. 2014. Peningkatan Akurasi Algoritma Backpropagation Dengan Seleksi Fitur Particle Swarm Optimizationdalam Prediksi Pelanggan Telekomunikasi Yang Hilang. Jurnal Pseudocode, Volume 1 Nomor 1.

Park, T. S., Lee, J. H., & Choi, B. , 2009, Optimization for Artificial Neural Network with Adaptive inertial weight of particle swarm optimization. Cognitive Informatics, IEEE International Conference , 481-485.

Prasetyo, H. R., Hadiani, R., Setiono. 2014. Analisis Data Runtun Waktu Debit Menggunakan Jaringan Saraf Tiruan Di Das Wuryantoro Pada AwlrKecamatan Wuryantoro. e-Jurnal Matriks Teknik Sipil Vol. 2 No. 2 : 64.

Riedmiller, M. and Braun, H. 1992. RPROP. A fast adaptive learning algorithm. In Proceedings of the 1992 International Symposium on Computer and Information Sciences, Antalya, Turkey, pp.279-285.

Siang, J.J. 2009. Jaringan Saraf Tiruan dan Pemrogramannya Menggunakan MATLAB. Yogyakarta.

Sari. D.R. 2015. Aplikasi Penerapan Metode Neural Network Menggunakan Algoritma Backpropagation Untuk Mengetahui Pembelian Dan Penjualan Bahan Bakar Industri. INFO TEKNIK Volume 16 No. 1. 47-60.

Sulistijanti. W. Peramalan Curah Hujan Wilayah Semarang Barat dengan Algoritma Resilient Backpropagation, Median Volume VI Edisi 20 Mei 2013

Performance Measurement Of *Mel Frequency CeptralCoefficient* (MFCC) Method In Learning System Of AlQur'anBasedIn*Nagham*PatternRecognition

Yesy Afrillia[1], Herman Mawengkang[2], [3]Marwan Ramli, [4]Fadlisyahand [5]RizkyPutra Fhonna

[1]Department of Informatics Engineering, Universitas Sumatera Utara, Medan, 20155.
[2]Department of Mathematics, Universitas Sumatera Utara, Medan, 20155.
[3]Department of Mathematics, Universitas Syiah Kuala, Aceh.
[4]Department of Informatics Engineering, Universitas Malikussaleh, Aceh.
[5]Department of Informatics Engineering, Universitas Sumatera Utara, Medan, 20155.

Email : [1]yesy.afrillia@gmail.com, [2]ramlimarwan@gmail.com,[5]rizkyputrafhonna@gmail.com.

Abstract. Most of research have used signal and speech processing in order to recognize*makhraj* pattern and*tajwid* reading in Al-Quran by exploring the *mel frequency ceptralcoefficient* (MFCC). However, to our knowledge so far there is no research has been conductedto recognize the chanting of Al-Quran verse using MFCC. This term is also well-known as*nagham* Al-Quran. The characteristics of *nagham* Al-Quran pattern is much more complexthen *makhraj* and *tajwid* pattern. In *nagham* the wave of the sound has more variation whichimplies the level of noice is much higher and has sound duration longer. The data testing inthis research was taken term by real-time recording. The evaluation measurement in the systemperformance of *nagham* Al-Quran pattern is based on true and false detection parameter withaccuracy 80%. To measure this accuracy it is necessary to modify the MFCC or to give moredata learning process with more variation.
Key words : speech recognition, *nagham* Al-Quran pattern, *mel frequency ceptral coefficient* (MFCC).

1. Introduction

Speaker awareness, from time to time referred to as speaker biometrics, comprises identity, verification (authentication), different types, and by extension, segmentation, Tracking and detection of speakers. It is a generic term used for any procedure which Involves knowledge of the identity or pattern recognition of a based on voice [1]. Biometrics technology is a self-recognition technique using body parts or human behavior. This technology fulfills two important functions of identification and verification. The identification system aims to solve one's identity. While the verification system aims to refuse or accept identity that someone claimed [2]. A speaker recognition system first attemptsto model the vocal tract traits of a person. This may be a mathematical model of the physiologicalsystem generating the human speech.However, the growing technology of many researchers build amore interesting and specific model on voice recognition systems [1].

Research on voice recognition has been largely done with various methods of signal processing, such as *Linier Prediction Coding* (*LPC*), *Mel Frequency Cepstrum Coefficients* (*MFCC*), *Neural PredictiveCoding* (*NPC*), etc. Which all of the above methods are based on *fourier* transform as a signalcharacteristic extraction method [3]. Feature and value extraction methods that have a direct effect inthe sound system [4]. The pattern recognition rate be effected by the method of extraction andclassification used. So far the method of extraction of Mel-Frequency Cepstral Coefficients or often called MFCC is more commonly used for speech recognition, as has been done by [5][6][7]. MFCC
uses discrete cosine transform (DCT) to calculate coefficients. by means of *mel filter* which is based on the human hearing system, MFCC Optimized for speech recognition.

IOP Conf. Series: Journal of Physics: Conf. Series **930** (2017) 012036 doi:10.1088/1742-6596/930/1/012036

2. Research purposes

The aim of this research isgive information how performance *mel frequecy ceptral coefficient*(MFCC) on the memorization system Qur'an based on pattern recognition *nagham* pattern to simplify *tahfidz* in memorization Al Qur'an to be more efficient so that minimizes the length of time learning inmemorizing the Qur'an.

3. Speech

Speech is useful for communication between human beings. Where each sound has the characteristics and levels of different frequency. Sound is a unique thing and have *range* specific frequency and the intensity of the sound which can and can't be heard by humans. The units of intensity to measure the intensity of the sound that is *desibel (dB)* taken from the name of the inventor, namely *Alexander Graham Bell.* Known as the inventor of the phone, While the unit of the sound frequency is *Hertz*, Taken from the name of a Physicist, *Heinrich Rudolf Hertz* to appreciate his contribution in the electromagnetic field [1].

4. Speech Recognition

Speech recognition is a process to recognize letters, words or sentences that are spoken. Speech recognition is better known by the term *Automatic SpeechRecognition* or *Computer SpeechRecognition* Where the use of a machine/computer for Recognize a person's voice or identificationfrom the spoken voice. Generally speaker talking in front of computer/machine then thecomputer/machine recognizes a person's voice exactly as it is spoken. Speech recognitionrecognizable into various task levels, recognition in the level of acoustic signal in the form of a level
test in the composition of sub-word units in the form of phonemes, words, phrases and sentences [8].

5. Feature Extraction

Speech feature extraction is tasked with changing voice signals to the coefficient vector feature that contains only the information needed for identity of the given phrase. Because speech has unique and different attributes contained in the spoken words of this attribute can be extracted from a lot of feature extraction recommendations and can be used for speech recognition tasks[9].
The purpose of feature extraction is to minimize the size of the data without change the characteristicsof the voice signal in each frame that can be used as a characteristic. Feature extraction is obtainedfrom converting the form of voice signal into parameterized representation [1].

6. *Mel Frequency Ceptral Coefficient* (MFCC)

The MFCC characteristic extraction process is a feature-taking process based on *discrete fouriertransforms*. MFCC is the most generally used method for the speech processing, because it isconsidered good enough to present feature and signal. The workings of MFCC in capturing sound and
frequency signals exactly as humans recognize sound characteristics [10]. The MFCC process diagramis shown in the following figure :

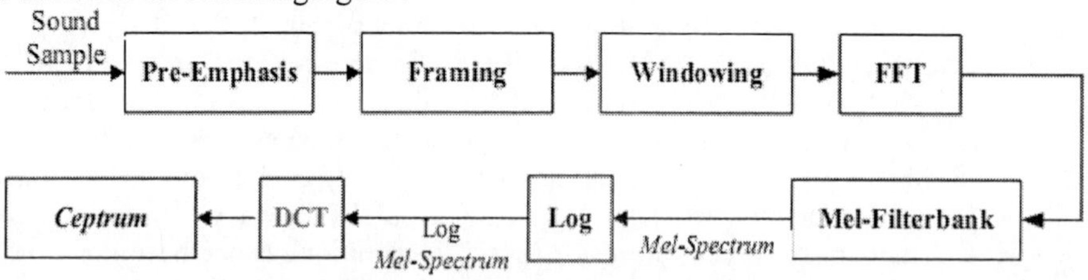

Figure 1. Diagram Process MFCC

6.1. Pre-Emphasis

Sound signal filtering process required recording or sampling process. The purpose of filtering is to get a smoother spectral form of the sound frequency. Filter pre-emphasis Is based on the input/output relation in the time-domain, expressed in the equation [1] :

$$y[n] = s[n] - \propto s[n-1]\ , 0.9 \leq \propto \leq 1.0$$

Where :
y [n] = result signal pre-emphasize filter
s [n] = before signal pre-emphasize filter

6.2. Framing

Sound signals are non-stationary signals, meaning their statistical properties are always changing with time. So it is not possible to extract the spectral features of speech at once. Therefore, the spectral character of sound signals is extracted through a window as a sound signal characterizing a particular sound part so that it can be made an assumption that the sound signals is stationary [1].

6.3. Windowing

The process of windowing is the process of filtering each frame by multiplying each frame with certain windows functions the same size as the frame. Windowing is also used to ensure continuous sound from the initial frame to the end [11]. A good window function should be narrowed to *the mainlobe*andextendstoits*side-lobe*.

6.4. Fast Fourier Transform (FFT)

The purpose of *Fast fourier transformation* (FFT) is to decompose the signal into a *sinusoidal* signal, Which consists of two units, namely real units and imaginary units. FFT Used for frequency analysis, thus simplifying voice processing as it corresponds to human hearing. FFT Is an algorithm that implements *discrete fouries transform* (DFT). DFT is the transformation of each *frame* with *N sample* From time domain to frequency domain.

$$X_n = \sum_{k=0}^{N-1} X_k\, e^{-2\pi j\, k n/N}$$

Defined :
N= The number of frame segments
X= Data value to *k*
N = 0,1,2,3,…, N-1 and k=0,1,2,3,…, N-1
k
j =-1

6.5. *Mel Frequency Wrapping*

Mel Frequency Wrapping is commonly carried out by using *filterbank. Filterbank* is one form of filter made with purpose to determine the energy size of a particular frequency-band in frequency, but for MFCC purposes, the *filterbank* should be applied in the frequency domain. *Filterbank* using convolution representation in doing filter to signal. *Convolution* can be done by multiplication between signal spectrum with filter bank coefficient.

6.6. Log

Logarithm values are obtained by converting DFT values into one value. Reduce the value of the mel filter bank by substituting each base log value. By using the *matlab* command "*log*" to extract the logarithm value of the mel-filter on the sound segment. human ear is less delicate in hearing sounds that have low amplitude and height [12].

6.7. Discrete Cosinus Transform (DCT)

DCT is the last step of the main process of MFCC feature extraction. The basic concept of DCT is correlating value of *mel spectrum* so as to produce a good representation of *property spectral local*. Basically the concept of DCT is the same as inverse *fourier transform*.

6.8. Ceptrum

Ceptrum is the inverse value of *spectrum.* Cepstrum Commonly used to obtain information from a voice signal spoken by humans. In this last step, *spectrum log mel* converted to ceptrums use Discrete Cosine Transform (DCT) Which is the value of the resulting mel frequency being converted into a time domain.

7. *Nagham*

Nagham is a word derived from Arabic which means song/rhythm. The popular term Nagham comes from the *qori* and *qoriah* of Egypt was teached his knowledge in Indonesia in the year 1973. Which mean chanting the Qur'an or make sounds in reading the Qur'an. Nagham then known as the Art of Read Al-Quran. Arts read Al-Quran is a science of sound art that apply the tone, tempo and rhythm inthe recitation of the Qur'an.

8. Results and Discussion

The scheme of the nagham al-qur'an pattern recognition system developed in this examine is illustrated in the following figure:

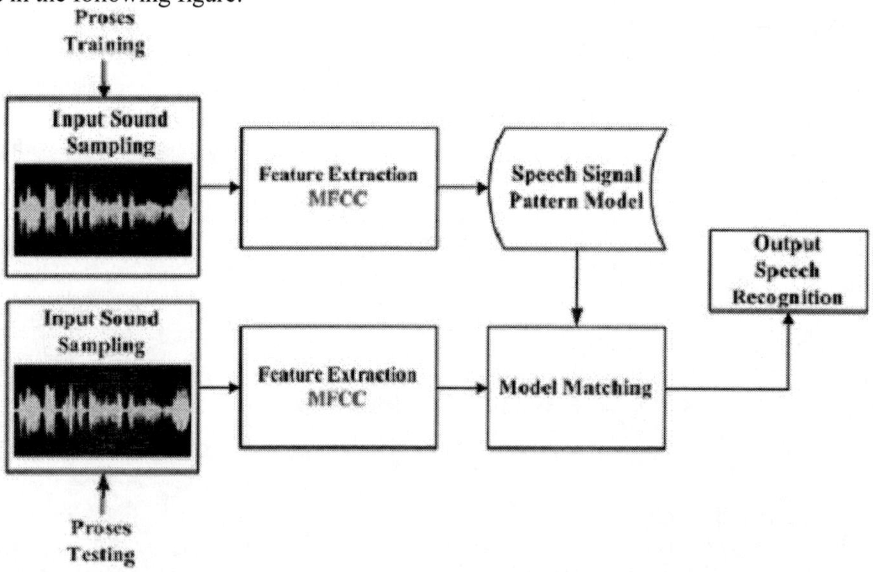

Figure 2. The performance scheme of the al-qur'an memorizing systemon the introduction of the nagham pattern4

There are two steps performed on al-qur'an memorizing system on nagham pattern recognition that is training process and testing process. In the training process, after the system receives voice input nagham al-qur'an, the next step is feature extraction using mel-frequency cepstral coefficient (MFCC), next model sound signal pattern will be stored in the database. In the testing process through the same

steps as the training process, but the sound will be processed taking voice in real-time then theprocess of matching the model that has been stored in the previous database in the training process to get the recognition of the sound pattern of *nagham* Al-Quran.
The sample test used in this research amounted to 10 sound an example of the pattern of the Qur'an *nagham* on the Al-Fatihah letter is an example of the characteristics of the 7 samples of the *nagham* pattern. The evaluation measurement in the system performance of *nagham* Al-Quran pattern is based on *true* and *false* detection parameter. *Detection rate* is the ratio between the number of successful *nagham* pattern recognition. While *false positive rate* Is the number of unrecognized sounds.
The evaluation measurement in the system performance of *nagham* Al-Quran pattern is based on true and false detection parameter with accuracy 80%. Table 1. illustrates some performance measurement results of the MFCC method in the al-qur'an recitation system based on the recognition of the nagham pattern.

Table 1. The results of the MFCC method show on therecitation system of al-qur'an based on the recognition of the nagham pattern

Data Speech Sampling Training	**Data Speech Sampling Testing**	**Detection True**	***False positive rate***	***Detection rate***
20	100	20	0.8	0.2
40	100	50	0.5	0.5
60	100	70	0.3	0.7
80	100	80	0.2	0.8

The results of the system evaluation show that the detection rate increase is intently associated to the expanding number of training. Figure 3 shows a graph of the results of the recitation system of AlQur'anbasedontherecognitionof the *nagham* pattern.

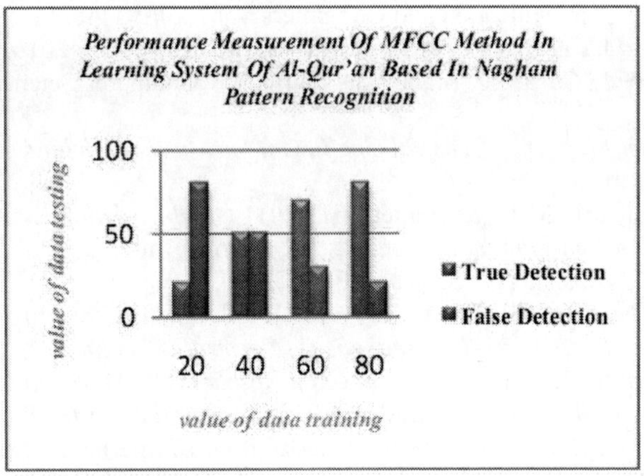

Figure 3. Graph of MFCC performance

Based on figure 3 the graph illustrates the measurement of performance performed after 2 samples of trained *nagham* Al-Quran pattern for the overall sound characteristics of the *nagham* pattern using the MFCC method. Amount of test sounds 10 samples of voice pattern *nagham* Al-Quran, at the beginning of the testing phase is done on a system that has been trained 2 samples of *nagham* pattern using MFCC method and obtained detection rate value = 0.2 or 20%. In the second step, the test is performed on a system that has trained 4 sound samples of *nagham* Al-Qur'an pattern and obtained detection rate value = 0.5 or 50%. In the third step, testing is done on a system that has trained 6 soundsamples of *nagham* Al-Qur'an pattern and obtained detection rate value = 0.7 or 70%. And the last stepof testing accomplished on a system that has been trained 8 sound samples pattern *nagham* Al-Qur'anand obtained detection rate value = 0.8 or 80%.

9. Conclusion and Future Work

In this research try to build voice recognition system based on *nagham* Al-Qur'an pattern using MFCCBased on signal processing techniques and sound. MFCC feature extraction applied in this research tocalculate a significant coefficient value for each input data in the form of sound. The results of thesystem evaluation show that the detection rate increase is closely related to the increasing number oftraining. Based on the detection rate of *nagham* pattern recognition on the recitation system of theQur'an using MFCC the accuracy rate reaches 80%. The future of this journal can still be developed that is by modifying the MFCC algorithm to optimizethe coefficient value in covering the weakness for sound input that has a more variable sound waveswith high noise level. In addition, other techniques for pattern matching should be attempted forsubsequent research.

10. Reference

[1] Beigi, H. 2011. *Fundamentals Of Speaker Recognition*. Spinger : New York.
[2] Putra, D. 2010. *Pengolahan Citra Digital.* Andi. Yogyakarta.
[3] I.Daubechies. 1988. *Ortonormal bases of Compactly Supported Wavelet*. Comm.Pure Apple. Math. PP : 909- 996.
[4] Al-Irhaim, Y.F. & Saeed, E.G. 2010. *Arabic Word Recognition Using Wavelet Neural Network*. Third Scientific Conference in Information Technology. PP: 416-420.
[5] Martinez, J., Perez, H., Escamilla, E. & Suzuki, M.M. 2012. *Speaker Recognition Using Mel-Frequency Ceptral Coefficient (MFCC) and Vector Quantization (VQ) Techniques*.International Confernce on Electrical Communication and Computers Vol.10. PP : 248251.
[6] Ismail, A., Idris, M.Y.I., Noor, N.M., Razak, Z. & Yusof, Z. 2014. *MFCC-VQ Approach forQalqalah Tajweed Rule Checking*. Malaysian Journal Of Computer Science Vol. 27 (4). PP: 275-293.
[7] Logan, B. 2000. *Mel Frequency Cepstral Coefficient for Music Modeling*. InternationalSymposium in Music Information Ritival.
[8] Tavanaei, A., Manzuri, M.T. & Sameti, H. 2011. *Mel-Scaled Discrete Wavelet Transform andDynamic Features for the Persian Phoneme Recognition*. Internasional Symposium onArtifical Intelligance and Signal Processing (AISP) (2). PP : 138-140.
[9] Kamarudin, N. Nematollahi, M.A. & Abd Rauf bin Hasan. 2014. *Feature Extraction UsingSpectral Centroid and Mel Frequency Cepstral Coefficient for Quranic Accent AutomaticIdentification*. IEEE Student Conference on Research and Development (SCOReD).
[10] Abro, B., Naqvi, A.B. & Hussain, A. 2012. *Qur'an Recognition for the Purpose of Memorisation Using Speech Recognition Technique*. International Multitopic Conference (INIMIC).
[11] Arshad, N.W., Sukri, S.M., Muhammad, L.N., Ahmad, H., Hamid, R., Naim, F. & Naharuddin, N.Z.A. 2013. *Makhraj Recognition For Al Quran Recitation using MFCC*. InternationalJournal Of Intelligent Information Processing (IJIIP) vol.4 (2). PP : 45-53.
[12] Ismail, A., Idris, M.Y.I., Noor, N.M., Razak, Z. & Yusof, Z. 2014. *MFCC-VQ Approach for Qalqalah Tajweed Rule Checking*. Malaysian Journal Of Computer Science Vol. 27 (4). PP: 275-293.

IOP Conf. Series: Journal of Physics: Conf. Series **930** (2017) 012037 doi:10.1088/1742-6596/930/1/012037

RSA KEY DEVELOPMENT USING FINGERPRINT IMAGE ON TEXT MESSAGE

Sayuti Rahman[1], Indah Triana[2], Sumi Khairani[3], Amru Yasir[4], Siti Sundari[5]

[1,2,3,5]Jurusan Teknik Informatika, Sekolah Tinggi Teknik Harapan Medan
[4]Universitas Dharmawangsa Medan
[1]masay.ram@gmail.com,[2] indahtriana46@gmail.com , [3]adek_sumi@yahoo.co.id, [4]cougara25@gmail.com, [5]sundaristth@gmail.com

Abstract. Along with the development of technology today , humans are very facilitated in accessing information and Communicate with various media , including through the Internet network . Messages are sent by media such as text are not necessarily guaranteed security. it is often found someone that wants to send a secret message to the recipient , but the messages can be known by irresponsible people. So the sender feels dissappointed because the secret message that should be known only to the recipient only becomes known by the irresponsible people . It is necessary to do security the message by using the RSA algorithm, Using fingerprint image to generate RSA key.This is a solution to enrich the security of a message,it is needed to process images firstly before generating RSA keys with feature extraction.
Keywords : RSA , Fingerprint , Cryptography , Image Processing.

1. Introduction

The development of information technology is currently increasing widely so that human needs for the latest information, accurate, fast and reliable is increasingly easy and more inseparable in human life. Such as through telecommunication network and internet.

With the development of information technology so many users communicate the Internet network, therefore many cases of interception of communication made by a handful of people who are not responsible for unnecessary purposes, because everyone has a secret, whether private secrets, family secrets, group secrets, Even state secrets. So therefore there is a need for security text messages when communicating over the Internet network.

Security issues are one of the important aspects of an information system. One of the important things in communication using computers and in computer networks to ensure the security of messages, data or information is encryption. Encryption is a process done to the original message to be encoded or made into another form so that it can't be seen the original form of the message, in other words the original message is often called the plaintext changed into a password or cipher text. To see the message again, the decryption process or change the message encoded into the original message again (Alvianto, 2015).

Encryption and decryption process requires an algorithm to be able to secure data well and can be used in communication and transaction information one of them is with RSA algorithm, RSA has been formulated as a practical algorithm to be implemented in cryptography system using public key designed by Ronald L. Rivest, Adi Shamir, and Leonard M. Adleman. The key pair used for the encryption process are the public key (e, n) and private key d as the decryption key where e, d and n are positive integers. The RSA algorithm is a block cipher algorithm that groups plaintext into blocks before encryption becomes cipher text. (Alvianto, 2015)

Information that has been encrypted with RSA algorithm can be sent and shared by distributing the public key on receiver and private key stored in file, but in this research how to use fingerprint image to generate public key as opening message. Image is a data that has many meanings that can be processed into complete information, the use of fingerprint image because it can represent the identity of the sender who can be recognized by the owner or called biometric identification. (S Rahman, 2017)

2. Methods

a. Data Security

Security and confidentiality of messages is a very important thing in communicating through the Internet because the network is prone to the crime by spreading the badness of someone to who is not entitled to know the information. Computer users that want the data unknown to unauthorized parties are always trying to get around how to secure the information to be communicated or to be stored. So the protection for data confidentiality increases, one way is data encryption. Encryption is a process of converting an original message into an unreadable character. There are several commonly used encryption algorithms such as Block Cipher, Stream Cipher, Data Encryption Standard (DES), Triple DES, Advanced Encryption Standard (AES), and so on.

b. Cryptography

The word cryptography comes from greek, "Kriptos" which means hidden and "graphien" which means write. So the word cryptography can be interpreted as "hidden text". This means cryptography is an art and science used to hide information or messages from unauthorized recipients. Cryptography is a science that relies on mathematical techniques to deal with information security such as confidentiality, data integrity and entity authentication. In a cryptographic algorithm, there are two elements:

1. Encryption, ie the process of converting the original message into an encoded message, and.
2. Decryption, ie the process of converting the encoded message to the original message.

c. RSA Algorithm Encryption

The next process after generating the key to the RSA is the Encryption process. Encryption is the process of converting the original message (plaintext) into a secret message (chipertext). In this process the original message is first converted into decimal form, then the decimal message are then divided into several decimal blocks of a regular basis. Each decimal block will have a value that must be less than the value of n called P. The formula used to perform the Encryption process of RSA is:

$C = P^e \pmod{n}$

where

P = Plain text
C = Cipher text
e = public key
n = public key (ab)

d. RSA Algorithm Decryption

he decryption process in RSA is similar to the encryption process, only in the decryption process using private key d. The formula for the decryption process of RSA is as follows:

P = Cd (mod n),

where

P = Plaintext
C = Ciphertext
d = private key
n = publik key (ab)

e. Image

Image is an important component in multimedia that has a deeper meaning compared to text data by displaying data visually and representing information that takes place when image retrieval, either object, condition, color, weather and many more meaning that can be extracted from an image. Viewed from a mathematical point of view, the image is a continuous function of the intensity of light in the field of two dimensions. Light sources illuminate objects, objects reflect light provided with different reflection levels of each object. This light reflection is captured by optical devices, such as human eyes, cameras, scanners, and so on. So the image of the object called the image is recorded. (Permadi, 2015)

f. Fingerprint

A fingerprint is formed by a combination of curves . The bright areas are called ridges while the dark areas are called valleys . Fingerprints are the result of reproduction of the skin of the finger surface either deliberately taken / stamped on ink or traces left to the object . The reproduction results in the form of paintings that consist of lines of papilair groove . Characteristics and features in this plot of papilair that distinguishes between people with each other . Human Fingerprints are divided into 5 main groups of fingerprints , namely : Whorl , Arch , Tanted Arch , Right Loop and Left Loop .

g. Problem analysis

The messaging security is very necessary when sending messages via electronic media. Then the message must remain safe during the delivery process. In the message delivery process, the public key can be known by anyone, but the key to the description is known only to the recipient of the message. For the sending of messages in secret then the need for security in the process of delivery. RSA cryptography is an asmetric key cryptography. The key to RSA cryptography consists of two keys: private key and public key. Given these two keys the process of sending messages does not require a lot of keys to sending many messages, simply made two keys, namely the public key to encrypting the message for the recipients and the description as a private key. Private keys are usually stored in a file, so it is easy about others to extract the private key. Then, to prevent theft of keys, it is necessary to use a fingerprint method as a key generator of the RSA algorithm.

In generating a key using fingerprint image. The process to do is to read the binary image of the fingerprint image. Here is an example of the pixel value of the fingerprint image.

1	0	0	0	1	1	1	1	0	0
1	0	1	0	1	0	1	1	1	1
1	1	1	0	0	0	1	0	1	1
0	0	1	0	1	0	1	0	1	0
1	1	0	0	1	0	0	1	1	0
1	0	1	0	1	0	1	1	1	1

Figure 1 Fingerprint Image Pixel

From the fingerprint image pixel data onto is known the number of pixels worth 1 is 34 , and the number of pixels is 0 is 26 . To get the value of a and b is taken from the first primary number of the number of pixels . Number of pixels values 1 becomes a reference value of variable a and pixel value 0 becomes reference variable value b . So the value of a obtained is 34 and the value of b is 26.

The flowchart of the key generator in RSA:

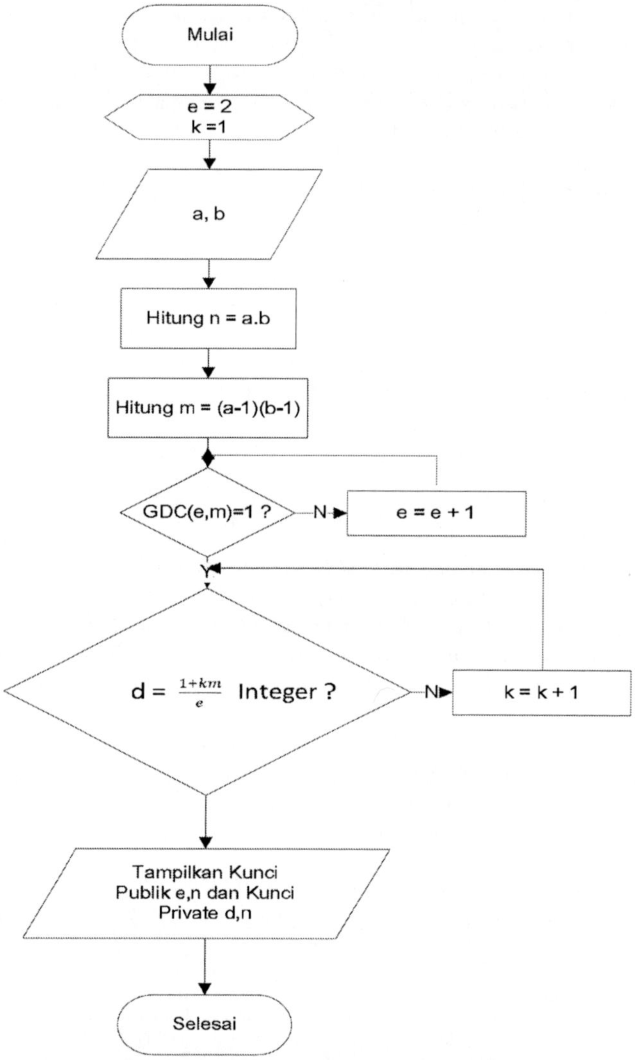

Figure 2 RSA Key Generation Flowchart

Figure 2 shows the RSA key generation process flowchart. The initial value of e is setting equal to two and k equals 1. The values of a and b are obtained from the input of the fingerprint image. Next the calculation of values of n and m. The value of e is obtained from the relative prime number of m that starts from 2 and increases one when it has not been obtained relatively prime. The value in is obtained from the formula $d = \frac{1+km}{e}$ where k starts from one and increases one onwards until the d value is obtained as an integer. Then get the public key e, n and private key d, n.

3. Results and Discussion

From the design result to generate RSA key is to use fingerprint image, this image as input variable a and b in RSA process, automatically generate both the public and private key, following the view of taking value of fingerprint image seen in picture 3.

Figure 3 Display of Key Generator Form

In Figure 3 it can be seen that the public e, n key has been obtained from the extract of the fingerprint feature to be used for encrypting text messages as shown in Figure 4.

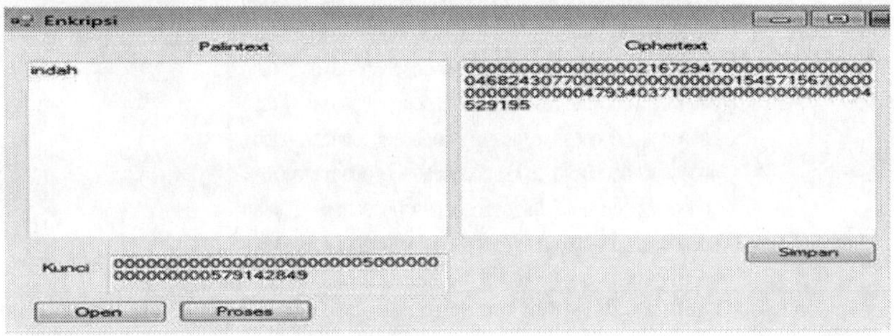

Figure 4 Encryption Process

The "indah" message on the image with the key generated from the fingerprint image succeeds to secure the message as shown in Figure 4, to decrypt the message using the stored private key.

The message decryption process can be seen in Figure 5 below:

Figure 5 Decryption Process

The following encryption process results are presented in table 1 :

Table 1 Encryption Results

No	*Plaintext*	*Ciphertext*	Citra	Keterangan
1	Nama saya Indah	000000000000000570602972000000000000000479340371 000000000000000328525475000000000000000479340371 000000000000000033554432000000000000000422715009 000000000000000479340371000000000000000455139245 000000000000000479340371000000000000000033554432 000000000000000335643046000000000000000468243077 000000000000000154571567000000000000000479340371 000000000000000004529195000000000000000033554432		Berhasil
2	Hari ini Hari Senin	000000000000000549473610000000000000000353650977 000000000000000236492201000000000000000403214886 000000000000000111317034000000000000000403214886 000000000000000554569632000000000000000403214886 000000000000000111317034000000000000000549473610 000000000000000353650977000000000000000236492201 000000000000000403214886000000000000000111317034 000000000000000081982662000000000000000033204660 000000000000000554569632000000000000000403214886 000000000000000554569632		Berhasil

The decryption process of the encrypted message is done successfully as shown in table 2.

Table 2 Description Results

No	Citra	*Ciphertext*	*Plaintext*	Keterangan
1		000000000000000570602972000000000000000479340371 000000000000000328525475000000000000000479340371 000000000000000033554432000000000000000422715009 000000000000000479340371000000000000000455139245 000000000000000479340371000000000000000033554432 000000000000000335643046000000000000000468243077 000000000000000154571567000000000000000479340371 000000000000000004529195000000000000000033554432	Nama saya Indah	Berhasil
2		000000000000000549473610000000000000000353650977 000000000000000236492201000000000000000403214886 000000000000000111317034000000000000000403214886 000000000000000554569632000000000000000403214886 000000000000000111317034000000000000000549473610 000000000000000353650977000000000000000236492201 000000000000000403214886000000000000000111317034 000000000000000081982662000000000000000033204660 000000000000000554569632000000000000000403214886 000000000000000554569632	Hari ini Hari Senin	Berhasil

4. Conclusion

From the design results RSA key generator using fingerprint image in the message security, obtained conclusion of which is as follows:

1. Application to secure text message using fingerprint image with RSA Algorithm Cryptography method for better security successfully done.
2. The fingerprint image is used as a prime number in RSA key generation.
3. Private key stored information in fingerprint image.
4. Public key is generated from fingerprint image in the form of numbers that can be stored or sent.
5. RSA algorithm method can be combined with other methods to improve the security of a file.
6. In terms of time, when the process of reading fingerprint image takes a long time.

5. Bibliography

[1] A. Fallis, 2013, Aplikasi Pengamanan File Menggunakan Kombinasi Algoritma C-RSA, Arcfour, Dan C-AES (CAC), J.Chem. inf. Model., vol. 53, no. 9, pp. 1689-1699.

[2] Nafi'iyah, Nur. 2015. Algoritma Kohonen Dalam Mengubah Citra Graylevel Menjadi Citra Biner, Vol.9 No.2, Agustus 2015.

[3] Hafizah, dkk. 2015. Penerapan Jaringan Syaraf Tiruan Dengan Algoritma Perceptron untuk Mendeteksi Karakteristik Sidik Jari, Vol.14, No. 2. 2015.

[4] Rosmala, dkk. 2012. Implementasi Mode Operasi Chiper Block Chaining (CBC) Pada Pengamanan Data, Vol. 3, No.2, Mei – Agustus. 2012.

[5] Sadikin, Rifki. 2012. Kriptografi untuk Keamanan Jaringan. Yogyakarta: ANDI.

[6] Hasibuan, Andi Hazri. 2013, Penambahan Chiner Reminder Theorem Untuk Mempercepat Proses Enkripsi Dan Dekrisi Pada RSA, Universitas Sumatera Utara Medan.

[7] Permadi, Yuda.2015, Aplikasi Pengolahan Citra Untuk Identifikasi Kematangan Mentimun Berdasarkan Tekstur Kulit Buah Menggunakan Metode Ekstraksi Ciri Statistik,Universitas Ahmad Dahlan.

[8] Hatta Muttaqin, Ismail, Risnandar. 2012, Modul Pengiriman Data Untuk Pemindai Barcode Nirkabel, Bandung: Politeknik Telkom.

[9] S Rahman, M Ulfayani. 2017, Perancangan Aplikasi Identifikasi Biometrika Telapak Tangan Menggunakan Metode Freeman Chain Code, CESS (Journal of Computer Engineering, System and Science) Vol.2 No.2, 121-130

Analysis Of Using Firewall And Single Honeypot InTraining Attack On Wireless Network

Tengku. Mohd. Diansyah[1], Ilham Faisal[2], Adidtya Perdana[3], Boni Octaviani Sembiring[4], and Tantri Hidayati Sinaga[5]

[123]Informatics Engineering and [45]Information System of SekolahTinggi Teknik Harapan
Jl. H. M. Joni No. 70 C Medan, Indonesia

[1]dia.10.22@gmail.com

Abstract: Security issues become one of the important aspects of a network, especially a network security on the server. These problems underlie the need to build a system that can detect threats from parties who do not have access rights (hackers) that are by building a security system honeypot. A Honeypot is a diversion of intruders' attention, in order for intruders to think that it has managed to break down and retrieve data from a network, when in fact the data is not important and the location is isolated. A way to trap or deny unauthorized use of effort in an information system. One type of honeypot is honeyd. Honeyd is a low interaction honeypot that has a smaller risk compared to high interaction types because the interaction with the honeypot does not directly involve the real system. The purpose of the implementation of honeypot and firewall, firewall is used on Mikrotik. Can be used as an administrative tool to view reports of Honeyd generated activity and administrators can also view reports that are stored in the logs in order to assist in determining network security policies.

Keywords: Honeypot, low interaction, firewallmikrotik, wireless

1. Network Security System

Network security systems are often at the moment still seen as the result of several factors. Factors related to the safety of this network vary, depending on the basic ingredients, but normally there are at least some things in the concept of security such as securities, integrity, and availability. In network security, there is also the risk of computer networks that are all forms of threats both physical and logic that directly or indirectly disrupts the ongoing activities within the network. The Risks in computer network caused by several factors such as weakness of network operating system, weakness of network communication system and weakness of computer hardware. This security can be combined with non- repudiation, authenticity, possession, utility.

The purpose of network security can be achieved by a network security method that can protect the system both from outside and outside the network, but not only protect but also must be able to overcome network attacks. If you want to determine what you want to protect then it should have a good and sound security plan from procedures to network security policies. Because if not planned then it will not be as expected in network protection.

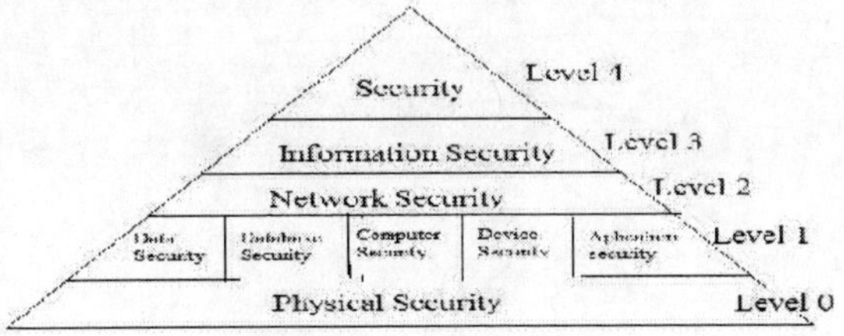

Figure1. Security Level 1

2. Honeypot mechanism

A Honeypot is a system or computer that is deliberately "sacrificed" to be the target of attacks from hackers. The computer serves every attack done by hackers in the penetration of the server. This method is intended for the administrator of the server to be attacked to know the penetration tricks that hackers do and can anticipate in protecting the real server. Any action was taken by an intruder trying to connect to the honeypot, then the honeypot will detect and record it. A Honeypot is a source of information systems that are usually designed to detect, trap, in an attempt penetration into the system. Generally, the honeypot consists of computers, data, and network segments that look Honeypot also have a monitoring feature to monitor attacker activity when Enter into the honeypot system. Known activities include ports being attacked, commands typed by attackers, and alterations by attackers on a fake server honeypot. This can be exploited by the Network Administrator as input to patch the actual system, configuring the original network segment for early prevention (Tambunan, et al, 2013). The picture of honeypot as follows.

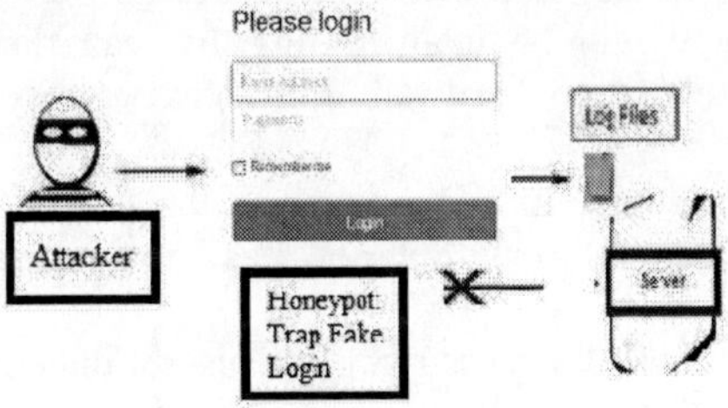

Figure 2. Attacker mechanism

A Honeypot is a computer system on the Internet that can manage to attract and trap people who are trying to penetrate other computer systems. A Honeypot is a server-installed forensic application specifically designed to monitor potential activities to attack and observe intruders how they get into the server computer system. (Aye Thu, 2013)

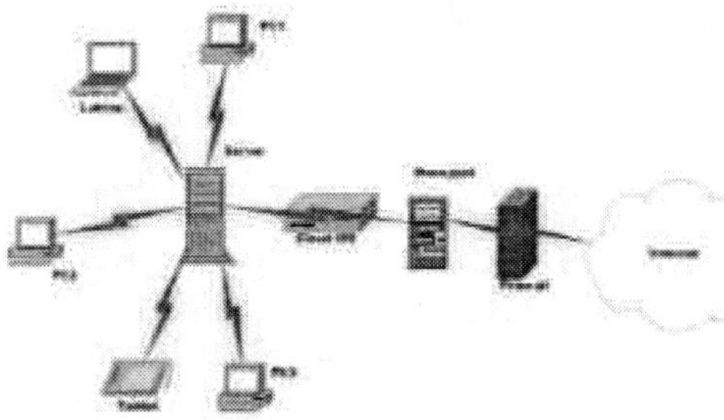

Figure 3. Firewall concept and using honeypot(Aye Thu,2013)

3. Honeyd

Honeyd is a type of Open Source Honeypot application written by NielProvos. Honeyd is a simple daemon that keeps virtual hosts on the network. The host can then be configured to run various services.

TCPnya Personality Can run as a particular operating system, to fool scanner fingerprints like Nmap or probe. Actually honeyd powerful enough and Provides a complete feature, but the configuration is not easy because it does not have a GUI (Mustofa&Aribowo, 2013). Honeyd is a low interaction honeypot type honeypot that performs network simulation as a whole like service FTP, SSH, HTTP, router in one machine / PC and can add multiple hops, packet losses, and latency (Tambunan, et al, 2013).

4. Firewall

In the computer network is very important for network security, especially related to applications involving various interests, there will be many things that can disrupt the stability of the computer network connection, whether related to hardware (physical security, power resources) and related to software (System, configuration, access system, etc.). Disruption of the system can occur due to accidental factors performed by the manager (human error), but not least also caused by a third party. Disturbances can include destruction, infiltration, theft of access rights, misuse of data or systems, to criminal acts through computer network applications. Security of the system should be done before the system is enabled. The use of the system should be done before the actual system is enabled. Overall.

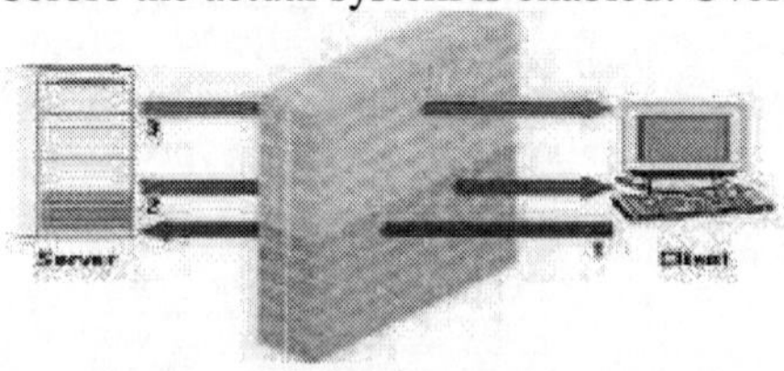

Figure 4. Firewall mechanism

5. Implementation

In this test the authors will conduct trials using low orbit applications as an attack testing application attack attack, this attack testing will conduct a direct attack to the server so that in the presence of this test will go into the trap honeypot.

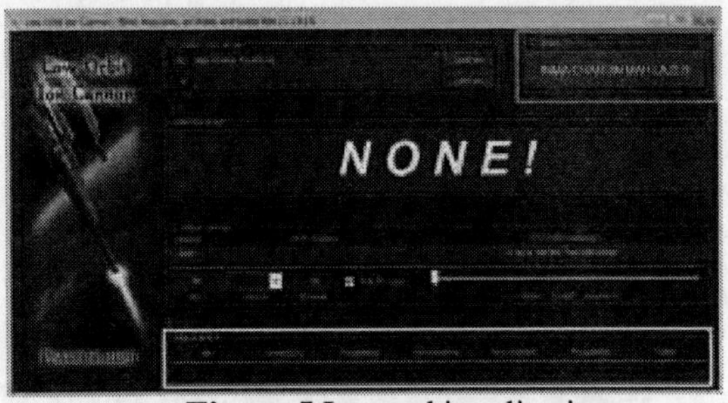

Figure 5.Low orbit aplication

Table 1. Honeyd implementation and firewall

IP Address	User Computer	Operating System	Port Applied
10.0.2.16	Real Server	Linux Ubuntu	135,139,445,1028
10.0.2.15	Server Shield Honyepot	Microsoft Windows SP1	135,139,445,1028
10.0.2.18	Client Intruder	Microsoft Windows Server	135,139,80,445,1028

The table is a scenario that will be tested in this honeypot test port used is port 135,139,445,1028 and windows SP 1 is acting as a shadow server used in this trial. The function of this shadow server is to perform manipulation so it is expected that the Microsoft server is hosted and active.

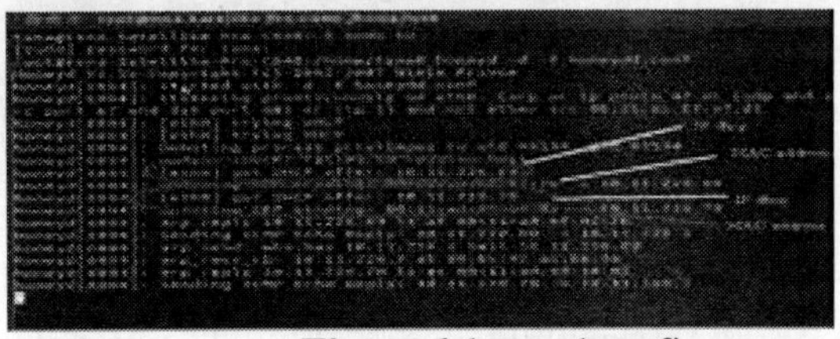

Figure 6. honeyd config

The above view is the result of the use of honeydconfig is the configuration file name that has been created before, this honeypot does not run in the background, in hoenyd it is automatically given information about IP DHCP, mac address of the shield already built from honeyd application.

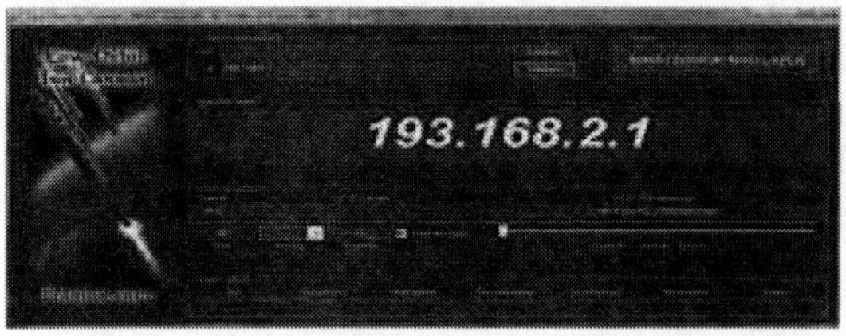

Figure 7. Attacker terting from orbit aplication

In this test directly to attack the server, the purpose of this attack is to make the server down so it can not be accessed, after this attack will not be able to directly attack the actual server but stuck in the fake serveror shield server so the server does not directly down But the attacker seems to be successful in attacking as shown below.

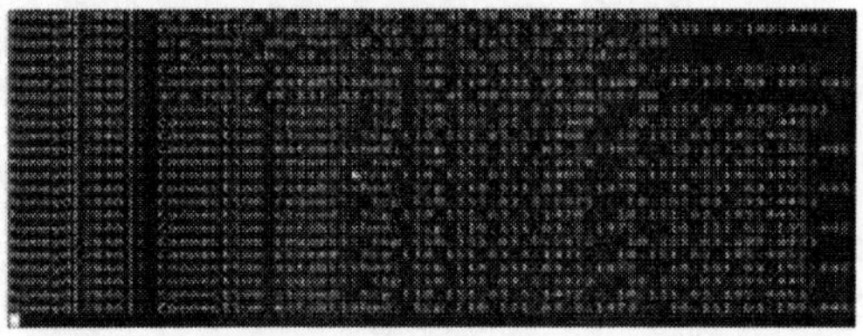

Figure 8. Attacker result that protected with honeyd

Display above after configuration and if port scanning successfully executed on at zenmaphoneyd will do the response from the result of scanning. In the above honeyd experiment will know the IP address that tries to communicate with the shadow host so that with the honeyd we can find the IP address of Intruder.

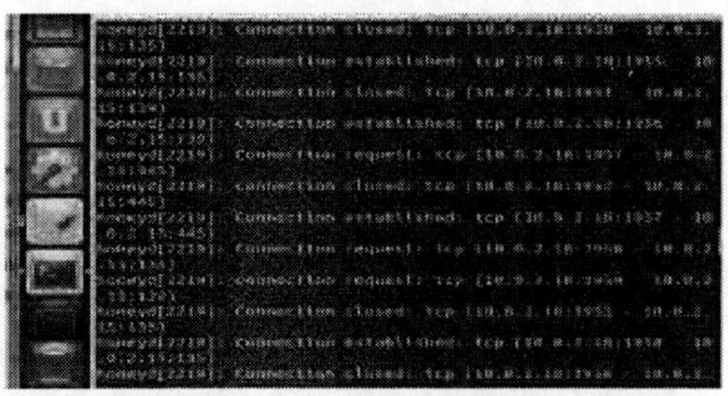

Figure 9. Testing port scanning

The attack is a type of port scanning attack so that the attacked port can be prevented following the attacker using IP 10.0.2.18 proved that honeyd is able to reduce attacks performed for the scanning port of the orbit application. Port scanning results can be seen as shown below In Implemetasi test scenario using this honeyd. Is done to determine the network condition of the gap of a network security testing using ubuntu 12.04 as a trial operating system and create a trap for honeyd application by utilizing the above configuration.

Table 2. Testing online

No.	Ip Address Scanned Detectted	
1	192.168.68.XXX	192.168.1.XXX
2	10.0.2.XXX	
3	172.192.85.XX	172.168.XX

In this test scenario using orbit tools as an application for port scanning and Flooding / in this study conducted experiments for 3 times so it can conclude new types of network than before, so that honeypot can accept new types of attacks and can learn the types of attacks that occur and stored In the folder ubuntuhoneydtmp / honeyd log file so will author can classify the type of attack.

6. Conclusion

At the time of this est the author get the conclusion that the honeypot and firewall can cooperate in restraining the incident that occurred so the attacker can't enter easily because the attacker into the trap honeypot that has been made, so the server can work safely, and honeypot is successful in Detects suspicious activity and captures the attacker's IP and is stored in a separate folder on the server trap honeypot.

REFERENCES

[1] Adam brown,ToddAndel 2016 "*What's Is Your Honeypot?*" (Universty of South Alabama), Mobile,USA

[2] Ariyus, Dony 2005. Computer Security Yogyakarta :PenerbitAndi

[3] AyeAyeThu, 2013. "*Integrated Intrusion detection and Prevention System With Honeypot on Cloud ComputingEnvirotment*" (International Journal of Computer Applications) Volume 64 Number 4

[4] Gray Plant Mooty, 2014. A. Legal guide to privacy and data security.

[5] Muh Masruri Mustofa, AribowoEko 2013 "*Penerapan system keamanan honeypot dan ids pada jaringan nirkabel (hotspot)*" (JurnalSarjana Teknik Informatika) e-ISSN : 2338-5197 Vol 1

[6] MohdDiansyah, Tengku 2015 "*Analisa Pencegahan Aktivitas Ilegal didalam Jaringan Menggunakan Wireshark*" (Journal Time)Vol 2 No 2

[7] PidieWiyanto. et al 2014 "*Aplikasi Monitoring KeamananJaringandenganmenggunakan IDS dan Router Mikrotik*" (JurnalJarkomInstitutSains&Teknologi AKPRIND Yogyakarta) e-ISSN : 2338-6312 Vol 2 No 1

[8] Project Honey Pot, 2015. *Project Honey Pot Statistic. Evailable at*: https:// www. projecthoneypot. org/statictic.php

[9] Zhan, Z., Xu, M. &Xu, S., 2013. *Characterizing Honeypot-Captured Cyber Attack: Statistical Framework and Case Study.(IEEE Transactions on Information Forensics And Security)*, 8 (11), pp. 1775-1789.

IOP Conf. Series: Journal of Physics: Conf. Series **930** (2017) 012039 doi:10.1088/1742-6596/930/1/012039

Analysis of Student Satisfaction in The Process of Teaching and Learning Using Importance Performance Analysis

P Sembiring[1], S Sembiring[2], G Tarigan[3], OD Sembiring4
[1,3,4] Department of Mathematics
Faculty of Math and Science
Universitas Sumatera Utara, Medan 20155 Indonesia
[2] Department of Computer Science
Faculty of Computer Science and Information Technology
Universitas Sumatera Utara, Medan 20155 Indonesia

Abstract. This study aims to determine the level of student satisfaction in the learning process at the University of Sumatra Utara, Indonesia. The sample size of the study consisted 1204 students. Students' response measured through questionnaires an adapted on a 5-point likert scale and interviews directly to the respondent. SERVQUAL method used to measure the quality of service with five dimensions of service characteristics, namely, physical evidence, reliability, responsiveness, assurance and concern. The result of Importance Performance Analysis reveals that six services attributes must be corrected by policy maker of University Sumatera Utara. The quality of service is still considered low by students.

1. Introduction

Extensive research in Education sector has been carried out studying the factors which can affect the satisfaction of students. Good quality of education provides better learning opportunities. The levels of satisfaction or dissatisfaction strongly affect the student's success or failure of learning.

Nowadays, researcher in higher learning institutions focused on identifying and satisfying the needs and expectations of their students. Such factors include student performance, faculty performance, classroom environment, learning facilities and institution reputation[2]. Education is a very important activity to develop the potential of Human Resources (HR)[1]. Educational activities in higher learning institution, in principle related to the two parties there are lecturers and students. The linkage of both parties will be harmonious if the clear position of each party in a professional manner, ie as a subject with rights and obligations [4].

Kotler and Clarke (1987) define satisfaction as a state felt by a person who has experience performance or an outcome that fulfill his or her expectation. Satisfaction is a function of relative level of expectations and perceives performance. The expectation may go as far as before the students even enter the higher education, suggesting that it is important to the researchers to determine first what the students expect before entering the university [12].

In the process of learning and teaching, the lecturer has a major role in determining the quality of teaching is implemented, is to give knowledge (cognitive), attitudes and values (affective) and skills (psychomotor) to students [12],[10]. In other words, the tasks and roles of the main lecturer is teaching field.

Teaching is a tool to achieve educational goals. Therefore, a teacher must be able to manage the classroom, using teaching methods and attitudes and characteristics of the faculty in managing the process of effective teaching and learning, developing the lecture materials, and improve the ability

of students to follow the course and master the educational goals they need to accomplish [3],[11].
Higher Learning Institutions in Indonesia including University of Sumatera Utara has been undergoing profound transformation during last decade. There are many higher education institutions have been established are trying to improve the quality of teaching and learning processes through improved various aspects. For instance facility of teaching and learning, a qualified lecturer, and a selection of new student enrollment to a higher standard is also on the higher side.
Ministry of Research Technology and Higher Education Republic of Indonesia, has provided various incentives for universities to be able to improve its quality. Such as by providing various financing schemes associated with research and community service, the provision of infrastructure facilities to support the educational process and so forth. So expect students who are taking their studies can also feel comfortable and satisfied with the services and facilities provided.
The paper is organized as follows. In Part II, we discuss research related to this study. While based on the evaluation of this system, we propose an evaluation method using SERQUAL in Part III. Section IV, describes the application of our method to empirical studies and discuss the value of this method. Finally, we provide some discussions and present our conclusions.

2. Literature Review

2.1. Student Satisfaction

There are some studies have been conducted to measure the level of student satisfaction at university. The concern for improving the University's understanding of the level of student satisfaction within the University System evolved from previous accountability survey results suggesting that graduating seniors and undergraduate alumni were less than satisfied with the quality of undergraduate instruction and non instructional services.
A definition of Satisfaction is a function of relative level of expectations and perceives performance. Satisfaction is a state felt by a person who has experience performance or an outcome that fulfill his or her expectation [5]. The expectation may go as far as before the students even enter the higher education. Palacio, Meneses and Perez [12], suggesting that it is important to the researchers to determine first what the students expect before entering the university. While, Carey, Cambiano and De Vore (2002), believe that satisfaction actually covers issues of students' perception and experiences during the college years.

2.2. Service Quality in Higher Learning Institution

A definition of quality is relating to how good a service is, and not necessarily how large or extensive. Quality can also be seen as relating to the fitness of a service or product to its intended purpose or use, subject to the expectations of the customer, user or public. The construct of quality as conceptualized in the services literature is based on the *perceived quality*. Zeithaml, (1987) define the perceived quality is as the consumer's judgment about an entity's overall experience or superiority. Similarly, Parasuraman, Zeithaml and Berry (1990) also stated that consumer perceptions of service quality result from comparing expectations prior to receiving the service, and their actual experience of the service.
Deshields et al (2005) state that higher learning institutions are focusing on identifying and satisfying the needs and expectations of their students. Mai (2005) studied the student satisfaction in higher learning institution and its influential factors. Aldemir and Gulcan (2004) examined the Turkish students' satisfaction in higher education. The results of study showed that for some Turkish university students, the quality of instructors, education, textbooks and being female and informed before attending university considered to important factors of satisfaction. Although a variety of assessment activities, including surveys and focus groups have been employed at the sector level to obtain student satisfaction data in support of program review requirements, there had not been a concentrated, comprehensive effort at a university-wide level to review and use those findings. The

purpose of the Student Satisfaction Study was to review these survey results and related literature, to pose and answer additional questions in order to understand more clearly students' perceptions of the quality of undergraduate instruction and non instructional services, and to make recommendations for needed improvements.

All these studies emphasis on certain factors of education offerings which determine the students' satisfaction with education and in turn loyalty to the institution. Therefore the objective of this study is to analyze the student satisfaction in the process of teaching and learning at higher education.

3. METHOD

This section describes the types of methods (qualitative, quantitative or mixed-method) along with the details of the methods of data collection and data analysis methods used. This section also describes the perspective that underlies the selection of a particular method.

This study examines the quality of service with five dimensions of services characteristics, namely; physical evidence, reliability, responsiveness, assurance and concern. For this purpose, responses were collected from students of 14 faculties in the University of Sumatera Utara Medan, Indonesia using questionnaire; the sample size consisted of 1204 students from different levels and disciplines.

3.1. Importance Performance Analysis (IPA) Method

Importance Performance Analysis (IPA) or the analysis of interest and performance is an easy application technique to measure the level of interest and the level of performance for the development of effective marketing programs, by analyzing the quadrants to show the relationship between the assessment of the level of interest and the level of performance (Sever, 2015)(Lai, *et al.* 2015). In this research there are two variables, the level of performance of the University North Sumatra that can give the students (denoted X) and the level of expectations of students (denoted Y). The formula used by are as follows (Martilla, *et al. 1997*) (Chu, *et al. 2012*):

$$Tk = \frac{X_i}{Y_i} \times 100\% \quad (1)$$

Description:
Tk = Level of concordance between performance and expectations
X_i = Total score assessment of the level of performance
Y_i = Total score assessment of consumer interests

In the analysis of Importance Performance Analysis mapping into four quadrants for all the variables that affect the quality of service, (Oh, 2001, Luo, *et al.* 2015) the horizontal axis (X) will be filled by the score level execution or performance, while the vertical axis (Y) will be filled by a score of importance. In simplifying the formula, then for each of the factors that affect the service satisfaction:

$$\overline{X} = \frac{\sum X_i}{n} \quad \overline{Y} = \frac{\sum Y_i}{n} \quad (2)$$

Description:
$\overline{X}$ = The average score of the level of performance
$\overline{Y}$ = The average score of the level of interest
n = Number of respondents

Cartesian diagram is a structure which is divided into four sections bordered by two lines that intersect perpendicularly at the points ($\overline{\overline{X}}$, $\overline{\overline{Y}}$) which $\overline{\overline{X}}$ is the average of the average score of the level of service performance all the factors or attributes and $\overline{\overline{Y}}$ is the average of the average rate score the entire interest rate factors or attributes that affect service. By using the following formula

$$\bar{\bar{X}} = \frac{\sum \bar{X}}{K} \quad \bar{\bar{Y}} = \frac{\sum \bar{Y}}{K} \tag{3}$$

Description:

$\bar{\bar{X}}$ = Average of the average level of performance

$\bar{X}$ = Score average level of performance

$\bar{\bar{Y}}$ = Average of the average rate of interest

$\bar{Y}$ = Score average rate of interest

K = The number of attributes that affect service

Importance Performance Analysis division can be seen in the following figure 1:

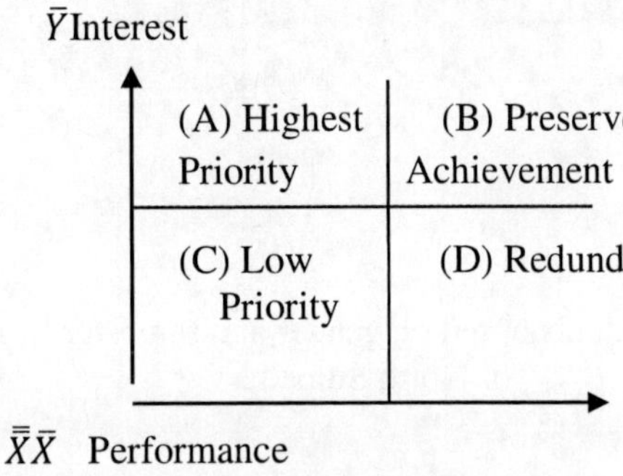

Figure 1. The Quadrants of Importance Performance Analysis
Source: (Liu, *et al.* 2012)

Description:

a. Quadrant I (highest priority) indicates factors or attributes that are deemed to affect customer satisfaction, including elements of the services that are considered very important, but the management has not done according to customer wishes. So disappointing or dissatisfied.
b. Quadrant II (preserves achievements) shows the basic service elements that have been successfully implemented enterprise, for it shall be retained. Considered very important and very satisfying.
c. Quadrant III (low priority) indicates several factors that are less importance for the customer, implementation by companies mediocre. Considered to be less important and less satisfying.
d. Quadrant IV (redundant) shows the factors that affect customer is less important, but implementation is excessive. Considered less important but very satisfying.

4. Results and Discussion

The main objective of this study is to find out the relationship between students' satisfaction and various education facilities and services in the process of teaching and learning at the University of Sumatera Utara.

Reliability of the data was checked using Cronbach Alpha which provides a value of 0.75 is more than the acceptable value of 0.70 (Nunnally [35]; Hair et al., [24]; Zikmund [53]) and a value ranges between 0 and 1 and the value close to 1 provides more reliability (Nunnally and Bernstein [36]). As reliability of the data plays a significant role in analyzing the results.

4.1. Characteristics of Respondents

The demographic information includes the following characteristic of respondents: gender, age, and semester of studies. From the 1024 respondents in this study435 (42.5%) are male and 589 (57.5%) are females. The calculated mean age of the respondents is 22 years old (66%) with the majority of the

students being 21 years old (34%). Most of the respondents are in the fifth semester of their study (65%), followed by seventh semester and above (35%).
The characteristics of respondents from 14 faculties at University of Sumatera Utara based on gender and semesters as shown on figure 2:

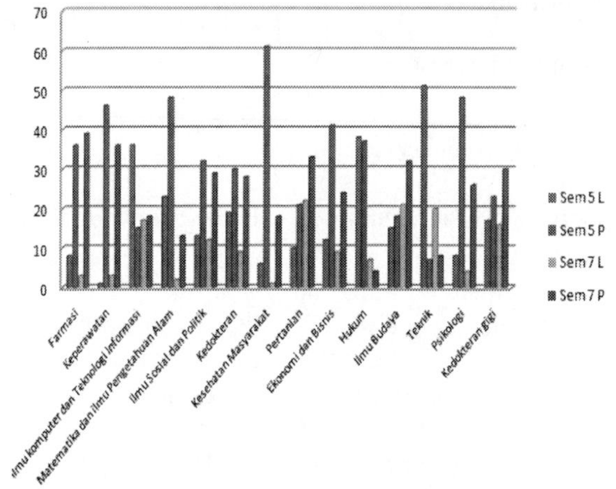

Figure 2. Characteristics of Respondents based on gender and semester from 14 faculties at University of North Sumatra

4.2. The Level of students Satisfaction
Based on respondents' answers, there is 11.77% respondents answer very unsatisfied, 20.74% students are lesser satisfied, 24.32% students state they are Satisfied and 9.15% of students answers Very Satisfied.

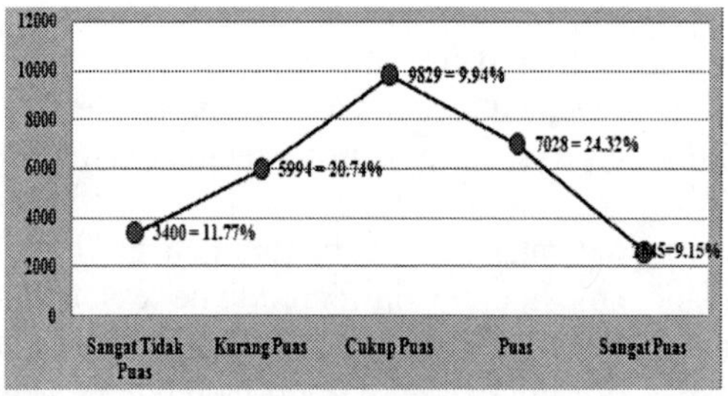

Figure 3. Distribution levels of students satisfaction at University on Sumatera Utara.

4.3. The Level of Importance
Importance Performance Analysis (IPA) method was used to measure the relationship between students' perceptions and priorities to improve the quality of products / services which is also known as a quadrant analysis. Quadrant analysis is divided into four quadrants covering the first quadrant main priority, keep achievement quadrant II, while Quadrant III and Quadrant IV low priority and redundant.
Quadrant analysis is a structure that is divided into four sections bounded by two lines that intersect perpendicularly at the points (x, y), where x is the average of the scores level implementation of all these factors and y is the average of score level interests of all the factors that affect the service user satisfaction. The data distributions by respondents felt in four quadrants as shown on figure 4.

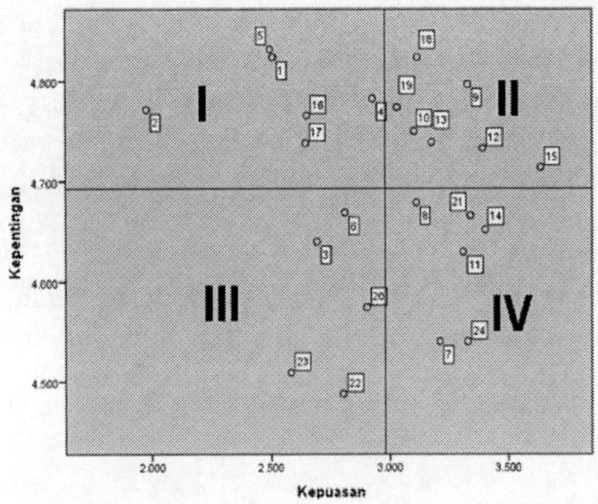

Figure 4: Quadrant Analysis Satisfaction vs Importance

1. Quadrant I is the low performance of high interest, service attributes that should be improved are:
   a. Convenience lecture rooms (chairs, ceiling fan, air conditioning, cleanliness, etc.).
   b. Availability region internet connection (WiFi) that support learning activities.
   c. Completeness of books, journals and literature lectures at the library.
   d. Hygiene and clean water in the toilet / bathroom.
   e. Availability of the department to respond and respond to student complaints.
   f. The teaching staff serve the maintenance of files quickly and professionally.

2. Quadrant II is a high performance high interest, the attributes of services that must be maintained are:
   a. Lecturer / Assistant lab deliver learning material clearly.
   b. Lecturer taught according to a set schedule.
   c. exam questions in accordance with the material provided.
   d. The system of values is done objectively.
   e. The ability of the faculty in the lecturer answered questions clearly.
   f. The security environment and their campus security officers.
   g. Ease in obtaining information about the educational system (curriculum, class schedules, scholarships, etc.).

3. Quadrant III is the low performance low importance, service attributes are not so expected is:
   a. The condition of the equipment in the lab practicum.
   b. Completeness equipment laboratory experiments.
   c. Clarity their administrative costs and the number of charges in the management of files.
   d. The lecturer substitute when absent lecturers teach.
   e. Discussion re-exam by professors and lab assistants.

4. Quadrant IV is a high performance low importance, service attributes that are less urgent and less expected are:
   a. Availability of adequate parking.
   b. Ease of finding books in the library.
   c. Supervisory exams begin and end with the exam on time.
   d. The system of payment of tuition
   e. Lecturers provide motivation for students while teaching in the classroom.
   f. The existence of student organizations on campus to support activities

The final section of the study is discussions on the important findings of the study will be reviewed in terms of its significance and support by other researches. This study attempts to examine the relationship between service quality dimensions (physical evidence, reliability, responsiveness, assurance and concern) and student's satisfaction.

This study indicates five-service quality (physical evidence, reliability, responsiveness, assurance and concern) that contributes most to satisfaction and has strong relationship with student's satisfaction. The results are consistent and also agree with Smith and Ennew (2001), Ham and Hayduk (2003) and Bigne et al. (2003) that found there is a positive relationship between service quality and student satisfaction.

This study agree with Soutar and McNiel (2003) in their research, they state that although all dimensions is service quality are actually useful in explaining student satisfaction, but that does not mean that all dimensions are significant.

5. CONCLUSIONs

Since all the variables are significantly and positively related to students' satisfaction it is concluded that teachers expertise, courses offered, learning environment and classroom facilities enhance the students satisfaction in higher education. Thus, it confirmed by other literature try to suggest here, which is by improving the service quality, it can be improve the students' satisfaction.

According to results teacher expertise and learning facilities' (class room comfort, internet connection, completeness of text books, journal, literature lectures in the library and hygiene and clean water in the toilet) are most influential factor on the student satisfaction. This means that teachers' expertise, courses offered and learning environment do a good job of enhancing students' satisfaction in higher education, especially at University of Sumatera Utara.

In the light of the research findings, there are some suggestions and recommendation for in the effort to improve and hence the level of student's satisfaction at University of Sumatera Utara. Institutions should pay special attention to induct, train and retain qualified and expert teachers for promoting the quality education, provide the conducive and favorable learning environment in the universities, to upgrade class room facilities', to establish the healthy and interactive communication between students and teachers and administrator.

REFERENCES

[1] Delahaye, B.. Student Resource Guide Human Resource Development Managing Learning and Knowledge Capital. Tilde University Press.2015. Prahran VIC 3181 Australia

[2] DeShields Jr., O. W., Kara, A. and Kaynak, E. 2005, "Determinants of business student satisfaction and retention in higher education: applying Herzberg's two factor theory", *International Journal of Educational Management,* 19(2), 28-139.

[3] Johnson, K.E. Golombek, P.R. Mindful L2 Teacher Education: A Sociocultural Perspective on Cultivating Teachers' Professional Development. Routledge. New York. 2016.

[4] Karatzoglou, B.. An in-depth literature review of the evolving roles and contributions of universities to Education for Sustainable Development. Journal of Cleaner Production : Learning for Sustainable Development In Regional Networks 49 : 44-53,2012

[5] Kotler, P., & Clarke, R. N. *Marketing for health care organizations*. Englewood Cliffs, NJ: Prentice-Hall,1987.

[6] Lai, I.K.W. Hitchcock, M.. Importance–performance analysis in tourism: A framework for researchers. Tourism Management. 48 : 242-267. ISSN 0261-5177,2015

[7] Liu, H. C., Mai, Y. T., Jheng, Y. D., Liang, W. L., Chen, S. M. & Lee, S. J. A novel discrimination index of Importance-Performance Analysis model," 2011 International Conference on Machine Learning and Cybernetics. Guilin. 2011. pp. 938-942. 2011

[8] Luo, C., Gao, H., Gao, C. & Ma W.. An importance-performance analysis of E-service quality. International Conference on Logistics. Informatics and Service Sciences (LISS). Barcelona. pp. 1-6. 2015

[9] Martilla, J.A. & James, J.C.. Importance-Performance Analysis. The Journal of Marketing. vol. 41. no. 2: 77–79,1977.

[10] Nunnally, J. C., "Psychometric Theory," 2nd ed., McGraw-Hill, New York, NY,. 1978

[11] Oh, H. Revisiting importance–performance analysis. Tourism Management. vol 22. no. 6: 617-627.. 2001

[12] Palacio, A. B., Meneses, G. D. & Perez, P. J. PThe configuration of the university image and its relationship with the satisfaction of students. . 2002. *Journal of Educational Administration*, 40(5), 486-505.

[13] Race, P.. The Lecturer's Toolkit: A Practical Guide to Assessment, Learning and Teaching Fourth Edition. Routledge. New York,2015

[14] Roberts, J. Language Teacher Education. Routledge. New York, 2016

[15] Scrivener, J. 2011. Learning Teaching: The Essential Guide to English Language Teaching. MACMILLAN Publishers Limited. UK.

[16] Sever, I. Importance-performance analysis: A valid management tool?. Tourism Management vol. 48 : 43-53, 2015.

[17] Smith, R. & Ennew, C. Service quality and its impact on word of mouth communication in higher education. Online: http://www.unim. nottingham.ac.uk /dbm/papers/ 2001

[18] Solomon, M. R. (1996). *Consumer behavior*. Englewood Cliffs, NJ: Prentice-Ha

[19] Wu, C. H., Lee, Y. C., Cheng, Y. C. & Tasi, S. B. The use of importance-performance analysis (IPA) in evaluating bank service. ICSSSM12. Shanghai. 2012. pp. 654-657.

Representative Model the Graph Theory in Calculations Kendall Correlation Coefficient

[1]Pasukat Sembiring, [2]Ujian Sinulingga, [3]Marihat Situmorang,[4]Sajadin Sembiring.

[1,2,3]Department of Mathematics, University of Sumatera Utara

[4]Department of Computer Science, University of Sumatera Utara

Abstract. Kendall correlation coefficient formula as an association or relationship size measure between two ordinal scale data variables, has been widely used in research in various fields of science. A study of the Kendall correlation coefficient formula associated with the graph theory through a complete asymmetric graph as an adjacency matrix, will be used in Kendall correlation calculations in the case of bivariate. The results of this study indicate that a complete asymmetric graph as an adjacency matrix can be used as an alternative method of Kendall correlation coefficient calculation.

Keyword: Graph Theory, Kendall Correlation Coefficient, adjacency matrix.

1. Introduction

Correlation is a study that discusses the degree of relationship between two or more variables [1]. Correlation is one of the statistical analysis techniques that are widely used by researchers because researchers are generally interested in the events that occur and try to connected. The degree of closeness of the relationship between two or more variables can be known by looking for the amount of correlation numbers commonly called the coefficient correlation.

Nonparametric statistical methods are often called distribution free methods because their statistical test models do not specify certain conditions about the distribution of population parameters [2]. The nonparametric statistics test specifies only the assumption that the observations should be independent and that the variables under study must essentially have continuity. Many of the nonparametric statistical tests are sometimes referred to as rank tests, since these nonparametric techniques can be used for scores that are not exact scores in the sense of scarcity, but rather scores that are solely ranks.

One method of measuring nonparametric correlation coefficient is the rank correlation coefficient of Kendall. The rank correlation coefficient of Kendall was first proposed by Maurice G. Kendall in 1938. Coefficient of rank correlation Kendall is denoted by τ. Kendall rank correlation coefficient (τ) is used to find the relationship and test the hypothesis between two or more variables, if the data is ordinal or rank.

IOP Conf. Series: Journal of Physics: Conf. Series **930** (2017) 012040 doi:10.1088/1742-6596/930/1/012040

The degree of closeness between the two variables can be shown by the ratio (comparison) between the actual score of +1 and -1 with the maximum score that can be achieved. +1 is assigned to a naturally arranged pair and -1 is assigned to a pair that is not naturally composed.
The rank correlation coefficient of Kendall (τ) has advantages when compared with spearman rank correlation coefficient (τ_s). τ is more general because it can be calculated as normal distribution and can be searched for the coefficient correlation [3].
Graph theory was first introduced by Leonard Euler in 1736 when he proved the possibility of passing through four areas linked to seven bridges over the Pregel river in Konigsberg, Russia in one pass through each bridge just once and returning to its original place [4]. The Konigsberg bridge problem can be expressed in graph by determining the four regions as vertices and the seven bridges as the edges connecting the corresponding vertex pairs.
The calculation τ using graph theory is done by forming complete asymmetric digraph with vertex is every object of research [5]. Complete asymmetric digraph and then poured into adjacency matrix. From the adjacency matrix formed can be calculated actually score. And for the maximum score that can be achieved is equal to the number of edges with n vertices on the complete asymmetric digraph. The number of arcs can be obtained from C_2^n. So that the correlation coefficient rank of Kendall (τ) is the ratio actual score with maximum score that can be achieved.

2. Methodology

The steps to determine the correlation coefficient of rank Kendall (τ) through graph theory approach that is:

1. Establishing a complete asymmetric digraph with vertices is every object of the research.
2. Establish adjacency matrix obtained from complete asymmetric digraph.
3. Calculating actual score of adjacency matrix formed.
4. Calculate the maximum possible score. Score maximum is equal to the number of arcs with n vertex on complete asymmetric digraph. The number of arcs can be obtained from C_2^n.
5. Subordinate actual score and maximum possible score into the correlation rank correlation coefficient formula (τ) with approach through graph theory.

2.1. Correlation Coefficient Rank Correlation (T)

The rank correlation coefficient of Kendall (τ) can be obtained by comparing actual score with maximum score possible [6]. Or in other words the actual score is the actual score +1 and -1. +1 is assigned to a naturally arranged pair and -1 is assigned to a pair that is not naturally composed. Whereas the maximum possible score is determined by an arrangement C_2^n that can be decomposed into $\frac{1}{2}n(n-1)$. So, Kendall rank correlation coefficient (τ) can be formulated:

$$\tau = \frac{\text{actual scorel}}{\text{maximum score}}$$

Furthermore the actual score is given the symbol S, and the maximum score is determined by the arrangement C_2^n, where n is the number of objects or individuals in the random variables X and Y. Can be mathematically written:

$$\tau = \frac{S}{C_2^n}$$

Or

$$\tau = \frac{S}{\frac{1}{2}n(n-1)}$$

Where
S = actual score (score count +1 and -1)
N = number of objects or individuals on random variables X and Y

There are times when random variables X and Y have the same object or often called the twin rank. If there are two or more observed values (either for the same random variable X or Y), then those values are given average rank. The effect of the twin rank values is to change the value of the denominator on the rank correlation coefficient formula of Kendall (τ)[7]. In this case the correlation rank correlation coefficient formula (τ) becomes:

$$\tau = \frac{S}{\left[\sqrt{\frac{1}{2}n(n-1)-T_x}\right]\left[\sqrt{\frac{1}{2}n(n-1)-T_y}\right]}$$

Where
S = actual score (score count +1 and -1)
N = number of objects or individuals on random variables X and Y
$T_x = \frac{1}{2}\sum t(t-1)$
T is the number of twin rank of each twin group for the variable X
$T_y = \frac{1}{2}\sum t(t-1)$
T is the number of twin rank for each of its twin groups for variable Y

2.2. Complete Asymmetric Digraph

Complete asymmetric digraph is an asymmetric digraph where there is exactly one between each pair of vertices. Complete asymmetric digraph with n vertices containing $\frac{1}{2}N(N-1)$ arc [8].

For evidence: Let D be a complete asymmetric digraph with n vertices v_1, v_2, ... v_n. Take any vertex (call v_1). Since D is a complete asymmetric digraph, then v_1 is associated with (n - 1) other vertices (v_2, v_3, ..., v_n). So, there is (n - 1) arc fruit.

Next, grab any second vertex (call v_2). Since D is complete asymmetric digraph, then v_2 is also associated with all remaining vertices (v_3, ..., v_n) so that there are (n - 2) arcs associated with v_2. The process is continued by counting the number of arcs associated with v_3, v_4, ..., v_{n-1} and which have not been calculated before. Many arcs obtained are: (n - 3), (n - 4), ..., 3,2,1.

So overall there is (n - 1) + (n - 2) + (n - 3) + ... + 3 + 2 + 1 = $\frac{n(n-1)}{2}$ arc

In the Figure 1 can be seen an example of complete asymmetric digraph.

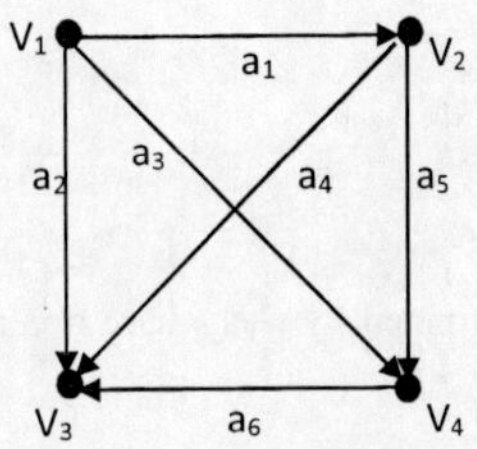

Figure 1. Complete Asymmetric Digraph

Complete asymmetric digraph can be represented in an adjacency matrix. The number of adjacency matrix rows and columns is the same as the vertex in the complete asymmetric digraph. The corresponding adjacency matrix is the matrix of square n x n, ie the matrix A = (a_{ij}) by:

$$(a_{ij}) = \begin{cases} +1 & \text{If any arc from } v_i \text{ ke } v_j \text{ and } (v_i, v_j) \text{ natural} \\ -1 & \text{If any arc from } v_i \text{ to } v_j \text{ and } (v_i, v_j) \text{ unnatural} \\ 0 & \text{If not a arc } v_i \text{ ke } v_j \end{cases}$$

3. Result and Discussion

3.1. Graph Form as An Adjacency Matrix in Determination of Rank Correlation Coefficient Kendall (τ)

The graph used in the determination of the rank correlation coefficient of Kendall (τ) is directed graph (digraph), because each object of observation in the given data is sequential paired data. Vertexes are the objects in the study. In the Kendall rank correlation method, the observed objects are linked to each of the next objects from the first to the second, then from the first to the third, and so on until next n object. After that proceed from second object to third object, second object to fourth object, and so on until object n. This continues through next n object to next *n* object. Or in other words, if the objects in the study is a vertex in the graph then from the above understanding can be written:

(v_i, v_j) with <j; ij = 1,2,3, ..., n.

v_i is the starting point and vj is the end point, so an arc with the vertex (v_i, v_j) represents the line from point i and point j. Because (v_i, v_j) is connected only once with i <j; Ij - 1, 2,3, ..., n so that there is no parallel line and no arc with end points (v_i, v_j) then the graph that is set is an asymmetric digraph. And because each pair of objects in the study must be connected or in other words there is exactly one arc between each pair of vertices (v_i, v_j) with i <j; Ij = 1, 2,3, ---, n then the graph form used in the T calculation is a complete asymmetric digraph.

Complete asymmetric digraph formed then poured into adjacency matrix. The number of bans and adjacency matrix columns is equal to the number of vertices in the complete asymmetric digraph. The corresponding adjacency matrix is the matrix of square n x n, in matrix A = (aij) by:

$$(a_{ij}) = \begin{cases} +1 & \text{if any arc from } v_i \text{ to } v_j \text{ and } (v_i, v_j) \text{ natural} \\ -1 & \text{If any arc from } v_i \text{ to } v_j \text{ and } (v_i, v_j) \text{ unnatural} \\ 0 & \text{If not have arc titik } v_i \text{ ke } v_j \end{cases}$$

So that the adjacency matrix is formed into:

$$\begin{matrix} & v_1 & v_2 & v_3 & v_4 & \dots & v_n \end{matrix}$$

$$A = (a_{ij}) = \begin{matrix} v_1 \\ v_2 \\ v_3 \\ v_4 \\ \dots \\ v_n \end{matrix} \begin{bmatrix} 0 & a_{12} & a_{13} & a_{14} & \dots & a_{1n} \\ & 0 & a_{23} & a_{24} & \dots & a_{2n} \\ & & 0 & a_{34} & \dots & a_{3n} \\ & & & 0 & \dots & a_{4n} \\ & & & & \dots & \dots \\ & & & & & 0 \end{bmatrix}$$

After the adjacency matrix is formed with the above conditions, then the actual score is determined by summing the +1 and -1 values obtained from the adjacency matrix. Then determine the maximum possible score. The maximum score may be equal to the number of arcs of complete asymmetric digraph, so the maximum score can be determined by C $ where n is the number of vertices of the complete asymmetric digraph.

Thus, the rank correlation coefficient of Kendall (τ) with the approach through graph theory can be determined by the formula:

$$\tau = \frac{S}{C_2^N} \text{ Or } \tau = \frac{S}{\frac{1}{2}n(n-1)}$$

For the rank Kendall correlation coefficient with twin rank then the formula used is:

$$\tau = \frac{2}{\left[\sqrt{C_2^n - T_x}\right]\left[\sqrt{C_2^n - T_y}\right]}$$

Where, S = actual score

$$T_x = \frac{1}{2}\sum t(t-1)$$

T is the number of twin rankings of each twin group for the variable x

$$T_y = \frac{1}{2}\sum t(t-1)$$

T is the number of twin rankings of each twin group for the y variables

Furthermore, we can develop null hypothesis and alternative hypothesis that can be tested by hypothesis test using formula:

$$Z = \frac{\tau}{\sigma_\tau}$$

Where
Z = statistical test for the correlation rank
τ = the rank correlation coefficient of Kendall
σ_τ = standard deviation, can be obtained by formulation:

$$\sigma_\tau = \sqrt{\frac{2(2n+5)}{9n(n-1)}}$$

3.2. Model Usage

As a test of Kendall rank correlation coefficient calculation model (τ) using graph theory, exemplified by the following data which is only a simulation data. The simulation data was taken from the second edition of nonparametric statistical book by Samsubar Saleh. The following is the observation data of the chief physician of the internal medicine section of 20 patients with high blood pressure.

Table 1. Observation Data of the Chief Medical Officer of 20 Disease Patients with High Blood

No.	Patient	Age	Tention
1	A	56	147
2	B	42	125
3	C	72	160
4	D	36	118
5	E	63	149
6	F	47	128
7	G	55	150
8	H	49	145

9	I	38	115
10	J	42	140
11	K	68	152
12	L	60	155
13	M	50	148
14	N	43	127
15	O	39	125
16	P	71	173
17	Q	65	169
18	R	73	176
19	S	44	132
20	T	26	115

To get the maximum score from the above data can be obtained from the maximum arc amount obtained from complete asymmetric digraph. 20 patients with high blood become vertex of complete asymmetric digraph. The maximum score is equal to the number of arcs of complete asymmetric digraph, so the maximum score can be determined by C_2^n where n is the number of vertices of the complete asymmetric digraph. From the data, it is known that the number of ***n*** is 20, so the maximum score can be obtained from the number of arcs of the complete asymmetric digraph using theorem 2, 4, 1, then the maximum score can be obtained by:

$$C_2^{20} = \frac{1}{2}20(20-1) = \frac{380}{2} = 190$$

Can be presented with digraph will be obtained graph with 20 vertexes and 190 arcs. And can be seen in the following figure:

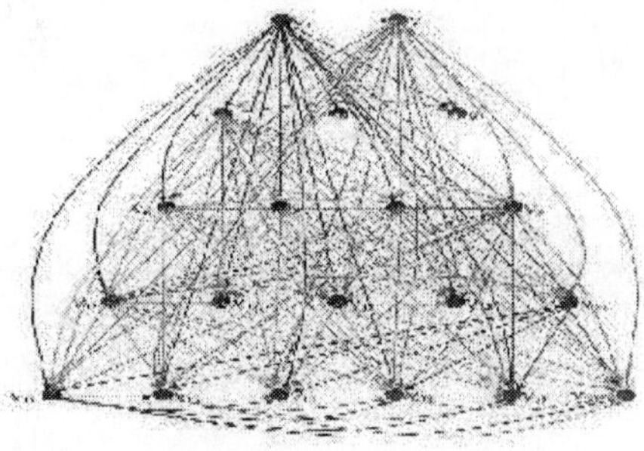

Figure 2. Graph G (20, 190)

So, the maximum score obtained is equal to the number of arcs of the complete asymmetric digraph of 190. To determine the actual score value can be obtained by using adjacency matrix in the form of upper triangular matrix.

Table 2. Ranking of Physician Observation of Internal Disease Section on 20 Patients with High Blood Patients

No.	Patient	Ranking of Physician Observation	
		Age	**Tention**
1	A	8	10

2	B	15,5	16,5
3	C	2	4
4	D	19	18
5	E	6	8
6	F	12	14
7	G	9	7
8	H	11	11
9	I	18	19,5
10	J	15,5	12
11	K	4	6
12	L	7	5
13	M	10	9
14	N	14	15
15	O	17	16,5
16	P	3	2
17	Q	5	3
18	R	1	1
19	S	13	13
20	T	20	19,5

Next rank is arranged in order.The results are presented in the following table:

Table 3. Ranking Arrangements Observation Doctors Head of Internal Disease Against 20 Patients Suffering from High Blood

No.	**Pasien**	**Ranking Pengamatan**	
		Umur	**Tekanan Darah**
1	R	1	1
2	C	2	4
3	P	3	2
4	K	4	6
5	Q	5	3
6	E	6	8
7	L	7	5
8	A	8	10
9	G	9	7
10	M	10	9
11	H	11	11
12	F	12	14
13	S	13	13
14	N	14	15
15	B	15,5	16,5
16	J	15,5	12
17	O	17	16,5
18	I	18	19,5
19	D	19	18
20	T	20	19,5

After the age rank is sorted naturally, the next step is to establish an adjacency matrix corresponding to the complete asymmetric digraph.

	V1	V2	V3	V4	V5	V6	V7	V8	V9	V10	V11	V12	V13	V14	V15	V16	V17	V18	V19	V20
V1	0	1	+1	+1	+1	+1	+1	+1	+1	+1	+1	+1	+1	+1	+1	+1	+1	+1	+1	+1
V2		0	-1	+1	-1	+1	+1	+1	+1	+1	+1	+1	+1	+1	+1	+1	+1	+1	+1	+1
V3			0	+1	+1	+1	+1	+1	+1	+1	+1	+1	+1	+1	+1	+1	+1	+1	+1	+1
V4				0	-1	+1	-1	+1	+1	+1	+1	+1	+1	+1	+1	+1	+1	+1	+1	+1
V5					0	-1	+1	-1	+1	+1	+1	+1	+1	+1	+1	+1	+1	+1	+1	+1
V6						0	+1	+1	+1	+1	+1	+1	+1	+1	+1	+1	+1	+1	+1	+1
V7							0	-1	+1	-1	+1	+1	+1	+1	+1	+1	+1	+1	+1	+1
V8								0	+1	+1	+1	+1	+1	+1	+1	+1	+1	+1	+1	+1
V9									0	-1	+1	-1	+1	+1	+1	+1	+1	+1	+1	+1
V10										0	+1	+1	+1	+1	+1	+1	+1	+1	+1	+1
V11											0	-1	+1	-1	+1	+1	+1	+1	+1	+1
V12												0	+1	+1	+1	+1	+1	+1	+1	+1
V13													0	-1	+1	-1	+1	+1	+1	+1
V14														0	+1	+1	+1	+1	+1	+1
V15															0	-1	+1	-1	+1	+1
V16																0	+1	+1	+1	+1
V17																	0	-1	+1	-1
V18																		0	+1	+1
V19																			0	+1
V20																				0

From adjacency matrix above calculated total value of +1 and -1.From here obtained actual score. The actual score can be seen from the table below. Each row in this table shows the number of +1 values (naturally arranged pair) and the number of values -1 (naturally incomplete pair) of each row in the adjacency matrix.

Table 4. Couples participating naturally (+1) and unbranded pair (-1) of each line in adjacency matrix

Many Natural Ordered Pairs (+1) Many Unnoticed Pairs Naturally (-1)

Many Natural Ordered Pairs (+1)	Many Unnoticed Pairs Naturally (-1)
19	0
16	-2
17	0
14	-2
15	0
12	-2
13	0
10	-2
11	0
10	0
9	0
6	-2
6	-1
5	-1
3	-1
4	0
3	0
0	-1
1	0
174	**14**

From the above table the actual score obtained by adding a lot of couples are sequenced naturally with many couples are not arranged naturally.

$$S = 174 - 14$$
$$= 160$$

Because that will be tested has twin rank, then the next search result of Tx and T_y. Where the observation of age is given by x and the observation of blood pressure is given by y. The rank of observation of age there is one group of twin values that is: rank 15,5 with t = 2. So, the value of T_x is:

$$T_x = \frac{1}{2}\sum t(t-1) = \frac{1}{2}[2(2-1)] = \frac{1}{2}[2] = 1$$

In the observation rank of blood pressure there are two groups of twin values: rank 16.5 with 1 = 2 and rank 19.5 with t = 2. So, the value of Ty is:

$$T_y = \frac{1}{2}\sum t(t-1) = \frac{1}{2}[2(2-1) + 2(2-1)] = \frac{1}{2}[2+2] = 2$$

Thus, the Kendall rank correlation coefficient value can be searched using the formula:

$$\tau = \frac{S}{\left[\sqrt{C_2^N - T_x}\right]\left[C_2^N - T_y\right]} = \frac{160}{\left[\sqrt{190-1}\right]\left[190-2\right]} = \frac{160}{\left[\sqrt{189}\right]\left[\sqrt{188}\right]} = \frac{160}{[13{,}7477][13{,}7113]}$$

$$= \frac{160}{188{,}4988} = 0{,}8488$$

Further testing against τ:

1. Determine the null hypothesis and alternative hypothesis, yaitu :
   H_o: there is no correlation between age and blood pressure
   H_1: there is a correlation between age and blood pressure
2. Criterion of decision making:
   H_o : accepted and H_1 is rejected if $-Z_{a/2} \leq Z_H \leq +Z_{a/2}$
   H_o : rejected and H_1 is rejected if $Z_H > Z_{a/2}$ or $Z_H < -Z_{a/2}$
3. The value of Z arithmetic is:

$$Z = \frac{\tau}{\sigma_\tau} = \frac{\tau}{\sqrt{\frac{2(2n+5)}{9n(n-1)}}} = \frac{0{,}8488}{\sqrt{\frac{2(2(29)+5)}{9(20)(20-1)}}} = \frac{0{,}8488}{\sqrt{\frac{90}{3420}}} = \frac{0{,}8488}{\sqrt{0{,}0263}} = \frac{0{,}8488}{0{,}1622} = 5{,}2330$$

4. The critical calue at $\alpha = 5\% = \pm Z_{1/2(1-\alpha)} = Z_{1/2(1-0{,}05)}$
   $Z_{1/2(0{,}95)} = Z_{0{,}475} = \pm 1{,}96$ (use normal curve)
5. Contribution
   H_o can not to process because $Z_H = 5{,}2330 > Z_{0{,}025} = +1{,}96$
   Meaning: there is a significant correlation between age with blood pressure from 20 patients with high blood pressure.

4. Conclusion

To determine the maximum score can be obtained from the accumulation of a*ii* arc of the complete asymmetric digraph and to obtain actual score can be obtained by using the adjacency matrix of complete asymmetric digraph in the form of upper triangular matrix. So, from the results of the above discussion can be concluded that the graph directed can be used to determine the value of correlation coefficient rank Kendall (τ).

Reference

[1]. Frey, L., Botan, C., & Kreps, G. (1999). Investigating communication: An introduction to research methods. (2nd ed.) Boston: Allyn & Bacon.
[2]. Siagian. 2000. *Metode statistika untuk bisnis dan ekonomi.* Jakarta: PT Gramedia Pustaka Utama
[3]. Djarwanto Ps. Edisi 2004. *Statistika Nonparametrik.* Yogyakarta: BPFE-Yogyakarta

[4]. Alavi, Y. 1985. *Graph Theory -with Applications to Algorithms and Computer Science.* USA: John Wiley & Sons, Inc

[5]. Beineke, Lowell W & Wilson, Robin J. 1983. *Selected Topics in Graph Theory 2.* New York; Academic Press Inc.

[6]. Wilson, Robin J. 1985. *Introduction to Graph Theory.* Third Edition. New York: Longman, Inc

[7]. Narsingh, Deo. 1984. *Graph Theory with application to Engineering and Computer Science.* New Delhi: Prentice Hall of India, Private Limited.

[8]. Narsingh, Deo. 1984. *Graph Theory with application to Engineering and Computer Science.* New Delhi: Prentice Hall of India, Private Limited.

[9] Saleh, Samsubar. Drs. M.Soc. 1996. Satistik Nonparametrik Edisi 2. Yogyakarta: Universitas Gajah Mada

IOP Conf. Series: Journal of Physics: Conf. Series **930** (2017) 012041 doi:10.1088/1742-6596/930/1/012041

Multithreading with separate data to improve the performance of Backpropagation method

Mulia Dhamma[1], Muhammad Zarlis[2], Erna Budhiarti Nababan[3]

[1, 2, 3] Department of Computer Science,
Faculty of Computer Science and Information Technology,
Universitas Sumatera Utara,
Medan, Indonesia, 20155

Email: muliadhamma@gmail.com[1], m.zarlis@yahoo.com[2], ernastp@yahoo.com[3]

Abstract. Backpropagation is one method of artificial neural network that can make a prediction for a new data with learning by supervised of the past data. The learning process of backpropagation method will become slow if we give too much data for backpropagation method to learn the data. Multithreading with a separate data inside of each thread are being used in order to improve the performance of backpropagtion method . Base on the research for 39 data and also 5 times experiment with separate data into 2 thread, the result showed that the average epoch become 6490 when using 2 thread and 453049 epoch when using only 1 thread. The most lowest epoch for 2 thread is 1295 and 1 thread is 356116. The process of improvement is caused by the minimum error from 2 thread that has been compared to take the weight and bias value. This process will be repeat as long as the backpropagation do learning.

1. Introduction

Artificial Neural Network(ANN) is an algorithm that can make prediction which is able to learn from non linear data and also learn data base on time series. ANN does prediction in two ways that is learning by supervised and learning without supervised. Backpropagation method is one of the method that implement the learning by supervised. This method has two phase learning cycle that are propagate forward and propagate backward. This method will propagate forward from input layer to output layer to compare the actual output with the desire output. The error value in output layer will be propagated back from output layer to get a new weight and bias value. The learning process will stop if the error value is lower than the expected error value. The performance of backpropagation method depend on the weight value, bias value, learning rate, expected error value and also data quantity. Learning with many data will reduce the performance of this method. The error that has been produced in each epoch will affect the performance. the smaller the error produced for each epoch then the learning speed of this method will become faster. Adjusting the right weight value will also improve the performance of the performance of backpropagation method. Khairani said that a backpropagation method can be speed up with many technique such as parallel training where the learning process done in parallel. Using single thread for backpropagation to learn many data will consume too much time. Researcher will use multithreading in order to improve the performance of backpropagation method.

2. Literature Review

2.1. *Thread*

Thread is the smallest unit in a process that can be scheduled by operating system. Multithreading is a type of model execution that can make multiple threads run at a time. The thread runs independently but they share a process that is origin from the main source. Multithreading can also be called threading.

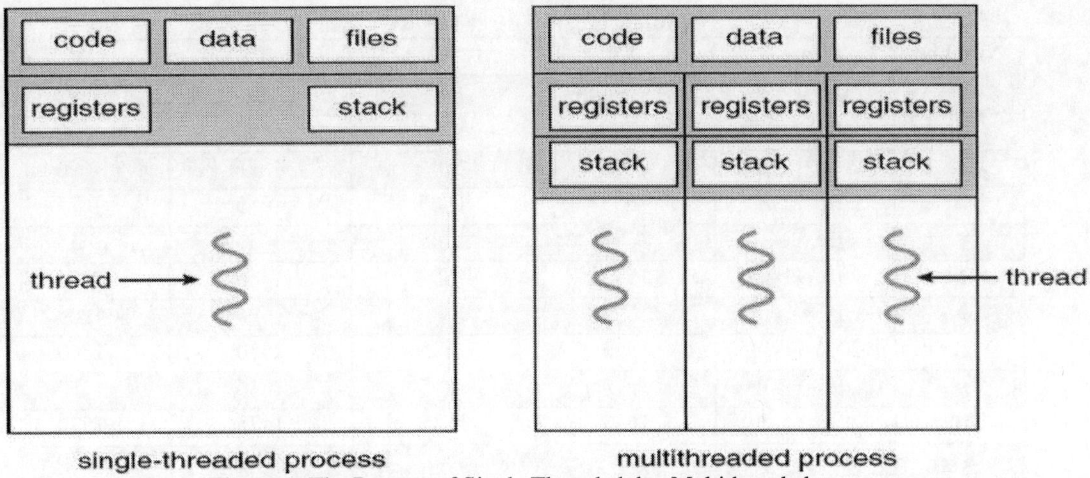

Figure 1. The Process of Single Threaded dan Multithreaded

The concept of multithreading consists of two that are concurrency and parallelism. Concurrency is a multithreading process that runs only on one processor where the processor can switch the execution on another thread in a very fast time and can be divided base on the number of cores that exist.

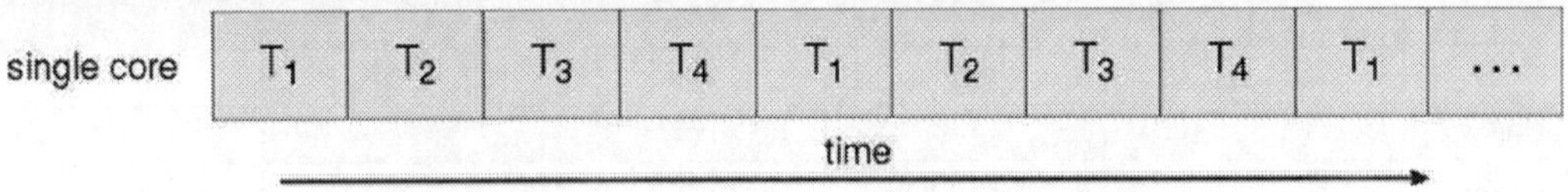

Figure 2. Execution Process on Single Core System

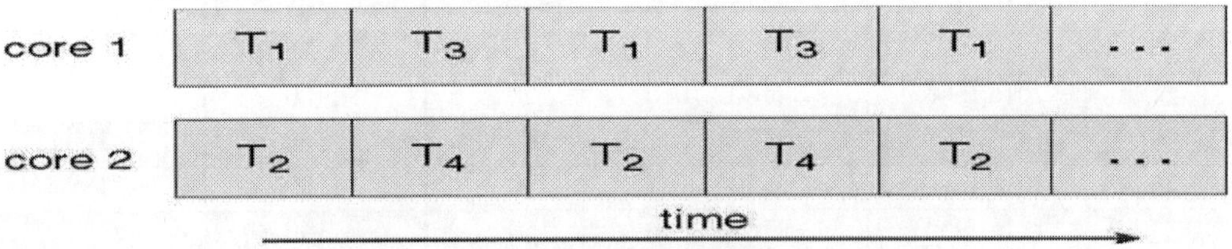

Figure 3. Execution Process on Mutli Core System

Parallelism is a multithreading process that can run more than one processor / multiprocessor simultaneously which can be called as parallel processing. Any thread on a different processor can run a concurrency process. Normally, parallel processing makes the program run faster because the performance of the processor is not split into each thread. Practically it would be quite difficult to divide the program on different cpu which is also run one data set without any connection between each parallel. Most computers have only one cpu, but there are also computers with multiple cpu, there is also the possibility to do the parallel processing by connecting computers on a computer network but it requires a pretty good software which is called distributed processing software.

3. Research Methodology

Backpropagaton method will consume so much time in learning many data so in order to improve the performance of this method the researcher will do multithreading with separating data into each thread. The total data that have been taken are 39 data from www.finance.yahoo.com for Adaro Energy Company(ADRO).

IOP Conf. Series: Journal of Physics: Conf. Series **930** (2017) 012041 doi:10.1088/1742-6596/930/1/012041

Table 1. Adaro Energy (ADRO) Data

Number	Date	X1	X2	X3	T
1	2016-12-30	1660	1705	1660	1695
2	2016-12-29	1700	1705	1680	1690
3	2016-12-28	1695	1700	1675	1690
4	2016-12-27	1655	1685	1655	1675
5	2016-12-23	1680	1705	1655	1655
6	2016-11-15	1595	1650	1465	1465
7	2016-11-14	1650	1680	1605	1605
8	2016-11-11	1590	1695	1550	1660
9	2016-11-10	1610	1685	1610	1650
10	2016-11-09	1635	1645	1475	1570
11	2016-11-08	1700	1720	1610	1630
12	2016-11-07	1685	1710	1675	1685
13	2016-10-26	1545	1595	1545	1555
14	2016-10-25	1540	1540	1520	1525
15	2016-10-24	1500	1540	1495	1525
16	2016-10-21	1510	1545	1450	1500
17	2016-10-20	1530	1540	1505	1510
18	2016-10-19	1485	1545	1485	1530
19	2016-10-18	1480	1520	1440	1440
20	2016-10-17	1435	1480	1435	1460
21	2016-10-14	1405	1425	1390	1425
22	2016-10-13	1415	1430	1405	1405
23	2016-09-06	1230	1245	1215	1225
24	2016-09-05	1180	1220	1180	1220
25	2016-09-02	1175	1190	1160	1170
26	2016-09-01	1150	1190	1135	1185
27	2016-08-31	1145	1190	1140	1150
28	2016-07-11	885	940	885	925
29	2016-07-01	860	895	860	885
30	2016-06-30	870	875	850	850
31	2016-06-29	865	885	845	855
32	2016-06-28	835	865	830	855
33	2016-06-27	830	840	820	835
34	2016-06-24	850	855	805	840
35	2016-06-03	760	825	760	815
36	2016-06-02	725	770	720	755
37	2016-06-01	710	725	705	720
38	2016-05-31	700	710	690	710
39	2016-05-30	700	710	695	695
39	2016-01-04	515	515	494	494

3.1. *Research Plan*

The researcher will compare the backpropagation with 2 thread learning process and the backpropagation with the normal learning process.

The value that will be used in backpropagation are :

- Weight and bias value = random value between (-0.5) – (0.5).
- Neuron input = 3.
- Hidden neuron = 4.
- Output neuron = 1.
- Error = 0.001.
- Max epoch = 100000.
- Learning rate = 1.

3.2. *Learning Process*

The data that has been randomized to be inserted into each thread will be normalized using the min-max normalization method. Chamidah said that the min-max method can increase the speed of convergence by scaling its value between 0 and 1.

$$x' = \frac{(x-a)}{b-a} \text{.............................(1)}$$

x' = normalization result, x = original value, a = lowest value, b = highest value.

The parameter will consist of :

- X1 = open price.
- X2 = highest price.
- X3 = lowest price.
- Y = target.

Each thread contains data in unorder sequence.

Learning process with normal backpropagation will run only on 1 thread. Learning process with 2 thread will start the learning of 2 thread in every epoch simultaneously. Every epoch in a thread can possibly stop the learning process if the error is lower than the expected. If not than the learning will continue to the next epoch again. The error in 2 learning thread will be compared in each epoch. The thread with the most lowest error will be selected. The selected thread will replace all its current weight and bias value to other thread except itself. This learning process of the thread will continue until the error is lower than the expected. If the learning process had done then the new data will be tested.

3.3. *Normalization process*

The normalization process using mix-max normalization will return a value between 0 – 1. The lowest value = 437 and the highest value = 1850.

Table 2. Data After Normalization

No	Date	X1	X2	X3	T
1	2016-12-30	$x' = \frac{(1660-437)}{1850-437}$	$x' = \frac{(1705-437)}{1850-437}$	$x' = \frac{(1660-437)}{1850-437}$	$x' = \frac{(1695-437)}{1850-437}$
2	2016-12-29	0.8903	0.8938	0.8762	0.8868
3	2016-12-28	0.8620	0.8832	0.8620	0.8762

4	2016-12-27	0.8797	0.8974	0.8620	0.8620
5	2016-12-23	0.8868	0.8903	0.8585	0.8797
			...		
39	2016-01-04	0.0552	0.0552	0.0403	0.0403

3.4. *Process of Separating Data*
The data that has been normalized will be shuffled and divided equally into each thread.

Table 3. Separating Data Into 3 Thread For 39 Data

P1	P2	P3
13 data	13 data	13 data

4. Result And Discussion

The research that has been done for stock predicting is used multithreading. The total data is 39 data which is took from www.finance.yahoo.com. The total testing are 5 times experiment with separate data into 2 thread.
The result of normal learning backpropagation for the smallest epoch are in the second testing = 356116 epoch and the biggest in the first testing = 587263. The average epoch is 453049.8
The result of 2 thread learning backpropagation for the smallest epoch are in the forth testing = 1295 epoch and the biggest in the second testing = 13600. The average epoch is 6490.

Tabel 4. Total Epoch for Normal Learning backpropagation & 2 Thread Learning

Model	Testing	Total Epoch
1 Thread / Normal Learning	1	587263
	2	356116
	3	364589
	4	572351
	5	384930
2 Thread	1	6650
	2	13600
	3	4970
	4	1295
	5	5935

5. Conclusion

Learning process in backpropagation using 2 thread with separate data can decrease the number of epoch. Data will be normalized before the learning process begin with min – max normalization which change the value between 0 and 1, then continue to separate the date into each thread. The result showed that the process of improvement is caused by the minimum error from 2 thread that has been compared to take the weight and bias value in each epoch for the most lowest error value and replace it on another thread.

6. Reference

[1] Gonzales J F 2012 *Java 7 Concurrency Cookbook* (Birmingham: Packt Publishing)
[2] Silberschatz A, Galvin P B & Gagne G 2013 *Operating System Concepts* (New Jersey: John Wiley & Sons)
[3] Khairani M 2014 *Improvisasi Backpropagation Menggunakan Penerapan Adaptive Learning Rate dan Parallel Training* (Medan: University of North Sumatera)
[4] Ghosh A W, Chakraborty M 2016 Hybrid optimized backpropagation learning algorithm for mutli-layer perceptron. International Journal of Computer Applications 57: 0975 - 8887
[5] Wagh S M & Pawar 2017 GPU parallelization of backpropagation neural network. International Journal of Engineering science and innovative technology 6: 2319 - 5967
[6] Chamidah N, Wiharto & Salamah 2015 Pengaruh Normalisasi data pada jaringan syaraf tiruan backpropagasi gradient descent adaptive gain(bpgdag). Jurnal Itsmart 1: 2301 - 7201
[7] Suresh S, Omkar S N & Mani V 2005 Paralel implementation of backpropagation algorithm in network of workstations. IEEE

Denni Algorithm An Enhanced Of SMS (Scan, Move and Sort) Algorithm

Denni Aprilsyah Lubis *, Opim Salim Sitompul, Marwan, Tulus, and M. Andri Budiman

*Planning and Development Agency Of Sibolga City Goverment Sibolga, Indonesia

*denni.lubis@gmail.com

Abstract. Sorting has been a profound area for the algorithmic researchers, and many resources are invested to suggest a more working sorting algorithm. For this purpose many existing sorting algorithms were observed in terms of the efficiency of the algorithmic complexity. Efficient sorting is important to optimize the use of other algorithms that require sorted lists to work correctly. Sorting has been considered as a fundamental problem in the study of algorithms that due to many reasons namely, the necessary to sort information is inherent in many applications, algorithms often use sorting as a key subroutine, in algorithm design there are many essential techniques represented in the body of sorting algorithms, and many engineering issues come to the fore when implementing sorting algorithms., Many algorithms are very well known for sorting the unordered lists, and one of the well-known algorithms that make the process of sorting to be more economical and efficient is SMS (Scan, Move and Sort) algorithm, an enhancement of Quicksort invented Rami Mansi in 2010. This paper presents a new sorting algorithm called Denni-algorithm. The Denni algorithm is considered as an enhancement on the SMS algorithm in average, and worst cases. The Denni algorithm is compared with the SMS algorithm and the results were promising.

1. Introduction

In 2010, Rami Mansi presents the SMS algorithm (Scan, Move and Sort), which is an enhanced of Quicksort algorithm [8]. Because the SMS algorithm is an enhanced of the Quicksort algorithm, the SMS algorithm categorized in Exchange Sort algorithm . In this paper, we presented a new sorting algorithm called Denni algorithm which is an enhanced of SMS algorithm. The study shows that the Denni algorithm is more efficient in terms of time complexity as compared to the SMS algorithm when dealing with a large size (n) of the input array, for the average and worst case.

2. Denni Algorithm

Denni algorithm consists of three procedures, Scan, Move, and Sort. The first procedure is (Scan), which scans the array and gives the values of the minimum of negative elements (minneg), the maximum of the negative elements (maxneg), the minimum of positive elements (minpos), the maximum of the positive elements (maxpos), the number of positive elements (NOP), and the number of negative elements (NON). Also, this procedure check whether the minimum value of negative equal to the maximum value of the negative and check whether the minimum value of positive equal to the maximum value of positive, if both conditions are true, then the input array is already sorted, otherwise, calls the procedure (Move).

The second procedure (Move) creates four temporary arrays, FreqPosArray of size (NOP), FreqNegArray of size (NON), PosArray of size (maxpos+1), and NegArray of size (|minneg|+1), and then initializes the PosArray, the NegArray, and the FreqArray with the value (minneg-1) to denoting the indices that will be skipped in the next phase. Then, this procedure distributes the elements on the four arrays, the positive elements are saved in the PosArray using the element itself as its index, the negative elements are saved in the NegArray using the absolute value of the element itself as its index, the frequent negative elements are saved in the FreqNegArray using a variable (j) as index and the frequent negative elements are saved in the FreqPosArray using a variable (i) as index, the variable I and j started from zero and incremented by one. To initialize and distribute positive elements, the iteration starts from (minpos) step incremented by one up to (maxpos), and iteration to initialize and distributing the negative elements, starts from (|maxneg|) step incremented by one up to (|minneg|).

The third procedure (Sort) copies the elements of the NegArray starting from the last index with ignoring the values of (minneg-1). Then it copies the elements of the PosArray starting from the first index with ignoring the values of (minneg-1). Iteration for copying the negative elements of NegArray starting from (|minneg|) step decremented by one up to (|maxneg|) and iteration for copying the positive elements of PosArray starting from (minpos) step incremented by one up to (maxpos). The copying is done on the original input array with overwriting the original values with the sorted values. After each copying operation of an element from the NegArray and the PosArray to the original array, the procedure searches the FreqPosArray and the FreqNegArray and then copies all element that are equal to the element that copied in the last copying operation (current element). if the current element is positive, the search will be performed on FreqPosArray and if the current element is negative then the search will be performed on FreqNegArray.

3. The Pseudocode Of Denni Algorithm

The pseudocode of the Scan procedure:

```
procedure scan(array, size)
if size > 1 then
    var a, maxpos, minpos, maxneg, minneg, NOP, NON
    maxpos:= 0
    minpos:=100000
    maxneg:=-100000
    minneg:= -1
    NOP:=0
    NON:=0
    for a:= 0 to size-1 do
        if array(a) >= 0 then
            NOP:= NOP+1
            if array(a) > maxpos then
                maxpos := array(a)
            end if
            if array(a) < minpos then
                minpos:=array(a)
            end if
        else
            NON:= NON+1
            if array(a) < minneg then
                minneg := array(a)
            end if
```

```
                if array(a) > maxneg then
                    maxneg:=array(a)
                end if
            end if
        end for
        if maxneg < minneg then
            maxneg = minneg
        if minpos > maxpos then
            minpos = maxpos
        if ((minpos ≠ maxpos)  and (minneg ≠ maxneg)) then
            Move(array, size, NOP, NON, maxpos, minpos, maxneg, minneg)
        end if
    end if
end procedure scan
```

The pseudocode of the Move procedure:

```
procedure move(array, size, NOP, NON, maxpos, minpos, maxneg, minneg)
var b,c,d,i,j
i:=0
j:=0
create a new array: FreqPosArray[NOP] and initialize by the value (minneg-1)
create a new array: FreqNegArray[NON] and initialize by the value (minneg-1)
if NOP > 0 then
    create a new array:PosArray[maxpos+1]
    for b:=minpos to maxpos do
        PosArray(b):= minneg-1
    end for
end if
if NON>0 then
    create a new array: NegArray[|minneg|+1]
    for c:= |maxneg| to |minneg| do
        NegArray(c):= minneg-1
    end for
end if
for d:= 0 to size-1 do
    if array(d) >= 0 then
        if PosArray(array(d))==minneg-1  then
                PosArray(array(d)):=array(d)
        else
                FreqPosArray(i):=array(d)
                i:=i+1
        end if
    else
        if NegArray(|array(d)|)==minneg-1  then
                NegArray(|array(d)|):= array(d)
        else
                FreqNegArray(j):= array(d)
                j:=j+1
        end if
    end if
end for
```

```
Sort (array, NegArray, PosArray, FreqPosArray, FreqNegArray, NON, NOP,
     maxpos, minpos, maxneg, minneg, i, j)
end procedure move
```

The pseudocode of the Sort procedure:

```
procedure sort(array, NegArray, PosArray, FreqPosArray, FreqNegArray, NON,
               NOP, maxpos, minpos, maxneg, minneg, i, j)
var index,x,y
index:=0
if NON > 0 then
   for x:= |minneg| downto |maxneg| do
      if NegArray(x) ≠ minneg-1 then
            array(index):= NegArray(x)
            index:= index+1
            for y:= 0 to j do
               if FreqNegArray(y)==array(index-1)   then
                  array(index):= FreqNegArray(y)
                  index:= index+1
               end if
            end for
      end if
   end for
end if
if NOP > 0 then
   for x:= minpos to maxpos do
      if PosArray(x) ≠ minneg-1 then
            array(index):= PosArray(x)
            index:= index+1
            for y:= 0 to i do
               if FreqPosArray(y)== array(index-1) then
                     array(index):=FreqPosArray(y)
                     index:= index+1
               end if
            end for
      end if
   end for
end if
end procedure sort
```

4. Comparison Denni Algorithm With SMS Algorithm

The main concept of the SMS algorithm and the Denni Algorithm exactly the same, which distributes the elements of the input array on additional temporary array. SMS Algorithm and Denni algorithm consists of three same procedures, Scan, Move and Sort. Differences of SMS algorithms and Denni Algorithm are as follows:

- The Scan procedure of the SMS algorithm useful only to get the values of the minimum (min), the maximum (max), the number of positive elements (NOP), and the number of negative elements of input array (NON) [8], however the Scan procedure of Denni algorithm useful to get minneg, maxneg, minpos, maxpos, NOP, and NON from the input array. The minneg, maxneg, minpos and maxpos useful to minimize the number of iteration on the Move procedure and Sort procedure in the SMS algorithm.

- The Move procedure of the SMS algortihm creates three temporary arrays, FreqArray of size (n) that contains all of frequent elements, PosArray of size (max+1) that contains positive elements, and NegArray of size (|min|+1) that contains positive elements [8], however The Move procedure of the Denni algortihm creates four temporary arrays, FreqPosArray size of (NOP) that contains frequent positive elements, FreqNegArray of size (NON) that contains frequent negative elements, PosArray of size (maxpos+1) that contains positive elements, and NegArray of size (|minneg|+1) that contains positive elements.
- On the procedure Move of the SMS algorithm, iteration to initialize PosArray starts from 0 (zero) step incremented by one up to (max), and iteration to initialize NegArray starts from 0 (zero) step incremented by one up to (|min|), but on the procedure Move of the Denni algorithm, the number of iteration is minimized, where to initialize PosArray, iteration starts from (minpos) step incremented by one up to (maxpos), and iteration to initialize NegArray starting from (|maxneg|) step incremented by one up to (|maxneg|).
- On the Sort procedure of the SMS algorithm, iteration for copying the elements of NegArray starting from (|min|) step decremented by one up to 0 (zero) and iteration for copying elements of PosArray starts from 0 (zero) step incremented by one up to (max). However, on Sort procedure of the Denni algorithm number of the iteration is minimized, iteration for copying elements of NegArray starts from (|minneg|) step decrement by one up to (|maxneg|) and iteration for copying elements of PosArray start from (minpos) step increment by one up to (maxpos).
- After each copying operation of an element from the NegArray and the PosArray to the original array, in the procedure sort of the SMS algorithm searches the FreqArray and copies all element that are equal to the element that copied in the last copying operation (current element), This resulted will be done iteration (n) times each found elements that not equal to (min-1). However, on the Sort procedure of the Denni algorithm number of the iteration is minimized, if the current element is positive, the search will be performed on FreqPosArray and if the current element is negative then the search will be performed on FreqNegArray, the number of elements in FreqPosArray (NOP) and FreqNegArray (NON) smaller than the number of elements on FreqArray (n), because FreqPosArray and FreqNegArray is simplification of FreqArray, where FreqPosArray contains the positive elements of FreqArray and FreqNegArray contains the negative elements of FreqArray (NOP+NON=n).
- The Denni algorithm is best used to sort an array of distinct elements. In this case, the value of (fneg and fpos) will be equals to 1, and the algorithm takes O(n+(maxpos-minpos)+(|minneg|-|maxneg|)) time.
- Denni algorithm is faster than SMS algorithm when dealing with a large size (n) of the input array, and when the values (maxpos/max) and (|minneg|/|min|) are much less than the value (n). In the best case, the SMS algorithm and the Dennni algorithm equally takes O(n) time, however in the average and worst case the Denni Algorithm takes O(n+(fpos*(maxpos–minpos))+(fneg*(|minneg|-|maxneg|))) time, while the SMS algorithm takes O(n+f*(nilai maksimum+|nilai minimum|)) time [8].
- In the other hand, SMS algorithm and Denni algorithm equally needs O(n+max+|min|+2) additional space.

5. Case Study

To prove that the proposed algorithm is faster than SMS algorithm when dealing with a large size of the input array, In this case study, the Denni and the SMS algorithms have been applied to sort a list of (100.000) elements for (10) sets of data and each set of data tested (10) times. Denni algorithm and SMS algorithm is implemented using Software (Bloodshed Dev-C + + version 4.9.9.2 GNU General Public License) and built-in functions (clock ()) is used to measure the elapsed time of both algorithms on the same computer using the same data set. The test results are as follows.

Table 1. Execution time for SMS and Denni algorithms for 100.000 elements of Input Array.

Set Of Data	Execution Time SMS Algorithm (ms)	Standart Deviation SMS Algorithm (ms)	Execution Time Denni Algorithm (ms)	Standart Deviation Denni Algorithm (ms)
1	8427	51	4303	62
2	7892	41	3995	52
3	7847	50	4006	49
4	7862	52	4033	67
5	7856	47	4016	46
6	7812	54	4010	66
7	7836	56	3986	81
8	7863	71	4001	54
9	7889	65	4049	71
10	7819	73	3994	71

6. Conclusions

In this paper presented a new sorting algorithm an enhancement of SMS algorithm is called Denni algorithm. Denni algorithm improve the efficiency of time complexity of SMS algorithm by minimizing iteration that occurs on the Move and Sort procedure. SMS algorithm and Denni algorithm equally takes O(n) time complexity in the best case. But in the average and worst case Denni algorithm is faster than SMS algorithm. In the average and worst case SMS algorithm takes O(n+f*(max+|min|)) time complexity, where f is the number of frequent elements, while Denni algorithm takes O(n+(fpos*(maxpos-minpos))+(fneg*(|minneg|-| maxneg|))) time complexity, where fpos is the number of frequent positive elements and fneg is the number of frequent negative elements.

In terms of space complexity, SMS algorithm and Denni Algorithm equally requires O(n+max+|min|+2) additional space to sort n elements.

References

[1] Alnihoud, J. & Mansi, R. Januari 2010. An Enhancement of Major Sorting Algorithms. *The International Arab Journal of Information Technology.* vol. 7. No. 1.

[2] Cormen, T., Leiserson, C., Rivest R. & Stein, C. 2001. *Introduction to Algorithms*. McGraw Hill.

[3] Dean, C. 2006. A Simple Expected Running Time Analysis for Randomized Divide and Conquer Algorithms. *Computer Journal of Discrete Applied Mathematics.* vol. 154. no. 1. pp. 1-5.

[4) Deependra, Kr. Dwivedi. 2011. Comparison Analysis of Best Sorting Algorithms. *VSRD-JCSIT.* vol. 1 (4). Pp. 261-267.

[5]. Hoare, R. 1962. Quicksort. *The Computer Journal*. pp. 10-15.

[6]. Kruse, R. & Ryba, A. 1999. Data Structures and Program Design in C++, Prentice Hall.

[7]. Knuth, E. 1998. *The Art of Computer Programming Sorting and Searching, 2nd edition.* Addison-Wesley.

[8]. Rami Mansi. April 2010. Enhanced Quicksort Algorithm. *The International Arab Journal of Information Technology.* vol. 7 no. 2.

IOP Conf. Series: Journal of Physics: Conf. Series **930** (2017) 012043 doi:10.1088/1742-6596/930/1/012043

FUZZY RULE SURAM FOR WOOD DRYING

Zakarias Situmorang

Faculty of Computer Science, Catholic University of Sain Thomas Medan Jl. Setiabudi No. 479-F Tanjungsari Medan 20132

zakarias65@yahoo.com

Abstract. Implemented of fuzzy rule must used a look-up table as defuzzification analysis. Look-up table is the actuator plant to doing the value of fuzzification. Rule suram based of fuzzy logic with variables of weather is temperature ambient and humidity ambient, it implemented for wood drying process. The membership function of variable of state represented in error value and change error with typical map of triangle and map of trapezium. Result of analysis to reach 4 fuzzy rule in 81 conditions to control the output system can be constructed in number of way of weather and conditions of air. It used to minimum of the consumption of electric energy by heater. One cycle of schedule drying is a serial of condition of chamber to process as use as a wood species.

Keywords : Fuzzy rule, look-up table, controller

1. Introduction

The wood drying process used the schedule of drying dependent for moisture of content in the wood, that condition of kiln in temperature and humidity of chamber. The controller used to control the actuators are heater, sprayer and damper, whenever the process used doing the optimal from time and energy and stability in wood schedule of drying. Main source of energy is solar energy from collector and alternative source energy by heater. Number of solar energy based of intensity of solar and alternative energy by heater is consumption of electric (Wang, X,G, et al, 2001)

The maximized use of solar energy in wood drying process is goal of control system. It's depended by a number solar energy and it's change and temperature of ambient. Responsibility of the change the solar energy in variable of temperature ambient and humidity ambient is the especial of goal the control system. The process of control is to hope maximized the use of solar energy and the minimized of consumption energy of electric. (Skuratov. N.V., 2003)

Process of wood drying is depended for schedule drying, which used to track of set point for temperature and humidity drying. The conditions of temperature and humidity drying in schedule are different for the each steps of the schedule drying of wood. Variable control the Wood drying of process kiln are temperature and humidity of air in chamber where dependent for moisture of content the wood. It's need to control of actuator system for heater, sprayer and damper, whenever the process used doing the optimal from time and energy and the actuator doing in the conditions data riel. (Lim P.K. and Natalie R.S,1995)

The Fuzzy rule suram implemented for schedule drying of Albasia Albizia wood and modification of membership function in range [0.5, 1]. (Situmorang,Z, et.all, 2009).

2. Wood Drying Process

Control variable of a solar energy wood drying process kiln is temperature and humidity to adapt variable of a drying schedule. Dimension of wood drying kiln has designed and built several type dry kiln for use lumber of housing structure. The Schedule drying is a cycle of drying and have the some level of process. A Level process doing at temperature and humidity variable are constant at set point any time. By the way need an

actuator control system (heater, sprayer and damper) then doing at effective the time and efficiency of energy (Garrahan.P., and Kidlark.J, 2001).
Air drying involves the open piling of freshly processed timber in stacks out of doors or in open sheds so that the wood surfaces are exposed to the surrounding atmosphere. During air drying there is little control over the factors that influence drying and hence the rate of drying is very much controlled by local atmospheric conditions. One guiding principle during air drying of timber is to ensure adequate air circulation through the stack. In practice this is achieved by separating each board within the stack using sticks or stickers
A drying schedule is a set of directions for the operation of a kiln during the drying period and is usually presented in the form of tables showing the temperatures and humidity to be used at various stages of the process. These schedules vary with species, size and grade of timber being dried; they are also influenced by local production and selling practices, by the degree of care in kiln operation and in kiln characteristics.

3. Solar Energy in a Prototyping of Wood Drying Kiln

Mearsurement of solar energy by Piranometer type MS-801 Chino, it has maximum voltage +50 mVDC, which to have a data from Agency of meteorology and geophysics Yogyakarta Indonesia , its shown in figure 1. with detail of a prototyping of wood drying

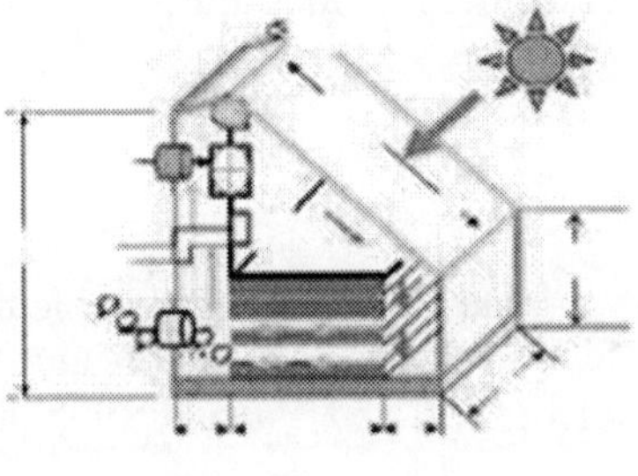

Figure 1. Detail of design the prototyping of wood drying kiln
(Situmorang, Z, et.all, 2009 and Haque.M.N., 2002)

Most important part of a dehumidifying kiln is a heat pump. Air in the kiln passes through the load of timber and absorbs moisture from the wood. Part of that air circulates through the heat pump where the moisture is condensed and drained out of the chamber. Dried, reheated air comes back to kiln chamber. Energy consumption is minimal, since there is almost no inner/outer air exchange. Air circulation in these kilns is smaller than in conventional kilns while the electrical power of flow fans and heat pump equals approximately the power of flow fans in conventional kilns. (Virginia Tech, 2007)

A number of variable solar radiation can be convert to variable temperature ambient and humidity ambient shown in table 1. (Dion,J.M.et.al, 1991)

Table 1. Convert solar radiation to variable temp. ambient and humidity for Albasia lbizia wood

No	Set Point Td_0 (°C)	Set Point Hd_0 (%)	MC (%)	Solar Radiation I_o (Watt/m²)	Temp. Ambient Ta (°C)			Hum. A mbient Ha (°C)		
					Min	Rate	Max	Min	Rate	Max
1	2	3	4	6	7	8	9	10	11	12
1	45	60	25.5-50	561.0	29.1	30.1	31.1	52	60	68
2	50	55	10.4-25.5	662.0	29.7	30.7	31.7	46	54	62
3	55	50	7.8-10.4	763.0	30.5	31.5	32.5	40	48	56
		45	7.3-7.8							
4	60	40	6.7-7.3	617.5	29.4	30.4	31.4	50	58	66
		35	6.4-6.7							
5	65 (F)	60	6.3-6.4	535.0	28.9	29.9	30.9	54	62	70

Note:

i. **For MC ≥ 25 %,(Td_0 = 45 s/d 55), The Solar Radiation** I_0 □ 460 □ 20.2(Td_o □ 40) **(watt/m²)**
ii. **For MC** □ **25 %,(Td_0 = 60 s/d 70), The Solar Radiation** I_0 □ 700 □ 16.5(Td_o □ 55) **(watt/ m²)**
iii. **Control priode for total used the energi from solar ,** Cp □ 0.1 □ 0.0003(I_0 □ 400)**(C/menit)**

F = Fakultatif

4. Design Membership Functions

Design of membership function for variables in fuzzy logic implemented to mapp of typical of triangle and trapezium as of a drying schedule as shown in figure 1. and table 1 It's needed to implemented control system of wood drying kiln (Klir, G.J, and Yuan.B, 1995 and Laurenzi. W., Popa.V, Comsa.Gh, 2003).

Fuzzification of membership function in range **[0.5, 1]**, and for weather (temperature of ambient Ta and change temperature ambient CTa) for set point temperature drying Td_0 = 45°C and Humidity drying Rd_0 = 60%) describe by figure 2. with variable M : over-cloudy ; B : cloudy; CB : bright-cloud ; C: clear; SC : clearest and for change of temperature ambient used -H = - High, -M = -Medium, -S = - Small, Z = Zero, S = Small, M = Medium, H = High

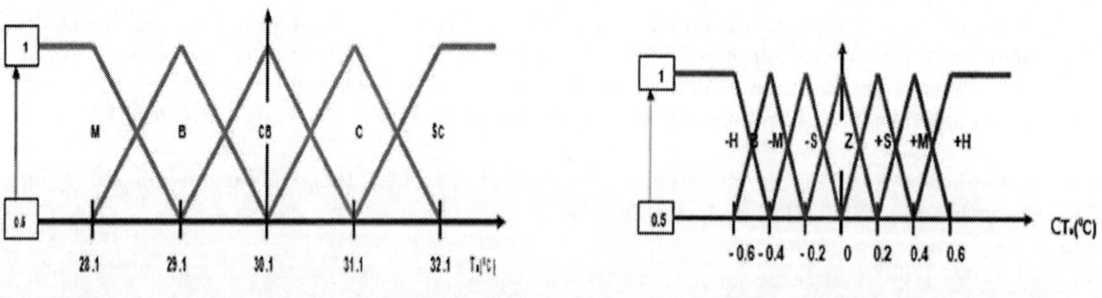

Figure. 2. Membership function of temperature ambient and change temperature ambient for Td_0 = 45°C and Rd_0 = 60%

Computation process of variable change of temperature ambient are given eq.1. with n □ 0 s/d □□

$$\mathbf{CTa\,[(n+1)T] = Ta[(n+1)T] - Ta[nT]}$$

Representative of weather is variable temperature ambient and this change used to maximize and to hoist responsibility of membership Function, with appropriate rule in a table look-up scheme, shown table 2.

Table 2. Look-up table for membership Function for temperature of ambient

CT \ T_a	**Over-cloudy (M)**	**Cloudy (B)**	**Bright-Cloud (CB)**	**Clear (C)**	**Clearest (SC)**
-H	Over-cloudy	Over-cloudy	Cloudy	Bright-Cloud	Clear
-M	Over-cloudy	Cloudy	Bright-Cloud	Clear	Clearest
-S	Over-cloudy	Cloudy	Bright-Cloud	Clear	Clearest
Z	Over-cloudy	Cloudy	Bright-Cloud	Clear	Clearest
+S	Over-cloudy	Cloudy	Bright-Cloud	Clear	Clearest
+M	Over-cloudy	Cloudy	Bright-Cloud	Clear	Clearest
+H	Cloudy	Bright-Cloud	Clear	Clearest	Clearest

Fuzzyfikasi of membership function in range [**0.5, 1**] of conditions of air (humidity of ambient Ha and change of humidity ambient CHa) for set point temperature drying ($Td_0 = 45^0C$) and humidity drying ($Rd_0 = 60\%$) describe by figure 3. with variable : P : Hot; AP : Rather-Hot; H : Swarm; S: Fresh; D : Cold, and for change of humidity ambient used -H = - High, -M = -Medium, -S = - Small, Z = Zero, S = Small, M = Medium, H = High.

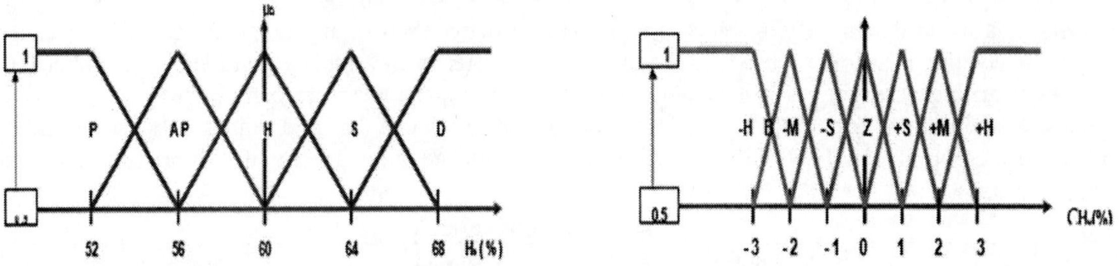

Figure 3. Membership Function of humidity ambient and change humidity for $Td_0 = 45^0C$ and $Rd_0 = 60\%$

Computation process of variable change of humidity ambient are given eq.2. with n □ 0 to □□ □

$$\mathbf{CHa\ [(n+1)T] = Ha[(n+1)T] - Ha[nT] \ldots \quad .(2)}$$

Representative of conditions of air in variable humidity ambient and this change used to maximum of membership function, with appropriate rule in a table look-up scheme, shown table 3. For other step will be to adjust at table 1.

Table 3. Look-up table for membership Function for humidity of ambient

H_a / CH_a	**Hot**	**Rather-Hots**	**Swarm**	**Fresh**	**Cold**
-H	**Hot**	Hot	Rather-Hot	Swarm	Sejuk
-M	**Hot**	Rather-Hot	Swarm	Fresh	Cold
-S	**Hot**	Rather-Hot	Swarm	Fresh	Cold
Z	**Hot**	Rather-Hot	Swarm	Fresh	Cold
+S	**Hot**	Rather-Hot	Swarm	Fresh	Cold
+M	**Hot**	Rather-Hot	Swarm	Fresh	Cold
+H	**Hot**	Swarm	Fresh	Cold	Cold

5. Implementation Control System

Automatic control unit has built-in programs for drying of all wood types regardless of a starting wood moisture content. It controls drying process automatically, so the operator presence is not needed during drying. Larger energy consumption is only during first day during heating stage when electrical heaters are turned on until working temperature is reached. Later, during drying stage these heaters are rarely turn on.

If a boiler installation exists in the drying complex, it makes sense to mount heat exchangers for hot water or steam (depending on the boiler) in the kiln. In that case, energy from the boiler would be used for heating of the drying chamber. That way kiln becomes combined (automatically uses boiler heat). This option is interesting only for a large capacity kilns and in the cases where electrical power supply is unstable (large voltage or current oscillation or frequent power cut-downs during winter periods). Drying is in that case performed in conventional way, heat pump is turned-off, and electrical energy is used only for flow fans. Combined kiln unites good characteristics of both dehumidifying and conventional kiln. That provides the greatest drying quality with the least energy consumption for the shortest time.

Drying quality in dehumidifying kilns is significantly better than in any conventional kiln, because the drying is equable in whole timber load, so there is no danger of developing degrade in the timber (cracks, checks, warping...). drying is fast enough since the working temperature can reach up 60 ^{0}C to. These kilns are environment friendly because all wastes that are produced during drying process are harmless for water and environment. Especially good results are achieved with drying of natural (untreated) wood (beech, ash,...).

Table 4. Input variable

No	Variable	Range	Describe
1	Temperature Drying Td	0 - 150 ^{0}C	
2	Temperature Ambient Ta	0 - 150 ^{0}C	Weather
3	Humidity Drying Rd	0 - 100 %	

4	Humidity Ambient	Ha	0 - 100 %	Conditions of air

Table 5. Output variable and implemented the fuzzy rule suram

No	Rule	Actuator			Conditions
		Heater	**Damper**	**Sprayer**	
1	**SUR-AM-1**	**off**	**off**	S	To drop of Temperature Drying Td and to hoist of Humidity Drying Rd by very suddenly [Heating- Process]
2	**SUR-AM-2**	H	**off**	**off**	To hoist of Temperature Drying by very suddenly and to drop of humidity drying Rd.
3	**SUR-AM-3**	H	D_3	**off**	To hoist of Temperature Drying Td and to stay of Humidity Drying Rd.
4	**SUR-AM-4**	H	D_2	**S**	To hoist of Temperature Drying and to drop of humidity drying Rd.
5	**SUR-AM-5**	H_1	D_3	**off**	To drop of Temperature Drying Td and to stay of Humidity Drying Rd
6	**SUR-AM-6**	H_2	off	S	To drop of Temperature Drying Td and to drop of Humidity Drying Rd
7	**SUR-AM-7**	off	off	**off**	To stay of Temperature Drying Td and to stay of Humidity Drying Rd
8	**SUR-AM-8**	H	D_1	**off**	To stay of Temperature Drying Td and to adjust Humidity Drying Rd with Humidity ambient Ha [Equalizing-Process]

Note: D_1 : Damper ON : 5 menit H : Heater ON : 15 meni
D_2 : Damper ON : 2 x 1 menit H_1 : Heater ON : 10 menit
D_3 : Damper ON : 1 x 1 menit H_2 : Heater ON : 5 menit
S : Sprayer ON : 1 x 1 menit

Result of Implementation the fuzzy rule suram for measurement of temperature drying for schedule Albasia Albizia wood drying, with specification : wood dimension : 4 x 6 cm, amount : 22 item. The figure 4 shown , that measurement of temperature drying can be to follow schedule drying.

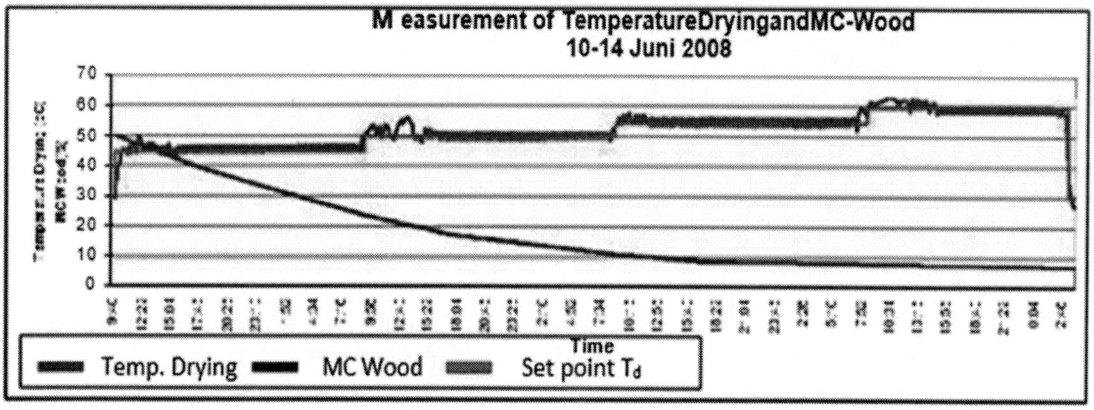

Figure 4. Measurement of temperature drying and MC-Wood at 10-14 Juni 2016

6. Conclusion

Rule suram based of fuzzy logic with variables of weather is temperature ambient andconditions of air is humidity ambient, and it can implemented for wood drying process. The membership function of variable of state represented in error value and change error with typical of triangle and trapezium. Result from analysis and evaluation, the8 fuzzy rule to control for output system can be constructed in a number of way of weather and conditions of air. It used to minimum of the consumption of electric energy by heater. The rule suram used to stability and equilibrium of schedule of drying in chamber by control of temperature and humidity. The result of implemented of fuzzy rule suram with the modification of membership function in range [0.5,1] represented approximate to he conditions riel are near enough and the actuator doing in the conditions data riel

Reference

[1] Dion,J.M., Dugard.L, Pranco.A, Tri N.M., Horwood. J.W., 1991, 'MIMO Adaftive Constrained Predictive Control Case Study : An Environmental Test Chamber', Automatica, 27, 1991,611-626, Pergamon-Press

[2] Garrahan.P., and Kidlark.J, 2001, 'Drying Hardwoods - Northern Hardwood Initiative :Industry Resource Manual', Forintek Canada Corp, 2665 East Mall, Vancauver, B.C. V6I 1W5.

[3] Haque.M.N., 2002, 'Modelling of Solar Kilns and The Development of An Optimised Schedule for Drying Hardwood Timber', Thesis Ph.D., Department of Chemical Engineering, University of Sydney., 354p.

[4] Klir, G.J, and Yuan.B, 1995, 'Fuzzy sets and Fuzzy Logic Theory and Applications', Prentice-Hall International, Inc, New York

[5] Laurenzi. W., Popa.V, Comsa.Gh, 2003, 'Computer Aided Design Of Drying Schedules In Order To Control The Drying Process', 8th International IUFRO Wood Drying Conference, p302 - 307.

[6] Lim P.K. and Natalie R.S,1995, 'Climate control of a storage chamber using fuzzy logic';

Proceedings of the 2nd ANNES IEEE.
[7] Situmorang, Z., Wardoyo.R, Hartati.S, Estiyanto.J.E, 2009,
[8] Skuratov. N.V., 2003, 'Computer Simulation and Dry Kiln Control', 8th International IUFRO Wood Drying Conference, 406 - 412
[9] Virginia Tech, 2007, 'Design Of Solar-Heated Dry Kiln',
[10] Wang, X.G. Liu, W. Gu, L. Sun, C.J. Gu, C.E. de Silva, C.W., 2001, 'Development of An Intelligent Control System for wood drying proceeses', Advanced Intelligent Mechatronics Proceedings.IEEE/ASME International Conference. Vol.I,, 371 - 376.

]IMPLEMENTATION ANALYSIS OF CUTTING TOOL CARBIDE WITH CAST IRON MATERIAL S45 C ON UNIVERSAL LATHE

Junaidi [1]Soni hestukoro [2]Ahmad yanie [3]Jumadi [4]Eddy [5]

[1]Junaidi Departemen Machine Engeneering Sekolah Tinggi Teknik Harapan
[2]Soni Hestukoro Departemen Machine Engeneering Poletecnic Negeri Medan
[3]Ahmad Yanie Departemen Electro Engeneering Sekolah Tinggi Teknik Harapan
[4]Jumadi Departemen Industri Engeneering Sekolah Tinggi Teknik Harapan
[5]Eddy Departemen Industri Engeneering Sekolah Tinggi Teknik Harapan

Email :Junaidi413@yahoo.com

Abstract. Cutting tool is the tools lathe .Cutting process tool CARBIDE with Cast Iron Material Universal Lathe which is commonly found at Analysiscutting Process by some aspects numely Cutting force ,Cutting Speed,Cutting Power,Cutting Indication Power ,Temperature Zone 1 and Temperatur Zone 2. Purpose of this Study was to determine how big the cutting Speed ,Cutting Power, electromotor Power,Temperatur Zone 1 and Temperatur Zone 2 that drives the chisel cutting CARBIDE in the Process of tur ning Cast Iron Material .Cutting force obtained from image analysis relationship between the recommended Component Cuting Force with plane of the cut and Cutting Speed obtained from image analysis of relationships between the recommended Cutting Speed Feed rate.
Keywords: Cutting Force,Cutting Speed,Cutting Power, Electromotor Power ,Temperatur Zone 1,Temperature Zone 2

1. Introduction

In the process of cutting, cutting tool moves relatively to workpiece and separates part of workpiece material, commonly called by chips. Part of cutting tool feeds into the workpiece material called by cutting element of the cutting tool. Turning process is the machining to produce parts of engine which is generally cylindrical(3). The basic principleof machining process outer and inner surface is cylindrical such as shafts, holes /drill, threaded and tapered. In the machining, the function of axis is to transmit power and circle, based on the function axis is designed to be strong in accepting the load. The axis has shaft strength and hardness so the material used is made of carbon steel S 45 C (2). Generally the process of making the axis is done in the lathe, using cutting tools on a rotating workpiece. Cutting tools is the most critical part of a machining process.

Material, parameter and geometry of cutting tool and cutting style will determine a machining process and affect the power of cutting tools. In a machining process cutting tool always changes. Cutting tool is a production component which can be wear and the price is relatively expensive. Cutting tool will be wear after being used for cutting. More wear, the cutting tool will be in critical condition. If it uses continuously the cutting tool wear will be faster, and someday cutting edge will be broken at all. The broken should be avoided to cutting tool, machine tool, workpiece because it can endanger to the operator, and also affect on the geometric and quality of production(3). Basically the wear will determine the limits of cutting cut power. The selection inappropriate type of cutting tool, workpiece material and cutting conditions can affect strength of the cutting tool. Therefore it is important to know the type of cutting tool, workpiece material and cutting conditions (cutting speed, depth of cut and feeding movement) on the cutting tool wear. Cutting speed can't be chosen randomly, if the cutting speed is low, it will take time a long to do it. If the speed is too high cutting tool will lose hardness it is one the heating effect soferates in fast time ,it will if the cutting tool operates in fast time,it will take the short life time.

Based on the background and the problems above, the formulations of the problems in this study are:

1. How do we know the cutting tool and the material used which relates to the counting formulation?
2. How do we choose a counting formulation for cutting tool and material?This paper presents the type of cutting tool and material that related the counting formulate to finish the cutting process corecttly and accurattly.

The deviation factors in cutting metal process is setting of machine tools, measurement methods, the movement of machine tools, the wear of cutting tool, temperature, and cutting styles. Rochim (2003) states that the wear of cutting tool is affected by the geometry of cutting tool, beside that it is influenced by all factors related to all machining processes, such as the types of workpiece material and cutting tool, cutting conditions (cutting speed, cutting depth and feeding movement), liquid cooling, and type of machining process. When cutting takes place, high temperature will occur in the cutting edge. This heat will flow in part to growl, workpiece and cutting tool. The heat caused by friction in transmission power system from machine tools (gears) will flow to the components so it will occur the difference temperature between the parts of the machine which are not the same size, the result from this is deformation. May be the spindle of the lathe is not the same level with the table or there is changing heath. Although the deformation is small but we should consider if we want to create the ideal product. Therefore to reduce geometric errors as the result of deformation from this temperature, it is usually warmed up the engine before production starts.

The strength and rigidity of the machine tool and workpiece are very important to reduce deformation caused by forces when cutting occurs. Bending occurs in the workpiece or other engine parts will reduce the accuracy of product. In the turning process, there is a cutting force such as radial force (the force on the depth of cut), tangesial force (force in cutting speed), and the longitudinal force (force funerals). Many factors affects the cutting forces like depth of cut, feed rate and the cutting speed, working forces can also be determined by empirical formulation such as specific cutting force. Specific cutting force (ks) is a numbers of force or energy required to move a unit volume of metal called specific cutting force or specific cutting energy. The relationship specific cutting energy with a workpiece material and tensile strength can be shown in table 1(1)

Table 1.Specific Cutting Force (1)

Material of workpiece	σb (Kg/mm^2)	*K*s	Material of workpiece	BHN	*K*s
Steel	30-40	132	Cast Iron	140-160	81
	40-50	145		160-180	86
	50-60	157		180-200	92
	60-70	170		200-220	98
	70-80	191		220-240	104
	80-90	200		240-260	108
	90-100	225			
	100-110	240			
	110-120	260			

From Table 1 above we can show the correlation between workpiece material, the tensile strength (σb) and specific cutting force (*Ks*). Tensile strength (σb) is close relation to carbon steel. Carbon steel is given every symbol which relate directly to the standard, heating performance and tensile strength as shown in Table 2 carbon steel for machine construction and steel rod is cold difinis to the shaft.(2)

Table 2. Carbon Steel For machine Contruction and difinis Steel rod to the exis.(2)

Standar and type	**symbol**	**Treatment Hot**	**Tensile strength (Kg/mm^2)**	**Information**
Carbon steel construction machinery (JIS G 4501)	**S30C**	**Penormalan**	**48**	
	S35C	,,	**52**	
	S40C	,,	**55**	
	S45C	,,	**58**	
	S50C	,,	**62**	
	S55C	,,	**66**	

Stainless steel rods	**S 35 C-D** **S 45 C-D** **S 55 C-D**		**55** **60** **72**	Pulled cold, digrinding, in-lathe, or connect between these things

The shape and size of the cutting cross affect the specific cutting force (ks), as well as the main cutting force Fz. A Specific cutting g style will decrease by increasing in cutting cross (see chart 10.1). The price of Ks will also be affected by the comparison of depth of cut, a, to feeding S.

Cutting speed affects little cutting forces. In cutting speeds below 75 m/ minutes, the cutting forces will come down with increasing rise and then a constant cutting speed if the speed is above 75 m/ minutes. It is the reason why the carbide tool has a constant cutting force unaffected by the cutting speed. The correlation between cutting speed and feeding S for tool life of 60 minutes, 240 minutes and 480 minutes as shown in .

The varieties of cutting tools determine tool life and final result from workpiece surface. The cutting edge angle in cutting tool edge is called cutting edge angle and the composition is also called cutting geometry. The composition of cutting edge angle and fingers of cutting edge are called by tool signature. Tool signature of a single point tool usually consists of seven elements, they are:

1. Back rake angle
2. The side rake angle
3. The end relief angle
4. The side relief angle
5. The end cutting edge angle
6. The side cutting edge angle
7. Nose radius

2. Research Methods

The implementation of this research will be conducted to purchase cast iron material and cutting tool HSS in selling place in Medan. And the implementation will be carried out in production process laboratory at STT Harapan Medan. It will be done in December 2016 until the end of February 2017. The process of making object test Cast iron which is larger than 3 x 140 mm and 20 mm diameter, cut into 3 parts. Then making process object test by using machine tools (milling, shaping, and drilling machine) to get shape and size object test is done by cold work, so it can be considered no changes in microstructure, deformation, plastic or residual stress (residual stress cause by the manufacturing process). Then 3 part of cutting tool is also geared up with grinder machine one by one. The next the biased of cutting tool is cut in turning machine.

Then it is installed by using a wrench, after that workpiece is installed by using dial indicator. Then turning machine is turned on by cutting the workpiece, try to move automatically and turn on for 1 hour or 60 seconds. In turning process we observe automatically how the condition of workpiece in turning process, whether cutting tool still have function or not, if it is not, we replace with the new sharpened cutting tool. Then we analyze several times to change cutting tool for 1 hour and how the condition of the axis whether it is smooth or rough, next we do the second test materials such as the first job by rotation and the same speed as the first job but different time that is 4 hours or 240 minutes. And the third job as the first job with different time that is of 8 hours or 480 minutes.

3. Data Analysis

Material Cutting Analysis (10)

$$Pc = \frac{Fc \times Vc}{4500}$$

Fc = Cutting Force (kg)
Pc=CuttingPower(Hp)
Vc = Cutting Speed (m/menit)
Pg = Pc + P idd

ηmk

Pg = Elektromotor Power (Hp)

ηmk = Mekanis Effesienci (%)

Pidd = Indication Power (Hp)

$\tan\theta = \frac{rc \times \cos\delta}{1 - rc \times \sin\delta}$

rc = cutting ratio 0,3

δ = in this from tool signatur.

$\theta + \beta - \delta = 45^0$

Tangencial Force (Ft)

$Ft = Fc \tan(\beta - \delta)$............ (Kg)

Scisories Force (Fs) $Fs = Fc \cos\theta - Ft \sin\theta$ ....(Kg)

Normality in Scisorse Force (Fns)

$Fns = Fc \tan(\beta - \delta + \theta) = Fc \tan 45^0$ ....(Kg)

Resultan Force (Fv) $Fv = \frac{Fs}{\cos(\beta-\delta+\theta)} = \frac{}{\cos 45^0}$

Friction Force (Ff)

$Ff = Fv \sin\beta$ ....(Kg)

Normal Force(Fn)

$Fn = \underline{Ff} / \tan\beta$

Friction Factore (η)

$\eta = \tan\beta$

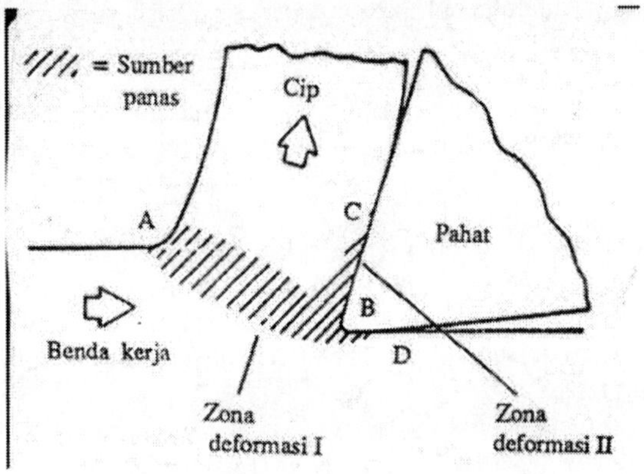

Figure 1. Tool Design
(**Fundamentals of Tool Design**:Syamsir A.Muin,page.151)

4. Temperatur in first Deformasi Zone 1 (1)

The amountof heat comes from first deformation zone is Qs and some of this heat is I (read gamma) conducted on the workpiece.The number (1-I) Qs transformed
with chip, so the increasing temperature from material through the first deformation zone are:

$$\Delta ts = \frac{(1-\Gamma)Qs}{\partial \,.\, Cp \,.\, V \,.\, ac \,.\, B}$$

Dimana :

B = Plane of the cut. (ft)

ac = dept of cut (mm)

Cp = Specifik Heat (Joule/kg ^{0}C]

Γ = From in Heatreatment

Γ = From Chart with looking for R tan Ø

R = Thermal Number

$$R = \frac{\rho \,.\, Cp \,.\, V \,.\, ac}{K}$$

K = Konduktivitas panas (Joule/m C)

ρ = Materials Weight (kg/m^3)

Wemercan do this equation including to determine some requirements to workpiece, he produces an equation which states (part of Qs is conditioned on the workpiece) as a unique function of R tan Ø
Ø = shear angle.

The theoreticalrelationship between I and R tan Ø in comparison with experimental data. it can be seen that a theory I ignore the price of a high R tan Ø is high speed and feed.In theory assumes plane heat source, heat can only flow in the workpiece by conduction; in fact the heat produced covers a large zone, most of it goes into the workpiece. The effects of spreading heat become very important in Maximum Temperature(Ø max)
Ø max = Ø m + Ø s + Ø0
Ø m = Temperature high up (^{0}c)
Ø m = it can with know lf/lo dan W0
Lo= Hotter long source Lo = lf / lo = $\frac{lf \times rc}{ac}$ ao = ac /rc
Wo = Konstanta 0,2
Ø0=HomeTemperature (27s/d30)^{0}C
In chart 6.10shows the effect of large variations in distribution of heat source which isn't uniform. if this curve is used, then Lo can be estimated from the wear on the tool face and wide from the heat source can be estimated from a micrograph photograph of chip.

5.Result

Known example data no.1,known

rc=0,3,Ff=674N,ρ=7800kg/m^3,
K= 43 J/m^{0}C ,Cp = 0,473 kj/kg^{0}C ,B = 2,5 mm , Lf = 7,5 mm ,ώ0 = 0,24 , θo = 28 ^{0}C
so ;

IOP Conf. Series: Journal of Physics: Conf. Series **930** (2017) 012044 doi:10.1088/1742-6596/930/1/012044

Varedire Force

Fv = 279,73 kg / 0,7071 = 395,6 kg

Resultan Energi (Pm)

From no1 Vc can = 190 m/menit = 3.16 m/detik.

Fc = 350 kg = 3430 Newton

Pm = Fc x Vc = 3430 x 3,16 m/detik = 10838,8 Nm/detik = 10838,8 j/detik

Pm = 45161,67 kal/detik = 45,16 kkal/detik.

Hotter in friction force (Pf)

Pf = Ff x V0 = Ff xVc x rc = 674 N x 3,16 m/detik x 0,3 = 638,952 Nm/detik

Pf = 638,952 j/detik = 2662,11 kal/detik = 2,662 kkal/detik

Hotter in Scisorse force(Ps)

Ps = Pm – Pf = 45,16 kkal/detik – 2,662 kkal/detik = 42,5 kkal/detik

$$R = \frac{\rho \times Cp \times V \times a}{K} =$$

$$= \frac{7800\ kg/m^3 \times 473\ J/kg^0C \times 3{,}16\ m/det \times 2{,}54.10^{-3} m}{43\ j/m^oC}$$

R = 688,6

$\tan \Theta = \tan 17{,}2 = 0{,}32 \rightarrow R \tan \Theta = 688{,}6 \times 0{,}32 = 213{,}952$

From grafic 4 ,with R tan 17,72 ,didapat Г =0

So :

$$\theta s = \frac{(1-\Gamma) \times Ps}{\rho \times Cp \times a \times V \times B} = \frac{(1-0) \times 42{,}5\ kkal/det}{7800\ kg/m^3 \times 473 j/kg^0C \times 2{,}54.10^3 m \times 3{,}16\ m/det \times 2{,}5.10^{-3} m}$$

$$\theta s = \frac{178.5\ J/det}{74\ J/det\ ^0C} = 2{,}41$$

$$\theta f = \frac{Pf}{\rho \times Cp \times a \times V \times B} = \frac{2{,}662\ kkal/det}{7800\ kg/m^3 \times 473\ j/kg^0C \times 2{,}54.10^{-3} \times 3{,}16\ m/det \times 2{,}5.10^{-3} m}$$

$$\theta f = \frac{2{,}662\ j/det}{74\ j/det} = 0{,}036\ ^0C$$

Result From Mathematic in calculating T= 1 Hour =60 minute and Cutting Force(Fc) = 350 Kg

Result From Mathematic in calculating live tool T= 1 hour =60 minutewith V= 190 m/minute, T= 4 hour = 240 minutewith V= 190 m/minutet ,T= 8 hour =480 minute withV = 190 m/minute Cutting Force (Fc) = 350 Kg can be show as :

Conclusion

1. Various types of cutting tool research can be applied to obtain the result of calculation analysis according to condition and industry
2. to obtain accurate calculation result analysis, it is necessary ketelitiaan read the picture on the diagram under study.

References

[1]. Ir.Syamir A.Muin (1989) *''Dasar-Dasar Perancangan dan Mesin-mesin Pe*rkakas '' CV.Rajawali,Jakarta.

[2]. Sularso dan Kiyokatsu Suga,(1983)*''Dasar Perencanaan Dan Pemilihan Elemen Mesin* '' PT,PradnyaParamita,Jakarta.

IOP Conf. Series: Journal of Physics: Conf. Series **930** (2017) 012044 doi:10.1088/1742-6596/930/1/012044

[3].Junaidi (2013) ''*HASIL PERHITUNGAN DAYA ELEKTRO MOTOR PADA PAHAT POTONG HSS DAN BENDA UJI BAJA KARBON S45 C DENGAN WAKTU 4 JAM YANG DIGUNAKAN PADA MESIN BUBUT UNIVERSAL* Journal AL Ulum Seri Saintek,Volume 1 Nomor 2.

[4].Dhaihatsu Training Centre (1987)*''Alat-alatukur''*

[5].ProsesBubut (*Turning*) *http://staf.uny.ac.id/sites/default/files*

[6].Junaidi (2010) *''Diktat Pratikum Proses Produksi* '' di Laboratorium Proses Produksi STT Harapan Medan.

[7].V.Arsinov,G.Alekseev (1970)'' *Metal Cutting Theory and Cutting Tool Design* ''

[8].Geoffrey Boothroyd (1975) School Of Engeneering Department Of Mechanical and Aerospace Engeneering University of Massacussetts, *''Fundamentals Of Metal Machining and Machine Tools''*

[9].M.Palay (1968) "*Metal Cutting Tool Production* ".10.P.Rodin. (1968) "*Design and Production of Metal-Cutting Tool Design* "

[10] .Junaidi(2015) '' *ANALYSIS CUTTING TOOL HSS WITH CAST IRON MATERIAL FROM UNIVERSAL LATHE'' IRSTC 2015 PROCEDINGS Book.*

Impact of GDP Information Technology in Developing of Regional Central Business (Case 50 Airports IT City Development in Indonesia)

Joko Suyono[1], Agus Sukoco[1], M Ikhsan Setiawan[1*], Suhermin[2] and Robbi Rahim[3]

[1]Narotama University, Jl. Klampis Semalang VII No.33, Surabaya 60117, Indonesia
[2]Sekolah Tinggi Ilmu Ekonomi Indonesia, Jalan Menur Pumpungan 30 Surabaya 60118, Indonesia
[3]School of Computer and Communication Engineering, Universiti Malaysia Perlis, Malaysia

*ikhsan.setiawan@narotama.ac.id

Abstract. Indonesia a great number of populations and demand of air transportation services keep increasing by the year in line with the increasing of population and welfare its people. Need for telematics solutions to support goods transport and distribution in cities is mainly due to the complexity of the processes taking place in urban transport systems and the importance of the optimisation of transport operations via ensuring adequate availability of linear and point infrastructure, while reducing the adverse impacts of the transport system on the environment. Efficient infrastructure supports economic growth, improves quality of life, and it is important for national security. Impact of GDP Information Technology in developing of Regional Central Business especially SME Business, are very large correlations and very significant supported by Passenger Arrival and Departure, Baggage Loaded and Unloaded, Cargo Loaded and Unloaded, Separated regional asset, Grant, Capital Expenditure, Investment of Regional Gov., GDP Agriculture-Forestry-Fishing, GDP Manufacturing, GDP Electricity-Gas, GDP Water supply-Sewerage-Waste Management-Remediation Activities, GDP Financial-Insurance Activities, GDP Business Activities, GDP Public Administration and Defense-Compulsory Social Security, GDP Education and GDP Other Services Activities

1. Introduction

Indonesia a great number of populations and demand of air transportation services keep increasing by the year in line with the increasing of population and welfare its people. Air transportation has dominant role, especially involved with the demand for fast transportation, which means the only one choice is air transportation. On the implementation of air transportation activities, there are two systems that need to be highlighted, they are airport as operators of supporting facilities and airline companies that operate the aircraft. Airport is the most vital part of air transportation. It is different with harbor as sea transportation support, or terminal as land transportation support. Without airport, air transportation activities cannot be done, especially for aircraft that has fixed wing and need landing pads that fulfil several requirements. Airport is an area that has necessary facilities and equipment to accommodate arrivals and departures of aircraft along with their passengers and cargo carried.

2. Literature Review

BPS (2017), GDP is defined as the total value added at basic price of all production units in a country during a given period (usually one year) plus net taxes on product (taxes less subsidies on products). The production units are grouped into 17 sections of industry, namely: (A) Agriculture, Forestry and Fishing, (B) Mining and Quarrying, (C) Manufacturing, (D) Electricity and Gas, (E) Water supply, Sewerage, Waste Management and Remediation Activities, (F) Construction, (G) Wholesale and Retail Trade; Repair of Motor Vehicles and Motorcycles, (H) Transportation and Storage, (I) Accommodation and Food Service Activities, (J) Information and Communication, (K) Financial and Insurance Activities, (L) Real Estate Activities, (M,N) Business Activities, (O) Public Administration and Defense; Compulsory Social Security, (P) Education, (Q) Human Health and Social Work Activities, and (R,S,T,U) Other Services Activities

Małecki, Krzysztof, Iwan, Stanisław, Kijewska, Kinga (2014) need for telematics solutions to support goods transport and distribution in cities is mainly due to the complexity of the processes taking place in urban transport systems and the importance of the optimisation of transport operations via ensuring adequate availability of linear and point infrastructure, while reducing the adverse impacts of the transport system on the environment. Chen, Xiaohong, Lin, Hangfei and Yang, Chao (2007), ITS (Intelligent Information System) plays an important role in promoting the performance of a city s overall transport system. Under the pressure of both urbanization and motorization, most cities in China require establishing an informational, integrated and intelligent transport system for vital homeland security. Considering the current conditions of China s infrastructure and the competence of transport technology priority should be given to technical skills and decision-making which can provide support for the scientific decision-making of transport planning and construction. Bin, Xu, Xiaohong, Chen, Hangfei, Lin and Chao, Yang (2013), To meet the requirements of present urban development in China, establishing ITS (Intelligent Information System) platform is mainly to support the planning decision making. The major function of this tool is by using abundant data resource and computing power, as well as Data Mining technique such as Fuzzy Analysis, Neural Network Study and Forecasting to analyze and dig the data in much deeper levels, which include the evolution laws of city traffic condition, the distribution characteristics of the contradiction between city traffic supply and travel demand clusters, the data calculation for Transport Monthly/Annual Report, the achievement of Automatic Transport Report, the evaluation of qualitative and quantitative research for current traffic situation, the identification of traffic problems and traffic regulation objects , and the support for developing the traffic improvement schemes. Miller, Harvey J (2011), Geographic Information Systems (GIS), provide an environment for managing, exploring, analyzing transportation data and communicating transportation information. Social computing can allow travelers to be active participants in the organization of the system rather than just passive recipients of transportation information. GIS – as well as the underlying geographic information science - have matured to the point that they can serve as a vital component of a smarter transportation system. Particularly relevant is the increasing ability of GIS to maintain and display spatio-temporal and moving objects data, improving capabilities for exploring and analyzing complex and massive spatio-temporal data, science and tools for simulating transportation, urban and other human systems from the "bottom-up" (at the level of the individual person, vehicle or object), and the development and adoption of data standards and information infrastructures for integrating and interoperating data.

Jimenez, Edgar, Claroa, João and de Sousa, Jorge Pinho (2014), According to the space available, or the ability of the airport to acquire surrounding land or partner with its tenants, the business approach may turn into the development of the airport city concept. Chung, Tae-won (2009), the particular business type such as aircraft chartering and sale, aircraft repairing and maintenance & assembling, and aircraft parts. It is needed to develop these businesses since world airport market is experiencing a consistent increase in airfreight demand, passenger demand and aircraft demand as the world agreed to open sky policy. From this point of view, if Incheon is able to attract the global firm that is doing business with aircraft chartering and sale related service, Incheon international airport will become a hub airport in Northeast Asia and this will stimulate activation of financial market that is related to aircraft intermediary so as to give the positive impact on the economy of Incheon. Wang, Ying, Chou, Chien-chang and Yeo, Gi-tae (2013), the development strategies for the Incheon and Taoyuan aerotropolises. It shows that the Incheon aerotropolis is primarily motivated by tourism, followed by the development of an airport-related industry zone and a residential zone. However, for the Taoyuan aerotropolis, the creation of an airport-related industry zone is the driving, followed by tourism and a residential zone

3. Methodology

Research sampling is 50 airports in Indonesia, there are: Sultan Iskandar Muda (Aceh Besar), Teuku Cut Ali (Aceh Selatan), Kualanamu (Medan), Binaka-Gunung Sitoli (Nias), Minangkabau (Padang Pariaman), Sultan Mahmud Badaruddin II (Palembang), Silampari (Musi Rawas), Raja Haji Fisabilillah (Tanjung Pinang), Ranai (Natuna), Halim Perdanakusuma (Jakarta Timur), Husein Sastranegara (Bandung), Nusawiru (Ciamis), Tunggul Wulung (Cilacap), Adi Sucipto (Sleman), Juanda (Sidoarjo), Abdul Rachman Saleh (Malang), Ngurah Rai (Badung), Bandara Internasional Lombok - BIL (Mataram), Muhammad Salahudin (Bima), Eltari (Kupang), Gewayantana-Larantuka (Flores Timur), DC Saudale Lekunik (Rote Ndao), A.A. Bere Tallo, Haliwen-Atambua (Belu), Komodo-Labuhan Bajo (Manggarai Barat), Frans Sales Lega Satar Tacik (Manggarai), Terdamu Sabu (Kupang), Umbu Mehang K-Waingapu (Sumba Timur), Supadio (Pontianak), Nangah Pinoh (Melawi), Iskandar - Pangkalan Bun (Kotawaringin Barat), Tjilik Riwut (Palangkaraya), H. Asan - Sampit (Kotawaringin Timur), Kuala Kurun (Gunung Mas), Syamsuddin Noor (Banjar Baru), Sepinggan (Balikpapan), Temindung (Samarinda), Bontang (Badak Bontang), Nunukan (Nunukan), Seluwing (Malinau), Naha-Tahuna (Kep. Sangihe), Mutiara(Palu), Kasiguncu (Poso), Syukuran Aminuddin Amir (Banggai Luwuk), Sorowako (Luwu Timur), Haluoleo (Konawe Selatan), Rendani (Manokwari), Wasior (Teluk Wondama), Moppah (Merauke), Nabire (Nabire) and Tolikara (Karubaga). Data collection from BPS (Statistics Institution) and Financial Ministry, to see impact of GDP Information Technology in developing of Regional Central Business especially SME Business. GDP Information Technology as Dependent variable and Independent/predictor variables are Passenger Arrival and Departure, Baggage Loaded an Unloaded, Cargo Loaded and Unloaded, Separated regional asset, Grant, Capital Expenditure, Investment of Regional Govt, GDP Agriculture-Forestry-Fishing, GDP Mining-Quarrying, GDP Manufacturing, GDP Electricity-Gas, GDP Water supply-Sewerage-Waste Management-Remediation Activities, GDP Wholesale and Retail Trade-Repair of Motor Vehicles and Motorcycles, GDP Accommodation-Food Service Activities, GDP Financial-Insurance Activities, GDP Business Activities, GDP Public Administration and Defence-Compulsory Social Security, GDP Education, GDP Human Health and Social Work Activities and GDP Other Services Activities. Institute for Digital Research and Education (2017), Statistic analysis use SPSS tools included Overall Model Fit, Anova Table and Parameter Estimates.

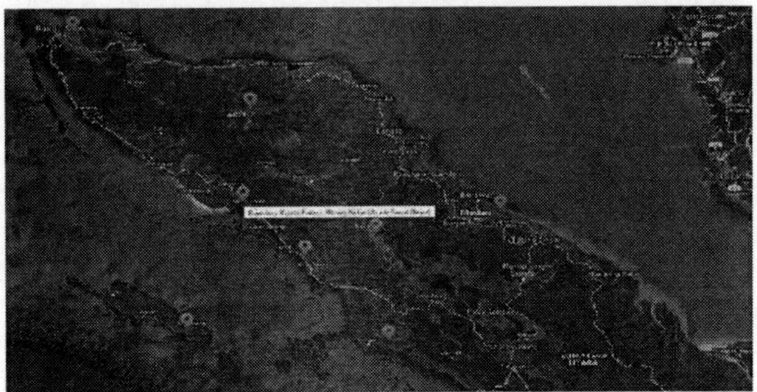

Figure 1. Airport Location in Aceh, Sumatera, Indonesia

4. Result and Analysis

The results of statistical analysis shows Overall Model Fit, ANOVA Table and Parameter Estimates of dependent variable GDP Information Technology and independent variables include Passenger Arrival and Departure, Baggage Loaded and Unloaded, Cargo Loaded and Unloaded, Separated regional asset, Grant, Capital Expenditure, Investment of Regional Gov., GDP Agriculture-Forestry-Fishing, GDP Mining-Quarrying, GDP Manufacturing, GDP Electricity-Gas, GDP Water supply-Sewerage-Waste Management-Remediation Activities, GDP Wholesale and Retail Trade-Repair of Motor Vehicles and Motorcycles, GDP Accommodation-Food Service Activities, GDP Financial-Insurance Activities, GDP Business Activities, GDP Public Administration and Defense-Compulsory Social Security, GDP Education, GDP Human Health and Social Work Activities and GDP Other Services Activities.

R value represents the simple correlation and is 0.996 (the "R" Column), which indicates a high degree of correlation. In this case, 99.6% can be explained, which is very large.

a. Dependent Variable: Informasi dan Komunikasi/ Information and Communication

b. Tolerance = .000 limit reached.

Model Summary

Model	R	R Square	Adjusted R Square	Std. Error of the Estimate
1	.999[a]	.998	.996	215.33423

a. Predictors: (Constant), Jasa lainnya/ Other Services Activities, Cargo Bongkar/ Unloaded, Pertanian, Kehutanan, dan Perikanan/ Agriculture, Forestry and Fishing, Pertambangan dan Penggalian/ Mining and Quarrying, Pengadaan Listrik dan Gas/ Electricity and Gas, Pengadaan Air, Pengelolaan Sampah, Limbah dan Daur Ulang/ Water supply, Sewerage, Waste Management and Remediation Activities, Industri Pengolahan/ Manufacturing, Penyediaan Akomodasi dan Makan Minum/ Accommodation and Food Service Activities, Cargo Muat/ Loaded, Administrasi Pemerintahan, Pertahanan dan Jaminan Sosial Wajib/Public Administration and Defence; Compulsory Social Security, Bagasi Bongkar/ Unloaded, Jasa Keuangan dan Asuransi/ Financial and Insurance Activities, Pesawat Datang/Arrival, Hibah, Belanja Modal, Penumpang Berangkat/ Departure, Jasa Pendidikan/ Education, Jasa Kesehatan dan Kegiatan Sosial/ Human Health and Social Work Activitie, Hasil pengelolaan kekayaan daerah yang dipisahkan, Perdagangan Besar dan Eceran; Reparasi Mobil dan Sepeda Motor/ Wholesale and Retail Trade; Repair of Motor Vehicles and Motorcycles, Pendapatan, Bagasi Muat/ Loaded, Penyertaan Modal (Investasi) Daerah, Jasa Perusahaan/ Business ...

Figure 2. GDP Model Analysis

ANOVA[a]

Model		Sum of Squares	df	Mean Square	F	Sig.
1	Regression	551029892.6	25	22041195.70	475.345	.000[b]
	Residual	1112851.938	24	46368.831		
	Total	552142744.5	49			

a. Dependent Variable: Informasi dan Komunikasi/ Information and Communication

b. Predictors: (Constant), Jasa lainnya/ Other Services Activities, Cargo Bongkar/ Unloaded, Pertanian, Kehutanan, dan Perikanan/ Agriculture, Forestry and Fishing, Pertambangan dan Penggalian/ Mining and Quarrying, Pengadaan Listrik dan Gas/ Electricity and Gas, Pengadaan Air, Pengelolaan Sampah, Limbah dan Daur Ulang/ Water supply, Sewerage, Waste Management and Remediation Activities, Industri Pengolahan/ Manufacturing, Penyediaan Akomodasi dan Makan Minum/ Accommodation and Food Service Activities, Cargo Muat/ Loaded, Administrasi Pemerintahan, Pertahanan dan Jaminan Sosial Wajib/Public Administration and Defence; Compulsory Social Security, Bagasi Bongkar/ Unloaded, Jasa Keuangan dan Asuransi/ Financial and Insurance Activities, Pesawat Datang/Arrival, Hibah, Belanja Modal, Penumpang Berangkat/ Departure, Jasa Pendidikan/ Education, Jasa Kesehatan dan Kegiatan Sosial/ Human Health and Social Work Activitie, Hasil pengelolaan kekayaan daerah yang dipisahkan, Perdagangan Besar dan Eceran; Reparasi Mobil dan Sepeda Motor/ Wholesale and Retail Trade; Repair of Motor Vehicles and Motorcycles, Pendapatan, Bagasi Muat/ Loaded, Penyertaan Modal (Investasi) Daerah, Jasa Perusahaan/ Business ...

Figure 3. Model Summary and ANOVA Analysis

Regression model predicts the dependent variable significantly well, at the "Regression" row and go to the "Sig." column. This indicates the statistical significance of the regression model that was run. Here, p is less than 0.05, and indicates that, overall, the regression model statistically significantly predicts the model (i.e., it is a good fit for the data).

Significant level below 0.05 included Passenger Arrival and Departure, Baggage Loaded an Unloaded, Cargo Loaded and Unloaded, Separated regional asset, Grant, Capital Expenditure,

Investment of Regional Gov., GDP Agriculture-Forestry-Fishing, GDP Manufacturing, GDP Electricity-Gas, GDP Water supply-Sewerage-Waste Management-Remediation Activities, GDP Financial-Insurance Activities, GDP Business Activities, GDP Public Administration and Defense-Compulsory Social Security, GDP Education and GDP Other Services Activities.

Coefficients[a]

Model	Unstandardized Coefficients		Standardized Coefficients	t	Sig.
	B	Std. Error	Beta		
Penyertaan Modal (Investasi) Daerah	-9.579E-9	.000	-2.164	-3.895	.001
Pertanian, Kehutanan, dan Perikanan/ Agriculture, Forestry and Fishing	-.091	.041	-.069	-2.228	.036
Pertambangan dan Penggalian/ Mining and Quarrying	.026	.014	.022	1.848	.077
Industri Pengolahan/ Manufacturing	-.008	.004	-.050	-2.111	.045
Pengadaan Listrik dan Gas/ Electricity and Gas	3.662	1.375	.366	2.663	.014
Pengadaan Air, Pengelolaan Sampah, Limbah dan Daur Ulang/ Water supply, Sewerage, Waste Management and Remediation Activities	17.940	3.282	.381	5.466	.000
Perdagangan Besar dan Eceran; Reparasi Mobil dan Sepeda Motor/ Wholesale and Retail Trade; Repair of Motor Vehicles and Motorcycles	.065	.060	.262	1.060	.300
Penyediaan Akomodasi dan Makan Minum/ Accommodation and Food Service Activities	-.240	.155	-.227	-1.548	.135
Jasa Keuangan dan Asuransi/ Financial and Insurance Activities	-.279	.125	-.162	-2.238	.036
Jasa Perusahaan/ Business Activities	2.675	.911	1.811	2.936	.007
Administrasi Pemerintahan, Pertahanan dan Jaminan Sosial Wajib/Public Administration and Defence; Compulsory Social Security	-.422	.116	-.267	-3.656	.001
Jasa Pendidikan/ Education	.456	.198	.500	2.305	.030

Figure 4. GDP Education Services 1

Coefficients[a]

Model	Unstandardized Coefficients		Standardized Coefficients	t	Sig.
	B	Std. Error	Beta		
Jasa Kesehatan dan Kegiatan Sosial/ Human Health and Social Work Activitie	-.198	.515	-.050	-.384	.704
Jasa lainnya/ Other Services Activities	1.407	.465	.654	3.026	.006

Figurre 5. GDP Education Services 2

5. Conclusion

A high degree of correlation, 99.6% is very large between dependent variable GDP Information Technology and independent variables Passenger Arrival and Departure, Baggage Loaded an Unloaded, Cargo Loaded and Unloaded, Separated regional asset, Grant, Capital Expenditure, Investment of Regional Gov., GDP Agriculture-Forestry-Fishing, GDP Mining-Quarrying, GDP Manufacturing, GDP Electricity-Gas, GDP Water supply-Sewerage-Waste Management-Remediation Activities, GDP Wholesale and Retail Trade-Repair of Motor Vehicles and Motorcycles, GDP Accommodation-Food Service Activities, GDP Financial-Insurance Activities, GDP Business Activities, GDP Public Administration and Defence-Compulsory Social Security, GDP Education, GDP Human Health and Social Work Activities and GDP Other Services Activities.

Impact of GDP Information Technology in developing of Regional Central Business especially SME Business, are very large correlations and very significant supported by Passenger Arrival and Departure, Baggage Loaded and Unloaded, Cargo Loaded and Unloaded, Separated regional asset, Grant, Capital Expenditure, Investment of Regional Gov., GDP Agriculture-Forestry-Fishing, GDP Manufacturing, GDP Electricity-Gas, GDP Water supply-Sewerage-Waste Management-Remediation Activities, GDP Financial-Insurance Activities, GDP Business Activities, GDP Public Administration and Defence-Compulsory Social Security, GDP Education and GDP Other Services Activities

Reference

[1] Agénor, P-R., Moreno-Dodson, B. (2006), Public infrastructure and growth: new channels and policy implications. Banca d'Italia, Italia.

[2] Baldwin, J.R., Dixon, J. (2008), Infrastructure Capital: What is it? Where is it? How much of it is there? Canadian Productivity Review. No 16. Ottawa: Statistics Canada.

[3] Bin, Xu, Xiaohong, Chen, Hangfei, Lin, Chao, Yang (2013), Decision Oriented Intelligent Transport Information Platform Design Research - Case study of Hangzhou City, 13th COTA International Conference of Transportation Professionals (CICTP 2013), Procedia - Social and Behavioral Sciences 96 (2013) 2230 – 2239

[4] BPS (2017), National Income Of Indonesia 2012-2016, BPS, Jakarta, Indonesia

[5] Chen, Xiaohong, Lin, Hangfei, Yang, Chao (2007), City Traffic, vol. 5, no.5, pp.28-32

[6] Chung, Tae-won (2009), A Study on Selections of Strategic Type of Business in Air-logistics Industry Clusters, The Asian Journal of Shipping and Logistics, Vol 25 No 1 June 2009, pp.83-102, email: logichung@idi.re.kr

[7] Grundey, D. (2008), Managing sustainable tourism in Lithuania: Dream or reality? Technological and Economic Development of Economy, 14(2), 118-129.

[8] Gu, W., & Macdonald, R. (2009), The Impact of Public Infrastructure on Canadian Multifactor Productivity Estimates. The Canadian Productivity Review. Research paper (21).

[9] Institute for Digital Research and Education (2017), https://stats.idre.ucla.edu/spss/output/regression-analysis/

[10] Jimenez, Edgar, Claroa, João, de Sousa, Jorge Pinho (2014), The airport business in a competitive environment, EWGT2013 – 16th Meeting of the EURO Working Group on Transportation, Procedia - Social and Behavioral Sciences 111 (2014) 947 – 954

[11] Laerd Statistics (2017), https://statistics.laerd.com/spss-tutorials/linear-regression-using-spss-statistics.php

[12] Macdonald, R. (2008), An Examination of Public Capital's Role in Production. Economic Analysis Research Paper Series. No. 50. Ottawa: Statistics Canada.

[13] Małecki, Krzysztof, Iwan, Stanisław, Kijewska, Kinga (2014), Influence of Intelligent Transportation Systems on reduction of the environmental negative impact of urban freight transport based on Szczecin example, 1st International Conference Green Cities 2014 – Green Logistics for Greener Cities, Procedia - Social and Behavioral Sciences 151 (2014) 215 – 229

[14] Mamatzakis, E.C. (2008), Economic performance and public infrastructure: an application to Greek manufacturing. Bulletin of Economic Research (60), 307-326.

[15] Martinkus, B., Lukasevicius, K. (2008), Investment environment of Lithuanian resorts: Researching national and local factors in the Palanga case. Transformations in Business & Economics, 7(2), 67-83.

[16] Miller, Harvey J (2011), Collaborative mobility: using geographic information science to cultivate cooperative transportation systems, International Conference: Spatial Thinking and Geographic Information Sciences 2011, Procedia Social and Behavioral Sciences 21 (2011) 24–28

[17] Snieska, V., Bruneckiene, J. (2009), Measurement of Lithuanian Regions by Regional Competitiveness Index. Inzinerine Ekonomika-Engineering Economics(1), 45-57

[18] Wang, Ying, Chou, Chien-chang, Yeo, Gi-tae (2013), Criteria for Evaluating Aerotropolis Service Quality, The Asian Journal of Shipping and Logistics, Vol 29 No 3 December 2013, pp.395-414, email: yingmickey@naver.com

Data Reduction Algorithm Using Nonnegative Matrix Factorization with Nonlinear Constraints

PasukatSembiring

[1]Department of Mathematics, University of Sumatera Utara, Medan, Indonesia
Email: pasukat_sembiring366@gmail.com

Abstract-Processing ofdata with very large dimensions has been a hot topic in recent decades. Various techniques have been proposed in order to execute the desired information or structure. Non- Negative Matrix Factorization (NMF) based on non-negatives data has become one of the popular methods for shrinking dimensions. The main strength of this method is non-negative object, the object model by a combination of some basic non-negative parts, so as to provide a physical interpretation of the object construction. The NMF is a dimension reduction method thathasbeen used widely for numerous applications including computer vision,text mining, pattern recognitions,and bioinformatics. Mathematical formulation for NMF did not appear as a convex optimization problem and various types of algorithms have been proposed to solve the problem. The Framework of Alternative Nonnegative Least Square(ANLS) are the coordinates of the block formulation approaches that have been proven reliable theoretically and empirically efficient. This paper proposes a new algorithm to solve NMF problem based on the framework of ANLS.This algorithm inherits the convergenceproperty of the ANLS framework to nonlinear constraints NMF formulations.

Keyword: Dimensional Reduction, Nonnegative Matrix Factorization, Sparse Matrix, Nonlinear Constraints

1. Introduction

In every second in the modern era, a lot of data sets are being generated. The number of people online write their blogs, their homepage design and disseminate their experiences through digital support, videos, images and others. Imagine also the data formed from living organisms and gene research data obtained from space or from our earth. Data of transaction's through e-banking and others. Be useful if the data has been processed. Associated with the boom in the amount of data so rapidly, there are several approaches for data processing, by using classical methods application, designing more powerful computing structures such as: distributed computing, multi-core processors, super computers and others. However, the amount of data growth and complexities tendency exceeds the increase computational capabilities [1]. One of very popular approach called model of reduction that attempts to reduce the complexity while maintaining the assumption of issues (preliminary data). The use of an appropriate model can save the time. In addition, different data types require different models to capture the data comprehension. Of course, a model considered appropriate so that the model can survive better met. An ideal model may not exist. For example, using sub-dominant space with the Singular Value Decomposition (SVD) has been proposed as the best model for reducing the complexity of data and complicated system [10]. This model offers the smallest error with the same complexity shrinkage, compared with other models. Partial-based analysis also referred to as additive model in the absence of the operating margin in the model. This is related to the formation of an object :

$$\text{Object}_i = \text{Part}_1\,(b_{i1}) \text{ with } \text{Part}_2\,(b_{i2}) \ldots,$$

Where, b_{ij} = present if part $i$is present in object j or , b_{ij} = absent if part $i$absent from object j. Their operating margin, which $b_{ij}<0$ for some $i.j$ result in each section can be interpreted as positive or negative quantity. If each part included in a vector space, this is true because orientation does not change the room establishment. But for other data types, such as: concentration, relative temperature, opportunities, and other sound spectrum - more negative quantity does not appear. Decomposition of non-negative objects with common methods such as SVD significantly alters the physical interpretation of the data. Other analyzes such as Principle Components Analysis (PCA) requires some features where the object is non-negative because nature is not owned, such sum to zero, orthogonality and others. So much so that became the main reference of this dissertation research. To focus on the data nonnegative through the study of non-negative matrix factorization (FMN). This technique explores for applying the non-negative object A by two other non-negative matrix W and H.
The rest of the paper is organized as follows. The concepts of Nonnegative Matrix Factorization (NMF) and related work are explained in section II. Proposed embedding and extraction algorithms are explained in section II. Experimental results are presented in section III. Concluding remarks are given in section IV.

2. Nonnegative Matrix Factorization

Nonnegative Matrix Factorization (NMF), was originally known as the factorization rank non-negative or matrix factorizations positive [11] have attracted much attention as a method of dimension reduction in machine learning and data mining [10,11]. NMF is used for high dimensional data in which each nonnegative valuable element. Due to the nature of this non-negative factor of the approach is much smaller give an interpretation. Each item of the data can be explained by a combination of linear additives from line items that are physically meaningful [10]. Various application of NMF has been reported in the field including text mining [7], bioinformatics [16].

Mathematically NMF can be formulated in the following way. Given an input matrix $A \in R^{m \times n}$, in which each element nonnegative, an integer k <min {m, n} NMF aims to acquire two vector $\in R^{m \times k}$ W x H $\in R^{k \times n}$ with non-negative elements, so

$$A \approx H \tag{2.1}$$

Measurement approach in Eq. (2.1) can be done in several ways, among other things, non Frobenius, divergence Kullbach - Leibber [10] and the divergence Bregaran [19]. The research was wearing a size selection of the most secure, ie non Frobenius. So,the factor W and H are obtained by solving the following optimization.

$$\min_{W \geq 0, H \geq 0} f(W,H) = \frac{1}{2} \| A - WH \|_F^2 \tag{2.2}$$

Where inequality W ≥0 and H ≥0 gives the sense that all of the elements W and H nonnegative. The issue in Eq. (2) is difficult.Various algorithms have been developed to solve Equation (2). At the first Paatero and Tappen [11] proposed NMF (or, more precisely positive matrix factorization) but their algorithm is not exactly related to the non-negative constraint. The most popular approach to solve Eq. (2) is the multiplicative update algorithm of Lee and Seung [10]. At each of iteration of this method, the elements W and H multiplied by a certain factor. Because the zero element is not in update, all components of W and H positive for all iterations. However, some issues about the performance [20], as well as the issue of convergence [21] reported. In [18], a simple algorithm, that resolve issues without the constraints of least squares at each iteration, has been used but the algorithm also has difficulty with convergence. Furthermore, some algorithms that are based on the framework of alternative non-negative least squares (ANLS) filed with the well's Performance [18].

Non-negative matrix factorization first introduced by Paatero and Tapper in [11]. But this term gained popularity thanks to Lee and Seung writing [10]. They propose that non-negativity important in human perception and also provide two simple algorithms for determining the representation of non-negative for non-negative data. Known $m \times n$ non-negative matrix **A** (ie. $A_{ij} \geq 0$) and reduced rank **r** ($r < \min(m,n)$), non-negative matrix factorization problem consists of the determination of the two non-negative matrix $\mathbf{U} \in \mathbb{R}_+^{m\times r}$ **dan** $\mathbb{R}_+^{n\times r}$ that approximates the matrix A, that is:

$$A \approx UV^T. \quad (2.3)$$

By paying attention to the columns of **A**, seen that each of these columns is approximated by a combination of non-negative r basis vectors which are the columns of **U.** Could be considered, the columns of U as a base of U, which is entirely in the non-negative. Each column A is approximated by an element of U, usually the element closest to the column. Also can exchange the role of U and V

to affirm that each row of A is approximated by an element of V, usually the element closest to the line, where V is generated by the columns of V. Where,

$$A_{:i} \approx \sum_{j=1}^{r} V_{ij} U_{:j}\,. \quad (2.4)$$

There are several ways to determine the difference between the data matrix A and matrix models UVT. But the most widely used measure is the Frobenius norm:

$$F(A,UV^T) = \frac{1}{2}\left\|A - UV^T\right\|_F^2 = \frac{1}{2}\sum_{i=1}^{m}\sum_{i=1}^{n}\left(A_{ij} - [UV^T]_{ij}\right)^2 \quad (2.5)$$

This is also referred to as distance Euclid (Euclidean Distance).

Suppose that U remains, the function $F(U, V) = \frac{1}{2}\left\|A - UV^T\right\|_F^2$ can be viewed as the composition of the Frobenius norm and linear transformation of V. Therefore, F is convex in V. Similarly, if V is fixed, F is convex on U.

In this study, non-negative matrix factorization problem will be studied with Frobenius norm. The main issues are as follows:

$$\min_{U\in R_+^{m\times r} V\in R_+^{n\times r}} \frac{1}{2}\left\|A - UV^T\right\|_F^2 . \quad (2.6)$$

The problem is Nonnegative Matrix Factorization. Finish it, when $m \times n$ non-negative matrix **A** and integer **r <min (m, n).** Where r is called the rank reduced. Henceforth, m and n will be used to denote the size of the matrix **A** and r is the rank reduced from a factorization. Nonnegative matrix factorization rewritten as a nonlinear optimization problem:

$$\min_{-U\leq 0 - V\leq 0} \frac{1}{2}\left\|A - UV^T\right\|_F^2 \quad (2.7)$$

Lagrange function associated is:

$$L(U,V,\mu,v) = \frac{1}{2}\left\|A - UV^T\right\|_F^2 - \mu \circ U - v \circ V, \quad (2.8)$$

Where μ and **v** are two equal-sized matrices, U and V containing multipliers Lagrange associated with restrictions nonnegativity $U_{ij} \geq 0$ dan$V_{ij} \geq 0$. Then the terms of Karush-Kuhn-Tucker, for nonnegative matrix factorization problem is, If (U, V) is a local minimum, then there $\mu_{ij} \geq 0$ dan$v_{ij} \geq 0$ such that:

$$U \geq 0, \quad V \geq 0, \tag{2.9}$$

$$\nabla L_U = 0, \nabla L_V = 0, \tag{2.10}$$

$$\mu \circ U = 0, \quad v \circ V = 0. \tag{2.11}$$

By developing (2.10) were obtained:
$AV - UV^TV - \mu = 0, \quad A^TU - VU^TU - v = 0$ or $\mu = -(UV^TV - AV), \quad v = -(VU^TU - A^TU).$
By combining this with $\mu_{ij} \geq 0$, $v_{ij} \geq 0$ and (2.11) provides the following terms:

$$U \geq 0, \quad V \geq 0, \tag{2.12}$$

$$\nabla F_U = UV^TV - AV \geq 0, \nabla F_V = VU^TU - A^TU \geq 0, \tag{2.13}$$

$$U \circ (UV^TV - AV) = 0, V \circ (VU^TU - A^TU) = 0, \tag{2.14}$$

Where the Lagrange multiplier is corresponding to the matrix **U** and **V** is also the gradient of **F** to matrix U and V. Since the distance Euclidean is not convex in the variables U and V, so that the terms of this is only a necessary condition. This is implied because of the existence of a saddle point and maximum

3. Non Linear Programming

In the section, it was found a certain optimization procedure to maximize or minimize a function of ***n*** variables. Worse, as noted earlier, this procedure cannot be used to solve problems that are very much exceeds its variable. Here will be presented a few, computationally feasible method to solve certain types of problems, non-linear programming. One problem, in nonlinear programming problem, which is must to maximize or minimize a function $f(\bar{x})$, the constraints of a set of equations or inequality constraints. Where $f(\bar{x})$, or at least one of the functions that appear in this set of constraints or both, is a non-linear function. The Problem of non-linear programming generally can be expressed as follows:

$$\text{Maximize } Z = f(\bar{x}) \tag{3.1}$$

$$\text{Constraints } g_i(x) \{\leq =, \geq\} b_i \tag{3.2}$$

Weir, one of the three relations $\{\leq =, \geq\}$ is designed against each of ***m*** constraints, the function $f(\bar{x})$ in equation (3.1) calls the objective function. In connection with the non-negativity constraint ($x_j \geq 0$) on some, or all, of the variables stated, separately or may be deemed to be included in the constraint. There is no known method of determining the global maximum of the problem, non-linear programming in general. Even, if the objective function and constraints in accordance with the properties of certain global maximum sometimes acceptable. For example proves that the global maximum of a function convex against a limited set convex and on, an extreme point of the set convex. Here will be downgraded "Kuhn-Tucker conditions" a set of conditions necessary, to make the maximum of a non-linear problems. These conditions in certain cases also produce global maximum. Kuhn-Tucker conditions are very useful indeed, in generating methods to resolve some types of non-linear problems. It can be seen how the use of Kuhn-Tucker conditions to develop an algorithm using the Simplex method to solve quadratic program. The Problem in the quadratic program is the objective function is quadratic and the constraints are linear

4. Proposed Algorithm

Mathematical formulation of NMF can be expressed in the following way. Known input matrices **A ε $\mathbf{R}^{M \times n}$** in which each element and a nonnegative integer ***k <min*** {***m***, ***n***}. NMF aims to determine two facts W ε $R^{m \times k}$ and H ε $R^{k \times n}$ with nonnegative elements, so:

$$A = WH \quad (4.1)$$

By using Fortenius norms, facts ***W*** and ***H*** is obtained by solving the following problems sptinuseri:

$$\frac{min}{N \geq 0, H \geq 0} f(N, H) = \frac{1}{2} \| A - WH \|_F^2 \qquad (4.2)$$

Where the inequality ***W ≥0*** and ***H ≱0*** means that all elements of ***W*** and ***H*** nonnegative.
Before proposing the algorithm to solve Eq.(4.2). This is necessary because the proposed algorithm starts from ANLS. Within the framework of the basic ANLS, the variable directly divided into two groups. Then both of groups updated. This basic framework can be summarized as follows:

1. Initialization H ε $R_{m \times l}$ with elements non negative
2. Complete the following problems until termination criteria are met:

$$\frac{min}{w \geq 0} (\| H^T W^T - A^T \|)_F^2 \qquad (4.3)$$

With ***H*** fixed, and

$$\frac{min}{H \geq 0} \| WH - A \|_F^2 \quad (4.4)$$

And **W** fixed

3. Column ***W*** is normalized into the unit L2 - norm and the corresponding rows of H scaled.

Alternatively, **W** can be just as initial iteration of data later in the Eq (4.4) and then Eq. (4.3). Note that each sub problems is a question of non-negativity least square (NNLS). Although, the initial problem (4.2) is non - convex, sub problem Eq. (4.3) is convex. NNLS issue in Eq (4.3) and has certain characteristics. NMF is an algorithm to reduce the dimensions. Very large initial dimensions, eg thousands, of small dimensions reduced. Because the matrix W ε $R^{m \times l}$ in Eq. (4.4) shape is very long and thin (m >> k) and the coefficient matrix H^T ε $R^{m \times l}$ in Eq. (4.3) is also long and thin (n >> k). Because the matrix W^T and H of the variables in Eq. (4.3) and (4.4) each of data and wide. To design the NMF algorithm based on ANLS framework, the need for a method to solve sub problem in Eq. (4.3) and (4.4). The algorithm is a kelaril method to NNLS active set by Lawson and Hanson [9]. The Set method actively seeking optimal set of active and passive by maintaining between two sets of variables. If the variable passive (ie, positive) solution is known, then the issue can be resolved by procedures of NNLS least square has no control over passive variable. The main limitation of this method is that the active set of the set of variables that meet non-negativity maintained precede vector settlement, while ensuring that the objective function decreases in each iteration. Means algorithm is slow when a large number unknown. Pivot major blocks method is proposed to identify the limitations of active set methods.

The first, this paper presents the block principal pivoting method for the NNLS problems with single right-hand side and then proposed the method developed to handle multiple right-hand sides. The NNLS problems with single right-hand side vector formulated as follow:

$$\frac{min}{x \geq 0} \| Cx - b \|_2^2 \quad (4.5)$$

Where, C ε $R^{P \times q}$, b ε $R^{q \times 1}$, and $p \geq q$. The sub problems in Eq. (4.3) and (4.4) can be decomposed into several independent instances of Eq.(4.5). Thus, an algorithm for Eq.(4.5) is a basic building block for an algorithm Eq. (4.3) and Eq.(4.4).
The Karush – Kuhn- Tucher optimality condition for Eq.(4.5) is written as follows:

$$y = C^T Cx - C^T b \qquad (4.6)$$
$$y \geq 0 \qquad (4.7)$$
$$x \geq 0 \qquad (4.8)$$
$$x , y = 0 , I = 1, \ldots \infty \qquad (4.9)$$

It is assumed that the matrix C has full column rank. In this case, the matrix positive $C^T C$ is positive definite and problems in Eq. (4.5) is strictly convex. Then, a solution x that satisfies the condition in Eq. (4.6) - (4.9) is the optimal solution of Eq. (4.5).

The set of indexes{ 1, ... ~ } are divided into two groups F and G to F ∪ G = {1, ... ~ } and $F \cap C = \emptyset$. Let's, X_F, X_C, Y_F and Y_G denote sub matrix C with corresponding indexes, and suppose C_F and C_G stated sub matrix C with corresponding column index initially: Take $X_G = 0$ and $Y_F = 0$, all the elements X_G and Y_F made zero.Thus, x and y have always satisfied the Eq. (4.1) for each price of X_F and Y_G. Now, calculate XF and YG using Eq. (4.6) and then check whether the computed value of X_F and Y_G meets the Eq. (4.7) and (4.8). The calculation of the X_F and Y_G is done by:

$$X_F = \arg min_{X_F} \|C_F X_F - b\|_2^2 \qquad (4.10)$$
$$Y_G = C_G^T (C_F X_F - b) \qquad (4.11)$$

Complete the Eq. (4.10) firstfor X_F then substitutes the result to the Eq. (4.11). Couples (X_F, Y_G) are called basic complementary settlement if they are obtained by Eq. (5.10) - (5.11). If the basic complementary settlement (X_F, Y_G) meets $X_F \geq 0$ and $Y_G \geq 0$, then it is called feasible. In this case X is the optimal solution of the Eq. (4.5) and BerLouthalgorithms. Otherwise, the complementary basic settlement (X_F, Y_G) is not feasible, and it needs to be updated F and G by exchanging variables which are the Eq. (5.7) and (5.8) do not apply. So, the definition of the following index set:

$$V = \{I \varepsilon F : x ; < 0 \} \cup \{ I \varepsilon G : y_i < 0\} \qquad (4.12)$$

The variable x with i ε v is called an improper variable. Now select an empty subset of $\hat{V} \subset V$ then, F and C are absolute with the following rules

$$F = (F - \hat{V}) \cup (\hat{V} \cap C) \qquad (4.13)$$

$$C = (G - \hat{V}) \cup (\hat{V} \cap F) \qquad (4.14)$$

Size $|\hat{V}|$ Presents how many variables are exchanged per iteration. If $|\hat{V}| = 1$, the algorithm is called the main single pivot algorithm. The algorithm repeats this procedure so that the number of unworthy variables is zero.

To speed up the search procedure, we can use the rule$V$ = V called the full exchange rule. This rule states that all variables F and G are exchanged that do not meet the Eg. (4.6) - (4.9), and the computational acceleration rule by reducing the number of iterations required until termination. However, such a rule can result cycles and fail to obtain optimal values, although these events are rare. To ensure the termination of the following exchange set for the Press (4.13), (4.14):

$$\hat{V} = \{ i : i = maks \{ i \varepsilon V \} \} \qquad (4.15)$$

Supporting rules, where only improper variables with the largest indexes are exchanged, there are the main single-pivot rule. This rule guarantees finite termination. By incorporating full exchange rules or support roles, the main pivot method block for NNLS issues with one segment is obtained. The algorithm is listed in Algorithm 1 as follow:

4.1 Algorithm 1

Input : $C \varepsilon R^{P \times q}$ and $b \varepsilon R^{P}$
Output : $x\ (\varepsilon R^{q \times 1}) = \arg\min x \geq 0 \|Cx - b\|_2^2$

1. Initialize $F = \emptyset$, $G = \{1, \ldots, q\}$, $x = 0$, $y = -C^T b$, $\propto = 3$, $\beta = q + 1$
2. Calculate X_F and Y_G by Eq (4.10) and (4.11)
3. While (X_F, Y_G) is not worth doing
4. Calculate V by Eq. (4.12)
5. If | V | <, Create β = | V |, α = 3 and $\hat{V} = V$
6. If | V | ≥β and α ≥1, make α -1 and $\hat{V} = V$
7. If | V | ≥β and α = 0, make $\hat{V}$ by Eq. (5.15)
8. Update F and G by Eq. (5.14), (5.14)
9. Update X_F and Y_G by Eq. (5.10) and (5.11)
10. End while

In the case support rules slower than full exchange rules, it is only in use if the full exchange rules do not work. Final Termination Algorithm 1 is achieved by controlling the number of undeserved variables. The variable α is used as a buffer against the full exchange rules that are tried. If the full exchange rules increase the number of improper variables, α minus one.

After α becomes zero, the supporting rule is used so that it makes the number of variables eligible less than the smallest value reached, filled in β. This occurs in a finite number of steps because the buffer rule has finite termination properties. As soon as the buffer rule reaches the latest low number of undeserved variables, return to the full exchange rule. There are used as default values for α, which means that the full exchange rule is tied up to three times until it has an undeserved number of variables. Since the variable is not feasible systematically reduced, the algorithm stops in the number of finite steps.

Suppose,if we want to solve the following NNLS issues:

$$\min_{x \geq 0} \|Cx - B\|_F^2 \qquad (4.16)$$

Where, $C \in R^{pxq}, B \in R^{pxr}$ *dan* $X \in R^{qxr}$.It can be done using Algorithm 1 for each vector of the right-hand side $b_1, \ldots, b_r$ with $B = [b_1, \ldots, b_r]$, since the X columns are not dependent on each other. But the computing is inefficient.
In Algorithm 1, the main computational load is more than the need to define X_F and Y_G as written in the Eq. (4.10) - (4.11). The Eq (4.10) can be solved by a normal equation:

$$C_F^T C_F X_F = C_F^T b \qquad (4.17)$$

And Eq.(4.11) can be written as:

$$Y_G = C_F^T C_F X_F - C_G^T b \qquad (4.18)$$

Required $C_F^T C_F$, $C_F^T b$, $C_G^F C_F$, and $C_G^T b$ to complete the Eq. (4.17) and (4.18) and these matrices and vectors vary throughout the iteration because F and G are cleared in the three iterations.
Note that for NNLS problems arising from the NMF, C matrix is very long and thin and calculates $C_F^T C_F$, $C_F^T b$, $C_G^F C_F$, and $C_G^T b$ is computationally expensive. However, $C^T C$ and $C^T B$ can be started

and reused later in the iteration. Since C_F^T C_F, $C_F^T b$, $C_G^F C_F$, and $C_G^T b$ can be taken so that the submatrix of C^T C and $C^T B$ is also small. This idea is used for the case of one right-hand side, but its influence is more observable in case many right-sided.

Other developments arise from the fact that X is flat and wide in the NNLS issue for NMF. Assuming, Algorithm 1 is run for vector with many right sides. In each iteration, there is an index set of F_j and G_j for each of the columns j ∈ {1, ..., n}, X_F, and Y_G since they are computed using Eq. (4.17) and (4.18). The idea is to obtain a group of columns that share the same set of indexes F_j and G_j and complete the Eq. (4.17) to the columns in the same group. If done so, recurrent computation of Cholesly factorization is avoided upon completion of the Eq. (4.17). So, if X is flat and wide, which is the case for the NNLS problem in NMF, more columns can share the set of F_j and G_j indexes, resulting in acceleration of the set process. The improved mainstream pivot method for many right-hand sections is listed in Algorithm 2.

4.2. Algorithm 2.

Input : $C \in R^{pxq}$, $B \in R^{px\,r}$
output : x ($\in R^{\,q\,x\,r}$) =arg $min_{x>0}\,\|Cx - B\|_F^2$

1. Calculate C^TC and C^TB
2. Start Fj = Ø and Gj = {1, ...q} ∀j ∈ {1, ...r}.
   Create X> 0, Y = - CTB, α (∈ Rr) = 3 and β (∈ Rr) = q +1
3. Calculate X_{Fj}andY_{Gj}; For all j ε {1, ..., r} by the Eq. (4.10) - (4.11) using column grouping.
4. While every (X_{Fj}, Y_{Gj}) is not worth doing
5. Specify a column index in the name of an ineligible settlement: I = {j: (X_{Fj}, Y_{Gj}) improper}
6. Calculate Vj∀ j ε I by the Eq. (4.12).
7. For all j ∈ I with | Vj | < B I, area β j = | Vj| , A j = 3 times $\hat{V}$= Vj
8. For all j ∈ I with | Vj | ≥ β j and α j ≥1, create, α i = α j - 1 and V j = Vj
9. For all j ∈ I with | Vj | ≥β j and α j¬ = 0, make $\hat{V}$ by the Eq. (4.15)
10. Update F_{jj} and C_j, ∀j ε I with the Eq. (4.14)
11. Update X_{Fj}dan$Y_{Ci}$$\forall_j$ ε I with the Eq. (4.10), (4.11) which using column grouping
12. End while

In NMF, discontinuation criteria can be reviewed from whether the point of stasiment, which is done with the following criteria:
In accordance with the terms of the KKT, (W, H) obtains the stasimer point of the Equation (4.2) if and only if

$$W \geq 0 , H \geq 0 \qquad (4.19)$$

$$\Delta fw = 0\, f(W, H) / 0W \geq 0 , \Delta fH (W, H) / 0H \geq 0 \qquad (4.20)$$

$$W * \nabla fw = 0 , H * \nabla f_H = 0 \qquad (4.21)$$

By * denoting multiplication of each element, define projected stadient $\nabla^P fw$ as
$(\nabla^P fw)_I = \int_0^{(\nabla fw)i}$ if $(\nabla fw)_I < 0$ or $W_i > 0$ if not and $\nabla^P fw$ by analog, condition in Eq. (4.14) - (4.21) can be written again as:$\nabla^P fw = 0$ dan$\nabla^P fH = 0$. Use the projected gradient number defined as:

$$\Delta = \sqrt{\|\nabla^P fw\|_F^2 + \| \nabla^P fh\|_F^2} \qquad (4.22)$$

By using this definition, the termination criterion is

$$\frac{\Delta}{\Delta_0} \leq \varepsilon \qquad (4.23)$$

Where Δ_0 is the value of Δ by marking the initial value of **W** and **H**, ε the tolerance value to be selected.

5. Conclusion and Future Work

This study proposes a new algorithm to complete FMN based on the ANLS framework as an alternative algorithm model for data reduction. Such an algorithm is built on the upper-pivot method for the nonnegative least square problem. Methods include the exchange of several variables between sets in the process with the aim of reaching the final partition into the active and passive set quicker. The method is improved so that the power applies the reduced gradient method to handle multiple right-hand case cases of nonnegative least square issues efficiently. This new algorithm has the same convergence properties as ANLS framework and can be extended to FMN formulation constrained such that FMN matrix power sparse.
Further research that can be done is to add Quasi-Newton and Variable -Matric methods in addition to the reduced gradient method. The addition of this method will be able to further accelerate convergence.

6. REFERENCES

[1] B. Corona, M. Nakano, H. Pérez, “Adaptive Watermarking Algorithm for Binary Image Watermarks”, *Lecture Notes in Computer Science,Springer, pp. 207-215, 2004.*
A. Reddy and B. N. Chatterji, "A new wavelet based logo-watermarking scheme," Pattern Recognition Letters, vol. 26, pp. 1019-1027, 2005.

[2] P. S. Huang, C. S. Chiang, C. P. Chang, and T. M. Tu, "Robust spatial watermarking technique for colour images via direct saturation adjustment," Vision, Image and Signal Processing, IEE Proceedings -, vol. 152, pp. 561-574, 2005.

[3] F. Gonzalez and J. Hernandez, " A tutorial on Digital Watermarking ", In IEEE annual Carnahan conference on security technology, Spain, 1999.

[4] D. Kunder, "Multi-resolution Digital Watermarking Algorithms and Implications for Multimedia Signals", Ph.D. thesis, university of Toronto, Canada, 2001.

[5] J. Eggers, J. Su and B. Girod," Robustness of a Blind Image Watermarking Scheme", Proc. IEEE Int. Conf. on Image Proc., Vancouver, 2000.

[6] Barni M., Bartolini F., Piva A., Multichannel watermarking of color images, IEEE Transaction on Circuits and Systems of Video Technology 12(3) (2002) 142-156.

[7] Kundur D., Hatzinakos D., Towards robust logo watermarking using multiresolution image fusion, IEEE Transcations on Multimedia 6 (2004) 185-197.

[8] C.S. Lu, H.Y.M Liao, “Multipurpose watermarking for image authentication and protection,” *IEEE Transaction on Image Processing*, vol. 10, pp. 1579-1592, Oct. 2001.

[9] L. Ghouti, A. Bouridane, M.K. Ibrahim, and S. Boussakta, “Digital image watermarking using balanced multiwavelets”, *IEEE Trans. Signal Process.*, 2006, Vol. 54, No. 4, pp. 1519-1536.

[10] P. Tay and J. Havlicek, "Image Watermarking Using Wavelets", in *Proceedings of the 2002 IEEE*, pp. II.258 – II.261, 2002.

[11] P. Kumswat, Ki. Attakitmongcol and A. Striaew, "A New Approach for Optimization in Image Watermarking by Using Genetic Algorithms", *IEEE Transactions on Signal Processing*, Vol. 53, No. 12, pp. 4707-4719, December, 2005.

[12] H. Daren, L. Jifuen,H. Jiwu, and L. Hongmei, "A DWT-Based Image Watermarking Algorithm", in *Proceedings of the IEEE International Conferenceon Multimedia and Expo*, pp. 429-432, 2001.

[13] C. Hsu and J. Wu, "Multi-resolution Watermarking for Digital Images", *IEEETransactions on Circuits and Systems- II*, Vol. 45, No. 8, pp. 1097-1101, August 1998.
[14] R. Mehul, "Discrete Wavelet Transform Based Multiple Watermarking Scheme", in *Proceedings ofthe 2003 IEEE TENCON*, pp. 935-938, 2003.
[15] Tu, J.H., Rowley, C.W., Luchtenburg, D.M., Brunton, S.L., Kutz, J.N.: "On dynamic mode decomposition: Theory and applications". Journal of Computational Dynamics, 2014. 1(2), 391– 421)

MODIFICATION OF RC6 BLOCK CIPHER ALGORITHM ON DIGITAL IMAGE

Mardiana[1] , Fajrillah[2] ,Yuyun Dwi Lestari[3],Ummul Khair[4]

[13]Department of Informatics, Sekolah Tinggi Teknik Harapan, Medan, Indonesia
[2]Department of Management, STIE IBBI, Medan, Indonesia

[1]mardiana.mcom@gmail.com

Abstrack. In the digital image of each pixel consists of three components, namely R, G, and B. If the three components are combined in a block then the block will have data with 24 bit or three characters. By default shortage of bits in each block will be filled by the next bit of the pixel so that each block will have the appropriate size is 32bit. The problem that happens is if the number of pixels is not in accordance with the division so it is necessary to block or adding padding bit 0 in register block, if padding occurs then the encrypted data size will change that affects the size of the image will change and will be hard pressed in preparing the return data encryption results in the form of images.

Keywords: RC6 Modification, Block Cipher, Digital Image, Encryption

1. Introduction

The importance of the security of digital images is increasing nowadays since the raise of communications using digital products over the network. Some experiments have been investigated over the encryption methods as the most important key to secure digital data [1-7]. DES (Data Encryption Data), RAS, and IDEA (International Data Encryption Algorithm) [8-10] have been studied to have a long computational time to process an image and need more power to compute the data. Besides, they also have some problems to secure the digital images [11]. Li *et al* that studied about a new chaotic key-based algorithm (CKBA) also found that CKBA method is not very secure, especially if it is seen from cryptographic viewpoint [12].

RC6 (Ron's Code 6) is an evolutionary method of RC5 block cipher. It is an encryption method which is easy to be applied for data security [14]. It has received much attention due to its simpler design. The difference between RC6 and RC5 is in the number of working registers used in the system. The addition of these working registers enhances the quality of security, decreases the rounds, and raises the throughput. Ahmed *et al* found that this method can be considered to be one of a real-time secure symmetric encryption for digital data such as digital images [11].

However, RC6 is found to have a problem with the RGB digital images which have an odd pixel size. The problem is happened if the number of pixels does not match the division of the block so that the padding is required. If padding occurs then the data size of the encryption will change. This condition will affect the image size, so it will be very to compile the data to returned it into the form of image.

Therefore, in this study, we proposed the modification of RC6 by changing the number of bits in every A, B, C, and D register and altering the constants P_{24} and Q_{24}. We studied the implementation of those two parameters in the digital images application. The comparation between the RC6 and the modified of RC6 is also given to evaluate the effectiveness of modified RC6.

2. Methodology

The encoding is done on each image pixel according to the width and height of the image. This is possible considering an image is a row of pixels that have components R (Red), G (Green), and B (Blue). These components are integers so they can be operated by converting them to ASCII form first.

RC6 is a chipper block method that uses four blocks of registers with each size of 32-bits. In other words each block will store four characters. In a digital image each pixel consists of three components: R, G, and B. If the three components are combined in a block then the block will have data of 24-bits or three characters. By default the lack of bits on each block will be filled by the bits of the next pixel so that each block will have a corresponding size of 32-bits.

In addition, RC6 algorithm is broadly divided into three algorithms, namely,

1 The Key Generation Algorithm. It is an algorithm used for key initialization that will be used in each round in both encryption operation and decryption operation.

2 Encryption Algorithm is an algorithm used for the changing operation of plaintext encryption into chipertext

3 Decryption Algorithm is an algorithm used for the changing operation of decryption to plaintext

The problem that occurs in RC6 algorithm is if the number of pixels does not match the division of the block so it needs a padding or the addition of bit 0 in the block register, if padding occurs then the data size of the encryption will change which affect the image size will change and will be very difficult in the rearrangement Encrypted data into image form. To overcome these problems then the modified RC6 algorithm was created.

The modified RC6 is done by changing the number of bits so that the constants of Pw and Qw are altered as well. For the RC6 algorithm, the value of Pw and Qw constants are 32 bits. The expansions of them have 8 digits with each digit containing 4 pixels. On the modified mode of RC6 algorithm, Pw and Qw constants only have 24 bits; so, there are 2 unused key expansions which means that the key expansion is reduced by 8 bits value. The difference in the registers is shown in the Figure 1.

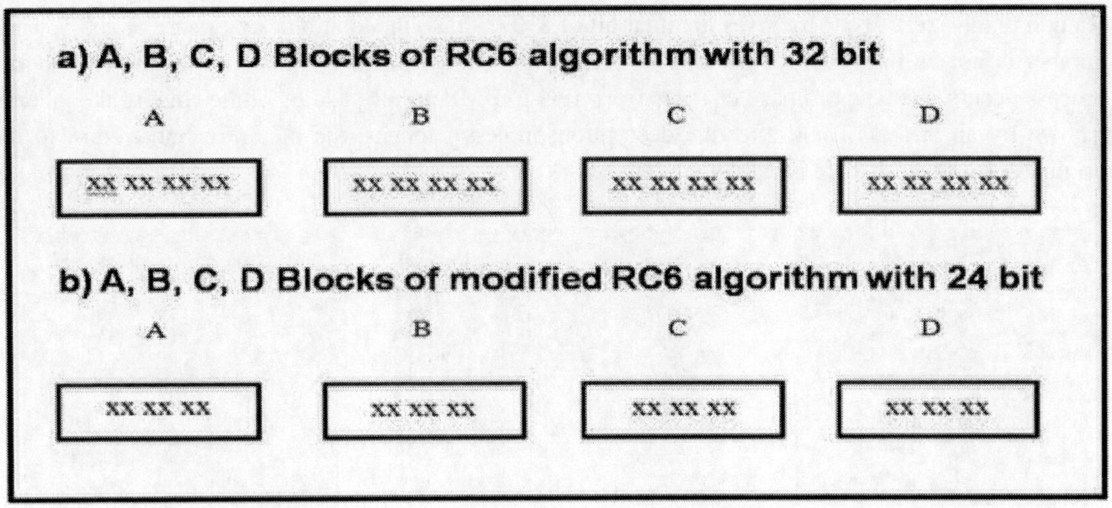

Fig.1 The difference between A, B, C, and D registers of RC6 and modified RC6

In this paper, to compare between RC6 and modified RC6, some investigations are done by applying RC6 as well as modified RC6 to several digital images. The images used in this investigation are in the format of BMP, JPG, or JPEG.

The First stage was the encryption process. In this stage, the pixel value of the images was then read and converted into RGB value. This value was converted again to ASCII, and being encrypted by whether RC6 or modified RC6. When the encryption process occured, the key was input into the data. After encryption process had done, the image was converted again into RGB value and then to pixel value, after that the image was saved in an encrypted form.

The second stage was the descryption process. The steps in this process are roughly the same as in the encryption process. After this process happened, the image had returned into its initial form.

3. Result and Discussion

3.1. Analysis of RC6 in the application of grayscale images

As have been reported by Ahmed *et al* that the use of RC6 in the grayscale digital images having size of 256 x 256 can be efficiently done [11]. It can be explained as the result of the limitation of the images size itself, i.e. 256 x 256. This number will ease the allocation of the value of every colors since every pixels has one value for each. That size limitation also makes the appearance of padding to be small or no padding sice the total value of the that size is calculated to be 4096; in other words, there are 16384 pairs of blocks, an even value. The same case also occurs when the images size is 512 x 512 since the total value of it is calculated to be 16384 as well.

3.2. Analysis of RC6 in the application of RGB images

In this study, RC6 algorithm is applied to two types RGB images, the first one is a picture that has an odd pixel size and the second one is an image that has even pixel size. The odd pixel being used is 591 x 616 in size and the even one is having size of 100 x 100.

An odd pixel size, 591 x 616, gives the total bit of 8737344, so the total block pairs are 68260,5. This number is not an integers, this padding is needed. As shown in the Figure 2, After the encryption process occurs, the type of image changes from JPG to BMP, it has padding at the edge of the picture (shown by an arrow). Then, after the decryption process happens, the picture cannot return to its original form because there is padding in it.

Fig.2 Application of RC6 algorithm to a chosen plainimage which has size 591 x 616

Therefore, an even pixel size is investigated, we use the size of 100 x 100 which gives the total block pairs to be 1875, an even value. Thus, after applying RC6 algorithm, the type of image changes from JPG to BMP. It does not have padding on the image because the total bit which exists in the picture meet the needs of the block; so, the original image can be returned. This process can be seen in the Figure 3.

Fig.3 Application of RC6 algorithm to a chosen plainimage which has size 100 x 100

3.3. Analysis of modified RC6 in the apllication of RGB images

Because of the problem of padding as mentioned above, we modify the RC6 algorithm that we called as modified RC6 (M-RC6). In this algorithm, we change the internal key generator based on the secret key 'K' to fill in the 'S' key table, where the size of the S key table is calculted as follows,

$$t = 2\,(r + 1) \text{ words} \tag{1}$$

The modification uses 2 magic constants, i.e., Pw and Qw which have function to generate the internal keys. The internal key generator of RC6 algorithm itself was altered from 32 bits to 24 bits; so the value of Pw and Qw also changes. In the RC6, the value of Pw is b7e15163, whereas the value of Qw is 9e3779b9. Due to the decrease of bit number from 32 to 24, the value of Pw becomes b7e151 and the value of Qw becomes 9e3779. Thus, the blocks A, B, C, and D are changed as well. As shown in the Figure, the modification of the bit number lets one pixel to be enough to fill in one block, so there is no padding needed since the RGB components can be borrowed from their neighbor blocks. Besides, modified RC6 can reduce the capacity or size of variables used during the process, thus making the process to be more efficient in consuming the memory as well as to be faster.

Fig.4 Application of modified RC6 algorithm to a chosen plainimage

We have attempted to apply this modified RC6 algorithm to some images and those images always can be returned to their original form as shown in the Figure 4.

IV. Conclusion

1. RC6 algorithm uses four block blocks with each size of 32 bits, while modified RC6 uses four registers block with each size of 24 bits. Thus, modified RC6 is more suitable for color images because each color image pixel has 24 bits.

2. RC6 algorithm is well applied to a greyscale image. However, it is not suitable to be used in the color images due to padding problems. It can only be used in some certain pixel sizes.

3. Modified RC6 can solve the padding problems that occur in digital images by changing the size of each block on the A, B, C, D blocks from 32 bits to 24 bits, so that one pixel is sufficient to fill a block. In the other word, no padding is required.

4. Modified RC6 can reduce the capacity or size of variables used during the process, thus making the process to be more efficient in consuming the memory and to be faster.

References

[1] Howard Cheng and Xiaobo Li, "Partial encryption of compressed images and videos," *IEEE Trans. Signal Processing,* vol. 48, no. 8, pp. 2439-2451,2000.

[2] Masaki Miyamoto, Kiyoshi Tanaka, and Tatsuo Sugimura, "Truncated Baker transformation and its extension to image encryption," in *Pmc. SPIE,* 1999, vol. 3814, pp. 13-25.

[3] Jiri Fridrich, "Symmetric ciphers based on two-dimensional chaotic maps," *Int. J. Bifurcation and Chaos,* vol. *8,* no. 6, "Fast encryption of image data using chaotic kolmogrov flows," *J. Electronic Imaging,* vol. 7, no.pp. 1259-1284, 1998.

[4] Philip P. Dang and Paul M. Chau, "Image encryption for secure internet multimedia applications," *IEEE Trans. Consumer Electronics,* vol. 46, no. 3, pp. 395403,2000.

[5] Josef Scharinger, 2, pp. 318-325, 1998.

[6] C. Alexopoulos, Nikolaos G. Bourbakis, and N. Ioannou, "Image encrytion method using a class of fractals," J. *Electronic Imaging,* vol. 4, no. 3, pp. 251-259, 1995.

[7] Henry Ker-Chang Chang and Jiang-Long Liu, "A linear quadtree compression scheme for image encryption," *Signal Processing: Image Communication,* vol. 10, pp. 279-290, 1997.

[8] National Bureau of standards. "Data Encryption Standard," Federal WXHL 255 Information processing standards Publication 46, US Government Printing Office, Washington, D.C., 1977.

[9] W. Stallings., "Network and Internetwork Security: Principles and Practice," Prentice-Hall, New Jersey, 1995.

[10] B. Schneier, "Applied Cryptography - Protocols, algorithms, and source code in C," John Wiley & Sons, Inc., New York, second edition, 1996.

[11] Ahmed, Hossam El-din H., Hamdy M. Kalash, and OS Farag Allah. "Encryption efficiency analysis and security evaluation of RC6 block cipher for digital images." In *Electrical Engineering, 2007. ICEE'07. International Conference on*, pp. 1-7. IEEE, 2007.

[12] Li, Shujun, and Xuan Zheng. "Cryptanalysis of a chaotic image encryption method." In *Circuits and Systems, 2002. ISCAS 2002. IEEE International Symposium on*, vol. 2, pp. II-II. IEEE, 2002.

[13] R.L. Rivest, "RC5 Encryption Algorithm," Dr Dobbs Journal, vol. 226, pp. 146-148, Jan. 1995.

[14] R.L. Rivest, M.J.B. Robshaw, R.Sidney, and Y.L. Yin, "The RC6TM Block Chiper," v.1.1, August 20, 1998.

The Time Window Vehicle Routing Problem Considering Closed Route

Nenna Irsa Syahputri[1] and Herman Mawengkang[2]

[1]Department of Informatics Technic, STTH, Jl. H.M. Jhoni No. 70 C, Medan
[2]Mathematics Department, University of Sumatera Utara

[1]nennairsa@stth-medan.ac.id [2]hmawengkang@yahoo.com

Abstract. The Vehicle Routing Problem (VRP) determines the optimal set of routes used by a fleet of vehicles to serve a given set of customers on a predefined graph; the objective is to minimize the total travel cost (related to the travel times or distances) and operational cost (related to the number of vehicles used). In this paper we study a variant of the predefined graph: given a weighted graph G and vertices a and b, and given a set X of closed paths in G, find the minimum total travel cost of a-b path P such that no path in X is a subpath of P. Path P is allowed to repeat vertices and edges. We use integer programming model to describe the problem. A feasible neighbourhood approach is proposed to solve the model
Keywords: Vehicle routing problem, closed path, integer programming, feasible neighbourhood search

.

1. Introduction

The Vehicle Routing Problem (VRP) is defined on a given graph $G = (V, A)$, where $V = \{v_1,v_2,\ldots,v_n\}$ is a set of vertices and $A \subseteq \{(v_i, v_j) : i \neq j, v_i, v_j \in V\}$ is the arc set. An optimal set of routes, composed of a cyclic linkage of arcs starting and ending at the depot, is selected to serve a given set of customers at vertices. The problem aims at minimizing the total travel cost (proportional to the travel times or distances) and operational cost (proportional to the number of vehicles used). This problem was first introduced by [1] to solve a real-world application concerning the delivery of gasoline to service stations. Due to its important role in the logistic system, the research domain of VRP has been widely explored. A comprehensive overview of the Vehicle Routing Problem can be found in [6] which discusses problem formulations, solution techniques, important variants and applications.

In reality, most of its operation, the logistic service faces difficulty to be in time to reach its customers, due to traffic congestion or other events. This paper discusses a variant of VRP in which there may be closed route. Forbidden sub-route involving pairs of edges occur frequently ("No left turn") and can occur dynamically due to rush hour constraints, lane closures, construction, etc. Longer forbidden subpaths are less common, but can arise, for example if heavy traffic makes it impossible to turn left soon after entering a multi-lane roadway from the right. If we are routing a single vehicle it is more natural to find a detour from the point of failure when a forbidden path is discovered.

Logically, a model whereby an algorithm identifies a potential path, and then this path is tried out on the actual network. In case of failure, further tests can be done to pinpoint a minimal forbidden subpath. Because such tests are expensive, a routing algorithm should try out as few paths as possible. In particular it is practically impossible to identify all forbidden paths ahead of time—we have an exponential number of possible paths to examine in the network. Therefore we impose an assumption of having no a priori knowledge of the forbidden paths, and of identifying forbidden paths only by testing feasibility of a path.

In terms of graph the problem can be defined as : given a graph *G*(*V*, *A*), and vertices *s* and *t*, and given a set *X* of forbidden paths in *G*, find optimal set of route *P* such that no path in *X* is a subpath of *P*. Routes in *X are called* exceptions, and the desired route is called a shortest exception avoiding route. We allow an exception avoiding route to be non-simple graph, i.e., to repeat vertices and edges. In fact the problem becomes hard if the solution is restricted to simple [2] . This problem has been called the Shortest Path Problem with Forbidden Paths by [3], [4].

2. Preliminary

We are given an directed graph *G*(*V*, *A*) with $n = |V|$ vertices and $m = |A|$ edges where each edge $e \in A$ has a positive weight denoting its length. We are also given a source vertex $s \in V$, a destination vertex $t \in V$, and a set *X* of route in *G*. The graph *G* together with X models a vehicle routing network in which a vehicle cannot follow any route in *X* because of the physical constraints .. We want to find a shortest route from s to t that does not contain any route in *X* as a subpath—we make the goal more precise as follows. A route is a sequence of vertices each joined by an edge to the next vertex in the sequence. Note that we allow a route to visit vertices and edges more than once. If a route does not visit any vertex more than once, we explicitly call it a simple route. *A* simple directed route from vertex *v* to vertex *w* in *G* is called a forbidden route or an exception if a vehicle cannot follow the route from *v* to w because of the physical constraints. Given a set *X* of forbidden route, a route (v_1, v_2, v_3, . . . , v_l) is said to avoid *A* if (v_i, v_i+ 1, . . ., v_j) $\notin A$ for all *i*, *j* such that $1 \leq i < j \leq l$. A route *P* from *s* to *t* is called a shortest A-avoiding route if the length of *P* is the shortest among all *A*-avoiding route from *s* to *t*. We will use the term "exception avoiding" instead of "*X*-avoiding" when *A* is equal to *X*, the set of all forbidden paths in *G*.

3. Vehicle Routing Problem

In the late fifties, [1] introduced the VRP, which can be viewed as an m-TSP with customer demands and vehicle capacity. An example of such a VRP is shown in Figure 1. The VRP introduced in [1] , strictly speaking, is called *Capacitated Vehicle Routing Problem* (CVRP), and is one of the simplest vehicle routing problems.

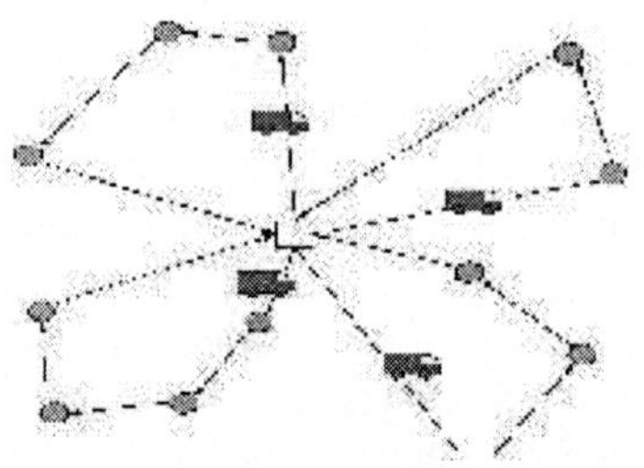

Figure 1. An example of the VRP.

3.1. Problem Formulation of VRP

Given a graph *G*(*V*, *A*) with nodes $V = C \cup \{0\}$ and arcs *A*, where *C* is the set of customers, and 0 is the depot. Moreover, we have a set *R* of resources which e.g. can be load and/or time. Each resource *r* $\in R$ has a resource window $[a_i^r, b_i^r]$ that must be met upon arrival to node $i \in V$, and a consumption $t_{ij}^r \geq 0$ for using arc $(i, j) \in A$. A resource consumption at a node $i \in C$ is modeled by a resource consumption at edge (*i*, *j*), and hence usually $t_{0j}^r \geq 0$ for all $j \in C$. A global capacity limit *Q* can be modeled by imposing a resource window [0, *Q*] for the depot node 0.

The VRP can now be stated as: Find a set of routes starting and ending at the depot node 0 satisfying all resource windows, such that the cost is minimized and all customers *C* are visited.

A solution to the VRP will consist of a number of routes

$$0 \to i_1^1 \to \cdots \to i_{k_1}^1 \to 0,$$
$$0 \to i_1^2 \to \cdots \to i_{k_2}^2 \to 0,$$
$$\vdots$$
$$0 \to i_1^n \to \cdots \to i_{k_n}^n \to 0$$

where n is the number of vehicles, and k_j is the length of the j'th route.

3.2. Model

In the following let c_{ij} be the cost of arc $(i, j) \in A$, x_{ij} be the binary variable indicating the use of arc (i, j) $\in$ A, and t_{ij}^r (the resource stamp) be the consumption of resource r $\in$ R at the beginning of arc $(i, j) \in A$. Let $\delta^+(i)$ and $\delta^-(i)$ be the set of outgoing respectively ingoing arcs of node $i \in V$. Combining the two index model from Bard et al. [3] with the constraints ensuring the time windows for the ATSP by Ascheuer et al. [1] a mathematical model can be formulated as follows:

$$\min \sum_{(i,j)\in A} c_{ij} x_{ij} \tag{1}$$

$$\text{s.t.} \quad \sum_{(s,j)\in\delta^+(i)} x_{ij} = 1 \qquad \forall\, i \in C \tag{2}$$

$$\sum_{(s,j)\in\delta^-(i)} x_{ij} = \sum_{(s,j)\in\delta^+(i)} x_{ij} \qquad \forall\, i \in V \tag{3}$$

$$\sum_{(s,j)\in\delta^-(i)} (T_{ji}^r + t_{ji}^r x_{ij}) \le \sum_{(s,j)\in\delta^+(i)} T_{ji}^r \qquad \forall\, r \in R, \forall\, i \in C \tag{4}$$

$$a_i^r x_{ij} \le T_{ji}^r \le b_i^r x_{ij} \qquad \forall\, r \in R,\ \forall\, (i, j) \in A \tag{5}$$

$$T_{ij}^r \ge 0 \qquad \forall\, r \in R,\ \forall\, (i, j) \in A \tag{6}$$

$$x_{ij} \in \{0,1\} \qquad \forall\, (i, j) \in A \tag{7}$$

The objective (1) sums up the cost of the used arcs. Constraints (2) ensure that each customer is visited exactly once, and (3) are the flow conservation constraints. Constraints (4) and (5) ensure the resource windows are satisfied. It is assumed that the bounds on the depot are always satisfied. Note, that no sub-tours can be present since only one resource stamp per arc exists and the arc weights are positive for all $(i, j) \in A : i \in C$.

For a one dimensional resource such as *load* a stronger lower bound of the LP relaxation can be obtained by replacing (4) to (6) with $\sum_{(i,j)\in\delta^+(S)} x_{ij} \ge r(S)$, where $r(S)$ is a minimum number of vehicles needed to service the set S. All though this model can not be directly solved it is possible to overcome this problem by only including the constraints that are violated. For more details on how to separate the constraint and calculate the value of $r(S)$ the reader is refered to [6].

3.3. Model For VRP With Forbidden Route

Given a set X of forbidden route, a route $(v_1, v_2, v_3, \ldots, v_l)$ is said to avoid A if $(v_i, v_i + 1, \ldots, v_j) \notin A$ for all i, j such that $1 \le i < j \le l$. A route P from s to t is called a shortest A-avoiding route if the length of P is the shortest among all A-avoiding route from s to t. We will use the term "exception avoiding" instead of "X-avoiding" when A is equal to X, the set of all forbidden paths in G.

It is necessarily to assume that costumers are not in the forbidden route.

The model can be written as :

$$\min \sum_{(i,j)\in A;(i,j)\notin X} c_{ij} x_{ij} \tag{8}$$

s.t. $\sum_{(s,j)\in\delta^+(i)} x_{ij} = 1$ $\quad \forall\, i \in C,\, i \notin X$ (9)

$$\sum_{(s,j)\in\delta^-(i)} x_{ij} = \sum_{(s,j)\in\delta^+(i)} x_{ij} \qquad i \notin X,\ \forall\, i \in V \tag{10}$$

$$\sum_{(s,j)\in\delta^-(i)} (T_{ji}^r + t_{ji}^r x_{ij}) \le \sum_{(s,j)\in\delta^+(i)} T_{ji}^r \qquad i \notin X,\ \forall\, r \in R,\ \forall\, i \in C \tag{11}$$

$$a_i^r x_{ij} \le T_{ji}^r \le b_i^r x_{ij} \qquad (i,j) \notin X, \forall\, r \in R,\ \forall\, (i,j) \in A \tag{12}$$

$$T_{ij}^r \ge 0 \qquad (i,j) \notin X, \forall\, r \in R,\ \forall\, (i,j) \in A \tag{13}$$

$$x_{ij} \in \{0,1\} \qquad (i,j) \notin X, \forall\, (i,j) \in A \tag{14}$$

In the above model every vehicle assigned will not be travelling along the forbidden route.

4. The Solution Basic Approach

Consider a mixed integer linear programming (MILP) problem with the following form

$$\text{Minimize } P = c^T x \tag{15}$$

$$\text{Subject to } Ax \le b \tag{16}$$

$$x \ge 0 \tag{17}$$

$$x_j \text{ integer for some } j \in J \tag{18}$$

A component of the optimal basic feasible vector $(x_B)_k$, to MILP solved as continuous can be written as

$$(x_B)_k = \beta_k - \alpha_{k1}(x_N)_1 - \cdots - \alpha_{kj}(x_N)_j - \cdots - \alpha_{k,n-m}(x_N)_{n-m} \tag{19}$$

Note that, this expression can be found in the final tableau of Simplex procedure. If $(x_B)_k$ is an integer variable and we assume that β_k is not an integer, the partitioning of β_k into the integer and fractional components is given by

$$\beta_k = [\beta_k] + f_k,\ 0 \le f_k \le 1 \tag{20}$$

suppose we wish to increase $(x_B)_k$ to its nearest integer, $([\beta]+1)$. Based on the idea of suboptimal solutions we may elevate a particular nonbasic variable, say $(x_N)_{j^*}$, above its bound of zero, provided α_{kj^*}, as one of the element of the vector α_{j^*}, is negative. Let Δ_{j^*} be amount of movement of the non variable $(x_N)_{j^*}$, such that the numerical value of scalar $(x_B)_k$ is integer. Referring to Eqn.(19), Δ_{j^*} can then be expressed as

$$\Delta_{f^*} = \frac{1 - f_k}{-\alpha_{kj^*}} \tag{21}$$

while the remaining nonbasic stay at zero. It can be seen that after substituting (21) into (19) for $(x_N)_{j^*}$ and taking into account the partitioning of β_k given in (20), we obtain

$$(x_B)_k = [\beta] + 1$$

Thus, $(x_B)_k$ is now an integer.

It is now clear that a nonbasic variable plays an important role to integerize the corresponding basic variable. Therefore, the following result is necessary in order to confirm that must be a non-integer variable to work with in integerizing process.

Theorem 1. Suppose the MILP problem (1)-(4) has an optimal solution, then some of the nonbasic variables. $(x_N)_j$, $j = 1, \ldots, n$, must be non-integer variables.

Proof.

Solving problem as a continuous of slack variables (which are non-integer, except in the case of equality constraint). If we assume that the vector of basic variables x_B consists of all the slack variables then all integer variables would be in the nonbasic vector x_N and therefore integer valued.

It is clear that the other components, $(x_B)_{i\neq k}$, of vector x_B will also be affected as the numerical value of the scalar $(x_N)_{j^*}$ increases to Δ_{j^*}. Consequently, if some element of vector α_{j^*}, i.e., α_{j^*} for $i \neq k$, are positive, then the corresponding element of x_B will decrease, and eventually may pass through zero. However, any component of vector x must not go below zero due to the non-negativity restriction. Therefore, a formula, called the minimum ratio test is needed in order to see what is the maximum movement of the nonbasic $(x_N)_{j*}$ such that all components of x remain feasible. This ratio test would include two cases.

1. A basic variable, $(x_B)_{i\neq k}$ decreases to zero (lower bound) first.
2. The basic variable, $(x_B)_k$ increases to an integer.

Specifically, corresponding to each of these two cases above, one would compute

$$\theta_1 = \min_{i\neq k|\alpha_{j*}>0} \left\{ \frac{\beta_i}{\alpha_{j*}} \right\} \tag{22}$$

$$\theta_2 = \Delta_{j*} \tag{23}$$

How far one can release the nonbasic $(x_N)_{j*}$ from its bound of zero, such that vector x remains feasible, will depend on the ratio test θ^* given below

$$\theta^* = \min(\theta_1, \theta_2) \tag{24}$$

obviously, if $\theta^* = \theta_1$, one of the basic variable $(x_B)_{i\neq k}$ will hit the lower bound before $(x_B)_k$ becomes integer. If $\theta^* = \theta_2$, the numerical value of the basic variable $(x_B)_k$ will be integer and feasibility is still maintained. Analogously, we would be able to reduce the numerical value of the basic variable $(x_B)_k$ to its closest integer $[\beta_k]$. In this case the amount of movement of a particular nonbasic variable, $(x_N)_{j*}$, corresponding to any positive element of vector $\alpha_{j'}$, is given by

$$\Delta_{f'} = \frac{f_k}{\alpha_{kj}} \tag{25}$$

In order to maintain the feasibility, the ratio test θ^* is still needed. Consider the movement of a particular nonbasic variable, Δ, as expressed in Eqns.(21) and (25). The only factor that one needs to calculate is the corresponding element of vectorα. A vector α_j can be expressed as

$$\alpha_j = B^{-1}a_j, j = 1, \ldots, n - m \tag{26}$$

Therefore, in order to get a particular element of vector α_j we should be able to distinguish the corresponding column of matrix $[B]^{-1}$. Suppose we need the value of element α_{kj*}, letting v_k^T be the k-th column vector of $[B]^{-1}$, we then have

$$v_k^T = e_k^T B^{-1} \tag{27}$$

subsequently, the numerical value of α_{kj*} can be obtained from

$$\alpha_{kj*} = v_k^T a_{j*} \tag{28}$$

in Linear Programming (LP) terminology the operation conducted in Eqns. (27) and (28) is called the pricing operation. The vector of reduced costs d_j is used to measure the deterioration of the objective function value caused by releasing a nonbasic variable from its bound. Consequently, in deciding which nonbasic should be released in the integerizing process, the vector d_j must be taken into account, such that deterioration is minimized. Recall that the minimum continuous solution provides a lower bound to any integer-feasible solution. Nevertheless, the amount of movement of particular nonbasic variable as given in Eqns. (21) or (25), depends in some way on the corresponding element of vector α_j. Therefore it can be observed that the deterioration of the objective function value due to releasing a nonbasic variable $(x_N)_{j*}$ so as to integerize a basic variable $(x_B)_k$ may be measured by the ratio

$$\left| \frac{d_k}{\alpha_{kj*}} \right| \tag{29}$$

where $|a|$ means the absolute value of scalar a.

In order to minimize the deterioration of the optimal continuous solution we then use the following strategy for deciding which nonbasic variable may be increased from its bound of zero, that is,

$$\min_j \left\{ \left| \frac{d_k}{\alpha_{kj*}} \right| \right\}, j = 1, \ldots, n - m \tag{30}$$

From the "active constraint" strategy and the partitioning of the constraints corresponding to basic (B), superbasic (S) and nonbasic (N) variables we can write

$$\begin{bmatrix} B & S & N \\ & & I \end{bmatrix} \begin{bmatrix} x_b \\ x_N \\ x_S \end{bmatrix} = \begin{bmatrix} b \\ b_N \end{bmatrix} \tag{31}$$

or

$$Bx_b + Sx_S + Nx_N = b \tag{32}$$

$$x_N = b_N \tag{33}$$

The basis matrix B is assumed to be square and nonsingular, we get

$$x_B = \beta - Wx_S - \alpha x_N \tag{34}$$

where

$$\beta = B^{-1}b \tag{35}$$

$$W = B^{-1}S \tag{36}$$

$$\alpha = B^{-1}N \tag{37}$$

Expression (33) indicates that the nonbasic variables are being held equal to their bound. It is evident through the "nearly" basic expression of Eqn. (34), the integerizing strategy discussed in the previous section, designed for MILP problem can be implemented. Particularly, we would be able to release a nonbasic variable from its bound, Eqn (33), and exchange it with a corresponding basic variable in the integerizing process, although the solution would be degenerate.

Currently, we are in a position where particular basic variable, $(x_B)_k$ is being integerized, thereby a corresponding nonbasic variable, $(c_N)_{j*}$, is being released from its bound of zero. Suppose the maximum movement of $(x_N)_{j*}$ satisfies

$$\theta^* = \Delta_{j*}$$

such that $(x_B)_k$ is integer valued to exploit the manner of changing the basis, we would be able to move $(x_N)_{j*}$ into B (to replace $(x_B)_k$) and integer-valued $(x_B)_k$ into S in order to maintain the integer solution. We now have a degenerate solution since a basic variable is at its bound. The integerixing process continues with a new set [B,S]. In this case, eventually we may end up with all of the integer variables being superbasic.

5. Conclusions

This paper presents a VRP model in which there are some forbidden route. The framework of the model stems from VRP with time windows. Then exclude the forbidden route from the previous assigned route. We solve the model using a feasible neighbourhood search.

6. References

[1] G.B. Dantzig, G. B., & Ramser, J. H., 1959. *The truck dispatching problem.* Management Science, Vol. 6, 80.

[2] Szeider, S., 2003. *Finding paths in graphs avoiding forbidden transitions.* Discrete Appl. Math., Vol. 126, No. 2-3, pp. 261-273.

[3] Villeneuve, D., & Desaulniers, G., 2005. *The shortest paths with forbidden paths.* European Journal of Operational Research, Vol. 165, No. 1, pp. 97-107.

[4] Ahmed, M., & Lubiw, A., 2009. *Shortest path avoiding forbidden subpaths.* Symposium on Theoretical Aspect of Computer Science, pp. 63-74.

[5] N. Ascheuer, N., Fischetti, M., & Grätschel, M., 2001. *Solving the asymmetric travelling salesman problem with time windows by branch-and-cut.* Mathematical Programming, Vol. 90, No. 3, pp. 475-506.

[6] Toth, P., & Vigo, D. (2nd Eds.) 2014 *Vehicle routing : Problems, Methods, and Applications, Society for Industrial and Applied Mathematics & Mathematical Optimization Society,* Philadelphia PA.

Modeling And Simulation Of Electrical Prevenion System Using Arduino Uno,Gsm Modem, And Acs712 Current Sensor

[1]Ummul Khair, [2]Abdul Jabbar Lubis, [3]Indra Agustha, [4]Dharmawati, [5]M. Zulfin

[123]Program Studi Teknik Informatika Sekolah Tinggi Teknik Harapan Medan
Jl. HM Jhoni No. 70 Medan , Indonesia

[4] University Of North Sumatera

[1]ummul.kh@gmail.com

Abstract. The current electricity needs is very primary, all objects including electronics require power, it encourages people not to be able to save electricity so the theft of electric power would be done. The use of ACS712 current sensor as the sensor with arduino uno would find out the power consumption continuously and prevent the theft of electricity because of the use of electricity which has been determined by PLN and the people fetl that it is not enough for every house, so the author made a tool for prevention of theft of electric power by using the arduino uno, buzzer, ACS712 current sensor, lcd, and relay then the power usage can be controlled according to the use to prevent the occurrence of theft of electricity so the use can be seen directly on the lcd 16x2and GSM modem to give information to employees of PLN so that it can reduceelectrical theft by the public.

Keywords : Electric theft prevention tools, ACS712 current sensor, Arduino uno.

1. INTRODUCTION

The development era increasingly, the current electrical needs are very primary. All electronic tools require electric power. Without realizing, the people make electricity as primary needs, so that we are very difficult to control and economize the use of electricity. The economies of people in Indonesia who are still low, the electricity fare is more expensive and it is difficult to control and save electricity usage, so there is a fraud done by the society to meet all the needs by making the theft of electric power without thinking big risks. So it losses the PLN (State Electricity Company) and it can dangerous to the crime of electricity theft such as fire because of short circuit.

The previous research related to the topic of discussion on electrical controller, which is done Temy Nusa, Sherwin R.U.A ST, MT. Dr.Eng Meita Rumbayan, ST, MT from UNSRAT Manado (2015). Using this tool we can save electrical energy on devices that are considered quite wasteful by utilizing microelectronics designed to monitor the consumption of electrical energy that utilizes a step-down transformer to measure power from PLN and At mega 328 microcontroller which function to process data from parameters needed to obtain the value of electrical energy consumption.

It is the same with the research was done by Mukhlas Arihutomo, Muhammad Rivai, Suwito from Institut Teknologi Sepuluh Nopember (2012). Using this tool, we can know the use electric current distribution at home, so if it is the excessive use of nets electric current, it is possible the electric current theft use electrical theories, that is if the use of the greater load while the voltage constant, the current consumption will also be greater. The current usage data on a line will be sent to the monitoring center through the media trasmisi nets.

Based on problem above, So the writer made a research about a device that can prevent electrical theft by containing a series of microelectronics such as *Arduino uno, Buzzer, Modem GSM, ACS712 current sensor, Power Supply, LCD, socket and plug.* This electrical theft prevention system uses arduino uno as a microelectronics that processes all data from other electronic components such as data from ACS712 sensors that connect directly to electric current and arduino uno. The system has 2 ways of working namely:

1. This tool measures the electrical current in the program by creating a maximum electric current limit that will be passed by the ACS712 current then LCD always displays the passing current from the ACS712 sensor when the electric current meets or exceeds the maximal current load so Buzzer rings in second as a reminder to prevent the occurrence of electricity theft so the people can reduce excessive electricity consumption and power back under maximal.
2. This tool measures the electric current programmed by making the limitation of maximal electric current or the continuation of the first workings when buzzer sounds and the person is not responding the buzzer ringing up to a few minutes then the electricity theft prevention system directly give a short message (SMS) that utilizes GSM modem communications installed on this tool to the nearest PLN with the requirement of PLN phone number is listed on the program of electricity theft prevention system so PLN can know where the theft of electricity and give warning to the user of electricity can be given sanctions from PLN.

2. RESEARCH METHOD

This method used an experimental method. Experiments were carried out on the design of circuit blocks to produce a tool. By conducting experiments on the design and manufacture of this tool, is expected to be obtained series and programs in accordance with the function and purpose of designing this tool. Block diagram can be seen in Figure:

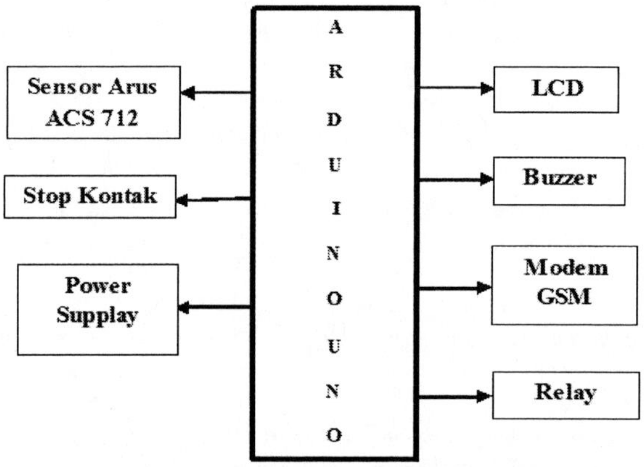

Figure 1 System Block Diagram

Explanation and function of each block in system block diagram above is as follows:

1. ACS 712 Current Sensor serves as a tool for measuring the electric current on this tool sensor. [2]
2. The socket has a function as the final terminal of the electrical installation
3. *Power Supply* has function as a charger on arduino uno. [1]
4. LCD (*Liquid Crystal Display*) has a function as an electrical current data. [3]
5. *Buzzer* is the sound signal when it exceeds a predetermined load [3]
6. GSM modem has a function as a modem to receive instruction *smartphone* android. [4]

2.1 Data Flow Diagram

The data flow diagram is a graphical representation that shows the flow of data from its source in the object through a process that transforms to another destination. Data *flow* diagram is a tool used to describe the flow of data through the system and the work or processing performed by the system, which exists on other objects.

1. Data Flow Diagram

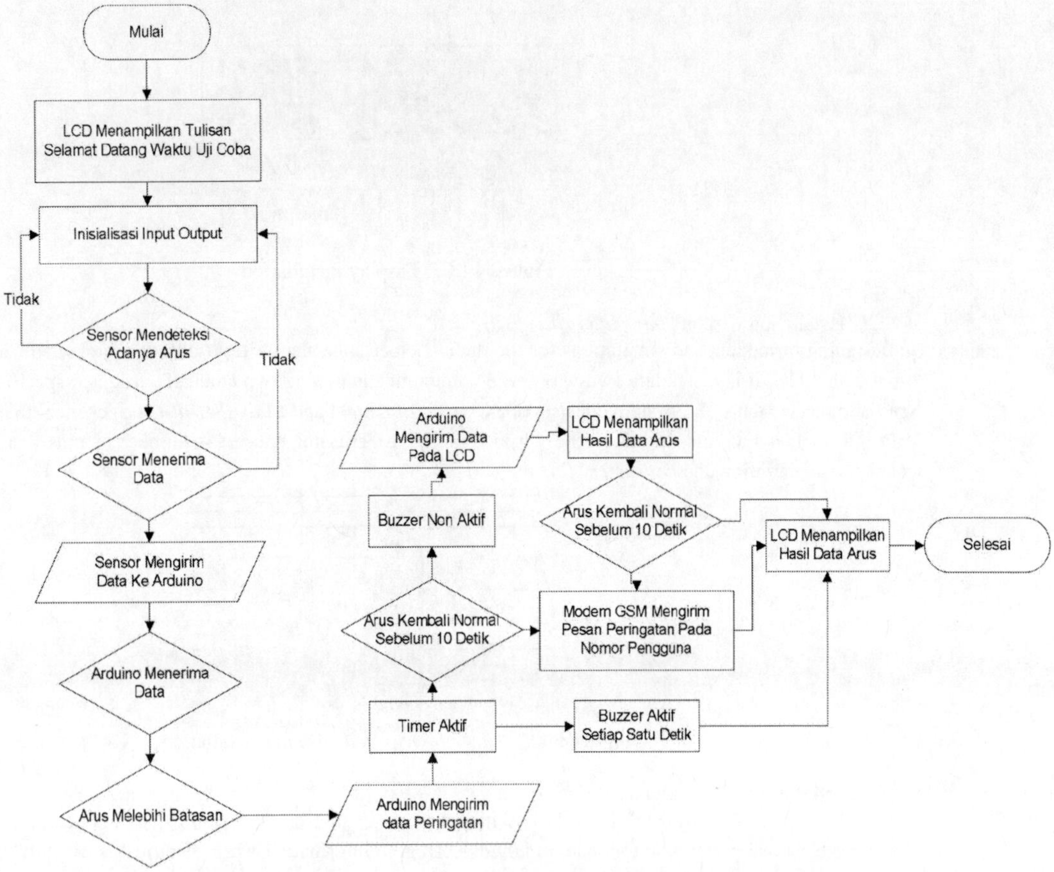

Figure 2 Data Flow Diagram Listrik Electrical Theft Prevention System

2. Arduino Uno Installation

In this research he writer used Arduino uno as a controller, Arduino uno is a *board* micro controller on the basis Atmega328 and a reset button. Arduino uno has a function as processing *input* and *ouput* data other components, While the series reset button has a function as a repeater of the arduino process if it does not work well when it runs. The Arduino uno series can be viewed in the image below:

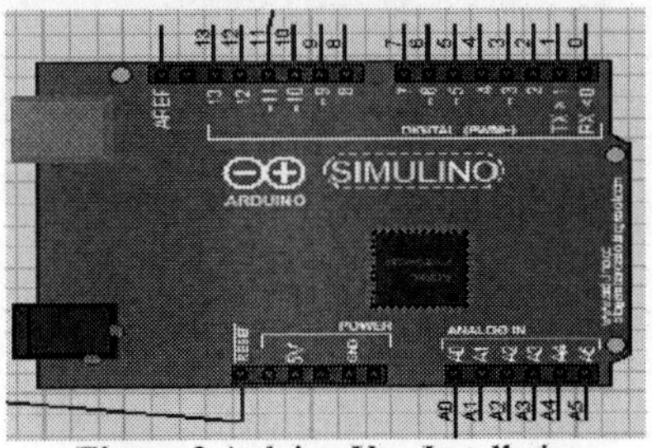

Figure 3 Arduino Uno Installation

3. LCD Display Installation

In this research the writer used LCD to perform information to electrical customers which use overload the electrical capacity that has been determined by arduino. The process of sending sms command if the usage of electricity exceeds that has been determined, and the process will run as before. The LCD circuit can be seen in the figure below:

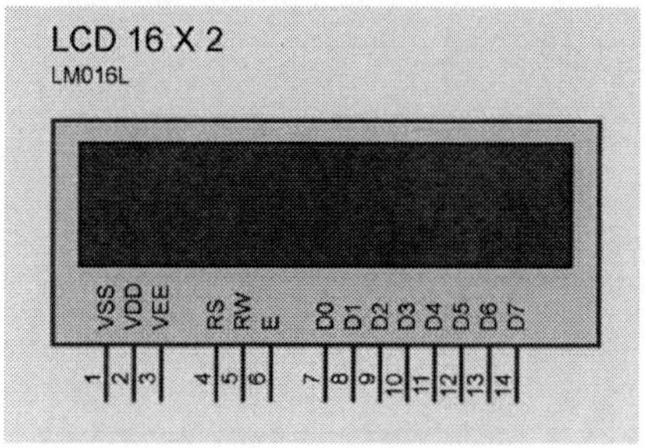

Figure 4 LCD Display Installation

4. I2C Installation Circuit (*Inter Intergrated Circuit*)

In designing of models and simulations for the theft of electricity, the writer used Inter Integrated Circuit or often called I2C, it is a standard two-way serial communication using two channels designed specifically to send and receive data. I2C system consists of SCL (*Serial Clock*) and SDA (*Serial Data*) channel that brings data information I2C and Arduino. The purpose of I2C LCD is to reduce the number of pins that use on connection on arduino and *character* LCD.

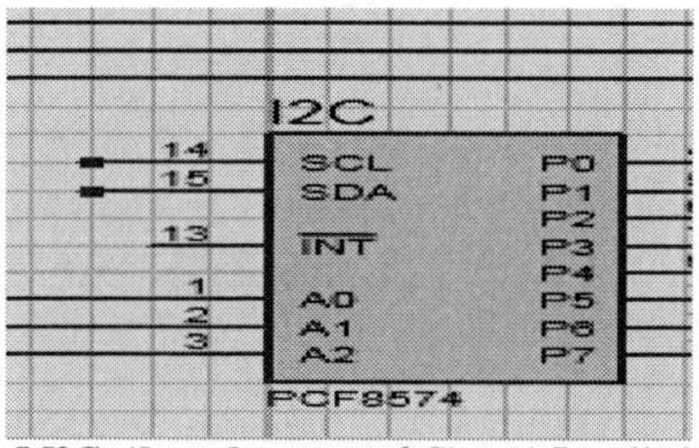

Figure 5 I2C (*Inter Intergrated Circuit*) Installation

5. GSM SIM900A Installation

This research was used SIM900A. SIM900A modul is a part that has function to communicate between a *prototype* electrical prevention tool and *handphone*. To give information when electrical is overload that has been determined and send from SIM900A ordered by arduino. AT Command is a command that can be given GSM / CDMA modem to send and receive data from SMS.

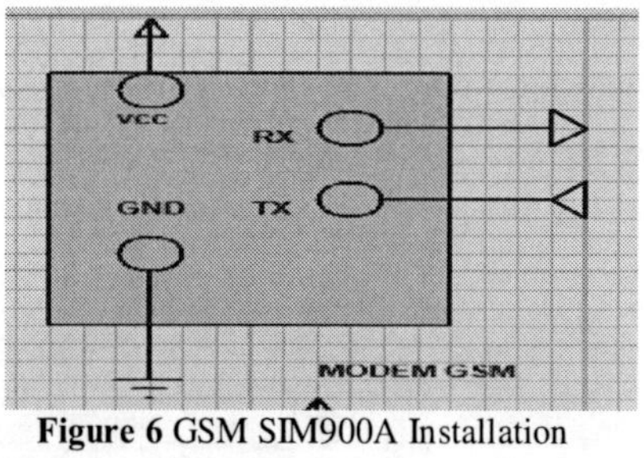

Figure 6 GSM SIM900A Installation

6. ACS 712 Current Sensor Installation

The use of ACS712 current sensor in designing *prototype* electrical prevention to measure the amount of electrical usage repeatedly has an output to the LCD as data read by ACS712 current sensor, principle of current sensor from ACS712 is ***Hall Effect*** to measure electric current. This sensor can measure Direct Current (DC) and alternating (AC). It can be used analog signal with low noise and lower error is 1,5% to TA = 25 °C, and 4% to -40 °C until 85 °C.

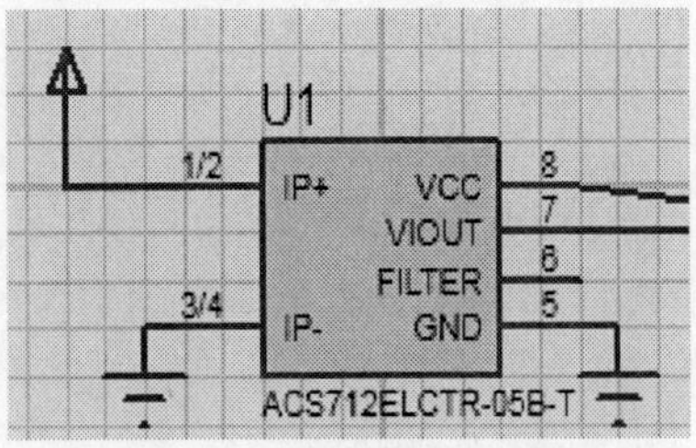

Figure 7 ACS 712 Current Sensor Installation Circuit

7. Socket Installation

Socket is an electrical installation material that has a function as a connecting between electrical current and electrical equipment. These sockets can not be used to turn on and off household appliances with power more than 2,000 watts or 16A.

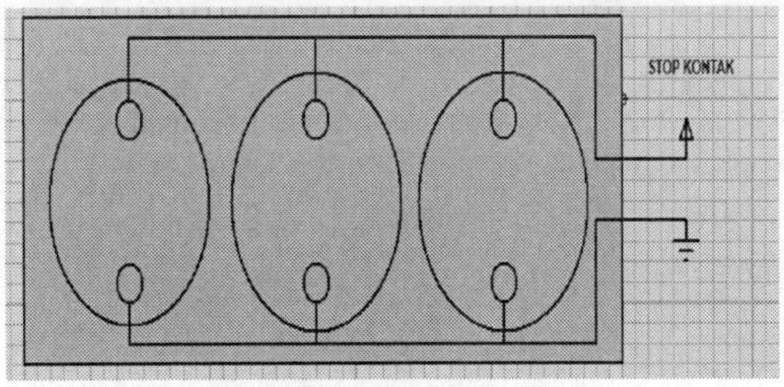

Figure 8 Socket Installation

8. *Buzzer* Installation

Buzzer is an audio device, the use of buzzer here as a warning from a Buzzer tool that uses the Piezoelectric effect to produce sound. The electric voltage applied to the Piezoelectric material, it will cause mechanical movement, the movement is converted into sounds that can be heard by the human ear by using the diaphragm and resonator.

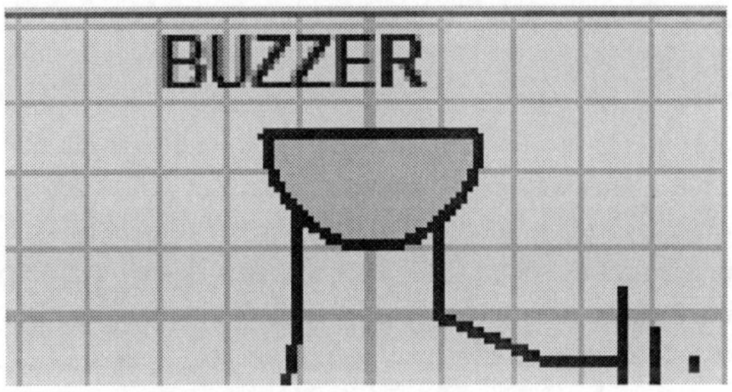

Figure 9 *Buzzer* Installation

9. Designing Relay

Relay is one of the electronic components consisting of a metal plate as a switch and a coil that has a function to produce a magnetic field. *Relay* used in this circuit is 250 *VAC*, it means that positive relay (coil 1) connected to voltage source 250 *VAC* and *negative* relay (coil 2) connected to *ground*, so the coil will produce magnetic field, it will pull the coil and make the switch (coil 3) connected to coil 4. Therefore, we can use coil 3 and 4 on *relay* as a switch to connect to *posistive* or *negative* voltage source.

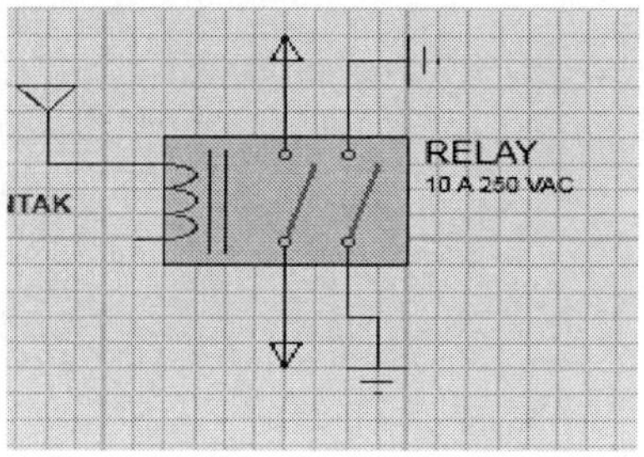

Figure 10 *Relay* Designing Circuit

10. Designing Overall Hardware
Designing hardware such as Arduino Uno, LCD (Liquid Crystal Display), GSM sim900a module, I2C (Inter Intergrated), ACS712 current sensor, Relay and Buzzer which is connected to a unity with Arduino Uno through ports which have been determined . At this stage, the hardware is transformed into a prototype so that it can be simulated directly.

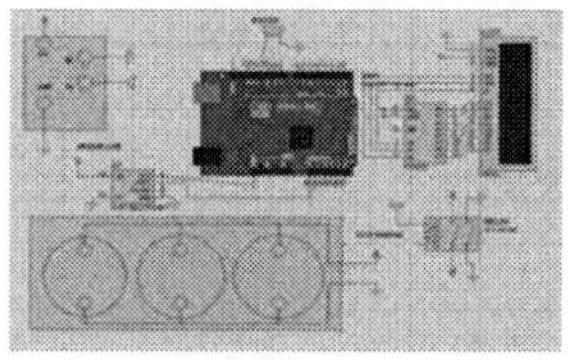

Figure 11 Designing Hardware

3. RESULT AND DISCUSSION

After doing test from the tool, the result are:

3.1 The Tool works

How the tool works are:

1. Active ACS712 current sensor to Arduino

Firstly all tools that have been built must be ready, then activate the arduino uno and wait a while to activate some other tools, after waiting for minutes the buzzer will be active and the sound. It means that tool is active then connect the socket to source of electrical current to ACS712 current sensor to give information like current data, then to check sensor, *user* must Connect some electronic devices to the socket, ACS712 current sensors will inform arduino, and arduino commands the LCD to display data that has been given from the ACS712 current sensor repeatedly without adding any previous data.

2. Active *Buzzer* as a warning

To active Buzzer, the User must provide an overload current that has been determined and remembered, it is only a simulation and checking, then if the current is overload, Buzzer is active to warn the user that electrical is overload than if Users respond it, the user will reduce the use of electricity and if the electricity consumption has decreased, the buzzer is not active and returns as before and it will appear on the LCD.

3. Active GSM 900A Modem and *Relay*

To activate the GSM 900A modem, the user must provide a predetermined overload current and it should be remembered, it is only a simulation and checking, then if the current is overload, Buzzer is active for 10 times respectively, if the user does not respond to the warning then arduino automatically give the command to send a short message (SMS) via GSM 900A modem to the registration number that has been determined such as user and it is displayed on LCD, then if the user does not reduce it, the relay disconnect the power automatically to prevent the theft of electricity during the process.

3.2 Testing Tool

The Test tool is done to test and know how the tool works according to previous planning.

IOP Conf. Series: Journal of Physics: Conf. Series **930** (2017) 012049 doi:10.1088/1742-6596/930/1/012049

1. Testing Arduino Uno Circuit with LCD

This test is done to determine whether the arduino is running well or not in accordance with planning. To find out if this arduino is running well then we must run the program Arduino IDE using visual language C. What must be done before running program is uploading the program on arduino.

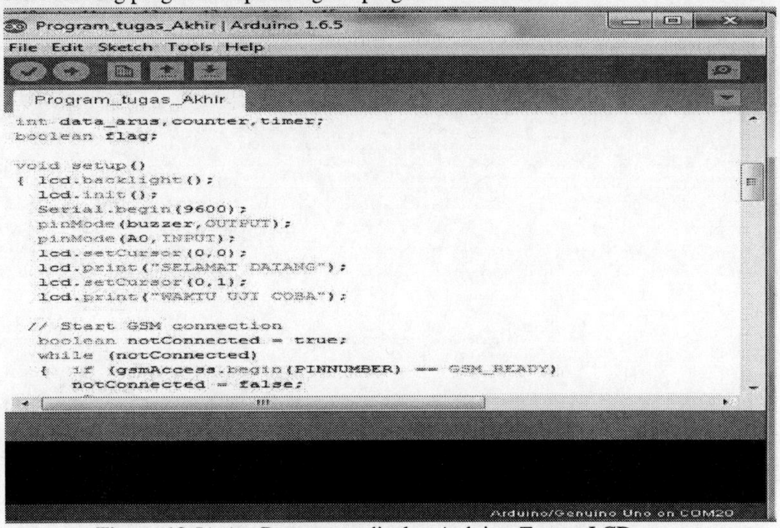

Figure 12 *Listing* Program to display Arduino *Text* on LCD

Program for arduino testing with LCD to display text as shown in Figure 4.4. In program lcd.*setCursor* (0,0); shows that first character " SELAMAT DATANG" is located on *cursor* x=0 dan y=0, it means that in the first column and raw. Then in program lcd.*setCursor* (0,1); shows that *cursor* x=0 and y=1 that is "WAKTU UJI COBA". Started from first column and second raw.

Figure 13 LCD Display

2. Testing Arduino with ACS 712 Current Sensor

Testing ACS712 current sensor circuit is done to find out the advantages and disadvantages of this current sensor in checking the electric current, and whether the sensor circuit is working well or not. The sensor circuit testing is done by installing ACS 712 current sensor in the circuit and contacted at socket. Similarly, in testing the arduino uno with the LCD, testing on the current sensor is also by using arduino IDE program. The measurements are made to measure current on electricity consumption, up and down data will appear on the LCD. After the program is uploaded to the tool set, This tool automatically runs according to the program, it can be seen in figure 15 with current sensor display.

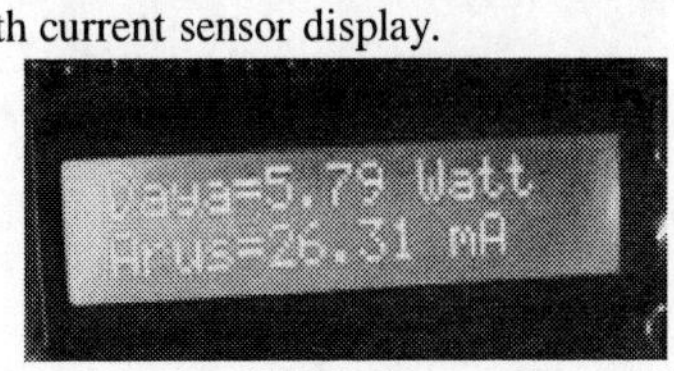

Figure 14 Display of ACS712 Current Sensor Value

3. Testing Arduino Uno Circuit with GSM SIM900A Modem

This test is done to find out whether SMS system runs well or not. To know it, we must run Arduino IDE program .

Figure 15 *Listing* Program at Arduino IDE for SMS instruction

4. CONCLUSION

The conclusion about modeling and simulation of electrical prevention system using arduino uno, GSM modem, and ACS712 current sensor are:

a. Using *buzzer* user get information such as little warning that is buzzer is active in seconds and it is not with different times to inform an excessive load on electricity.

b. Using GSM 900 automatically, it sends message like a warning to inform *overload*.

c. Using *relay,* the electricity will be disconnected automatically when SMS sends for the certain time in order electricity theft.

5. References

[1] Arduino. 2016. Tersedia Online – www.arduino.cc. Di akses Mei 2016.

[2] Hilman, HR. Jufri. 2016. ***Rancang Bangun Alat Ukur Daya Arus Bolak – Balik Berbasis Mikrokontroler Atmega 8535.*** Jurnal Tugas Akhir. Jurusan Fisika Fakultas MIPA Universitas Sumatera Utara. Medan.

.[3] Permadi, Muhammad Teddy. 2015.***Rancang Bangun Smart Home Berbasis SMS Gateway Menggunakan ATMEGA16 (Fokus Hardware).*** Tugas Akhir. Teknik Informatika STT-Harapan Medan.

[4] Wiharto, Yudi. 2011. ***Sistem Informasi Akademik Berbasis SMS Gateway***. Jurnal Palembang : Politeknik PalComTech Palembang.

Weighted hybrid technique for recommender system

Suriati[1], **Meisyarah Dwiastuti**[2], **Tulus**[3]

[1]Department of Informatics, STT Harapan Medan, Medan, Indonesia
[2]Department of Computer Science and Electronics, Universitas Gadjah Mada, Yogyakarta, Indonesia
[3] Department of Mathematics, Universitas Sumatera Utara, Medan, Indonesia

[1]suriati_19@yahoo.com, [2]meisyarah.dwiastuti@gmail.com, [3]tulus@usu.ac.id

Abstract. Recommender system becomes very popular and has important role in an information system or webpages nowadays. A recommender system tries to make a prediction of which item a user may like based on his activity on the system. There are some familiar techniques to build a recommender system, such as content-based filtering and collaborative filtering. Content-based filtering does not involve opinions from human to make the prediction, while collaborative filtering does, so collaborative filtering can predict more accurately. However, collaborative filtering cannot give prediction to items which have never been rated by any user. In order to cover the drawbacks of each approach with the advantages of other approach, both approaches can be combined with an approach known as hybrid technique. Hybrid technique used in this work is weighted technique in which the prediction score is combination linear of scores gained by techniques that are combined.The purpose of this work is to show how an approach of weighted hybrid technique combining content-based filtering and item-based collaborative filtering can work in a movie recommender system and to show the performance comparison when both approachare combined and when each approach works alone. There are three experiments done in this work, combining both techniques with different parameters. The result shows that the weighted hybrid technique that is done in this work does not really boost the performance up, but it helps to give prediction score for unrated movies that are impossible to be recommended by only using collaborative filtering.

1. Introduction

Recommender system becomes an important component in an information system or webpages nowadays. In e-commerce and social networking websites, recommender system has substantial role in developing their business [1,2,3]. The recommender system can help user to find things that he might not expect to look for before, for example a book that matches user preference but not quite popular [4]. A good recommender system can make the user trust and rely on the recommendation given so he keeps using it in the future. This is an essential reason to escalate a business.

Recommendation given by a system can be either personalized or non-personalized. Non-personalized recommender system gives general recommendation for all users. For example, it recommends the 10 top movies for ones whose average ratings are the highest. A personalized recommenderbuilds a profile for each user by predicting his score of preference towards the item. The preferences can be retrieved by user's action on the system, either implicitly or explicitly. Implicit action is activity user does on the system that unintentionally influences the preference score because the system record and compute it. For instance, clicking, opening a product page, closing a product page, etc. Explicit action is activity that user intentionally does to show his preference towards a product. For instance, giving a rating for a product.

There are some familiar approaches to build a recommender system, namely content-based filtering[5] and collaborative filtering[6]. Content-based filtering analyzes relation between user preference and item description. It builds a profile for user to show his preference based on characteristics of the items. Opinions from other users are not involved in making the prediction. A drawback of this approach is that it cannot distinguish the quality of item. Two items with the same characteristics will be considered have the same quality even though human can easily categorize them as different items. As the number of items increases, the number of items in the same category grows. It makes ineffectiveness of content-based approach.

Collaborative filteringgives recommendation to a user by examining rating behavior of other users [6]. It assumes that other users' opinions can be aggregated in such a way as to predict the preference of a user. Thus, the prediction is not influenced by item features. Since human's opinion is involved, the quality of information gained becomes better and more accurate if it compares to approach that only depends on computer intelligence. However, this approach also has drawbacks. Pure collaborative filtering cannot deal with early rater problem. It cannot provide a prediction for a new item or an item that has not been rated by any user since there is no users' ratings to compute the prediction.

There are two known types of collaborative approach: user-based and item-based collaborative filtering. In user-based collaborative filtering, the prediction for a certain user is influenced by opinions from other users similar to him. As the number of users increases, the amount of system work also rises. However, in a small community, there are users whose opinions do not consistently agree with any group of people so they do not get benefit from collaborative filtering [7]. Item-based collaborative filtering, which analyzes items relationship instead of users relationship, can be the alternative approach since it provides a better quality than user-based collaborative approach [8].

In order to balance the advantages and drawback of each recommendation approach, hybrid approach can be implemented [9]. Hybrid approach is a combination of two or more approaches in building a recommender system. There are some hybrid techniques, such as weighted, switching, mixed, feature combination, cascade, feature augmentation, and meta-level technique [9].

Since collaborative approach uses ratings as information, it cannot recommend item which has never been rated by any user. On the other hand, content-based approach involves item's features to compute rating prediction. Even if the item has not been rated, but it has feature that a user likes, it can be recommended by the system. The combination of both approaches using hybridization, hopefully, can increase the quality of recommendation. Weighting technique is used to combine those recommendation approaches.

In this paper, we implementthe combination of content-based and item-based collaborative approaches using weighted hybrid technique. The objective of this work is to show, firstly, how an approach of weighted hybrid technique can work in a movie recommender system and, secondly, the performance comparison when both approaches are combined and when each approach works alone.

2. Related Work

Researches concerning recommender system have shown a great interest since last two decades. In [5], Pazzani did some surveys on implementation of content-based recommender system, including the possible techniques and algorithms used. Tapestry [10], known as the earliest recommender system using collaborative filtering, was designed to recommend documents to a collection of users in which user's annotations on documents are used for filtering. The GroupLens research system [11, 12]introduced a collaborative filtering solution for Usenet news and movies.

Burke [9] did surveys and experiments on possible methods to perform a hybrid recommender system. He mentioned 6 known methods of giving recommendation and 7 possible techniques to combine two or more of them. The summary shows that 6 out of 7 hybrid techniques can be implemented to combine content-based filtering and collaborative filtering and there are some researches have been done for those techniques. His work also shows combinations which are impossible and redundant.

Claypool et al. [7] has ever worked on combining collaborative and content-based filtering through a linear combination for an online newspaper recommender system. They consider collaborative filtering as a good approach to make accurate prediction which is personalized by combining opinions by humans and content-based filtering can make a fast prediction. Therefore, combining both of them will hopefully result a fast and accurate prediction.

Another work on combining those techniques is done by Melville et al. [13]. They introduced a framework where a sparse user ratings matrix can be converted into a full ratings matrix by implementing content-based predictions, and collaborative-filtering technique is used to provide recommendations.

3. Methods

3.1. Content-based filtering

Content-based filtering (CB) technique gives recommendation to user by computing similarity between user and item based on item features. Supposed that we have item matrix**A**with element $a_{i,j}$representingrelationship between item i and item feature j, and rating matrix**R** with element $r_{u,i}$representing ratings that user u give to item i, we want to build user profile matrix **B**with element $b_{u,j}$representing relationship between user$u$ and item feature j. The user profile matrix is obtained by computing matrix multiplication of rating matrix and item matrix, as shown in equation (1).

$$\boldsymbol{B} = \boldsymbol{R} \times \boldsymbol{A} \tag{1}$$

Since it is too large to compute the whole user profile matrix at once, we get$b_{u,j}$ by computing dot product of vector $\boldsymbol{a}_j$and$\boldsymbol{r}_u$.Before applying dot product, the ratings in user vector$\boldsymbol{r}_u$ is subtracted by a threshold. Threshold is a number or point that distinguishes range of rating considered as 'like' and 'unlike'. We experiment different values for threshold that will be explained in the subsequent section. Rating that is lower than the threshold will have minus value after subtraction. After applying dot product, the result must be normalized in order to have the same scale for all features. Normalization can be done by dividing the result by product of norm of the vectors. In other word, it is the same as we apply cosine similarity formulation over those two vectors, the subtracted user vector$\boldsymbol{r}_u$ and feature vector$\boldsymbol{a}_j$.

Rating prediction can be computed then by finding similarity between a user and items that he has not rated. If both item profile $\boldsymbol{a}_i$and user profile$\boldsymbol{b}_u$ are represented as vector with item features as the indices, cosine similarity can be used to compute how close both vectors are, that represents how likely the user and the item are. the range of the score is between -1 and 1. Rating prediction of user u to item i can be computed using equation (2). In the equation, x is the highest rating value on the system.

$$p_{u,i} = (x - t)sim(\boldsymbol{b}_u, \boldsymbol{a}_i) + t \tag{2}$$

3.2. Item-based collaborative filtering

The concept is: if a user likes items, the system assumes that he will like items similar to those he likes. Unlike CB technique, this technique does not use item features as information. It is interested in other users' opinion to find similar items. This technique predicts rating given by user $u$to item$i$ by examining similarity between item i to items that have been rated by user u, selecting k similar items, and computing the weighted average of user's rating on these similar items. The computation of prediction pu,i can be done as follows:

$$p_{u,i} = \frac{\sum_{j \in S} sim(i,j) r_{u,j}}{\sum_{j \in S} |sim(i,j)|} \quad (3)$$

in which S is a set of k similar items to i, j is a member of S, and $r_{u,j}$is the rating user u has given to item j. The k similar items are selected using k-nearest neighborhood (k-nn) algorithmwith $k = 30$. The similarity metric used is cosine similarity.

3.3. Weighted hybrid technique

Weighted technique computes the prediction score as results of all recommendation approaches by considering them as variables in a linear combination. This technique gives each of them weight and summing up the weighted results. Supposed that there are c recommendation approaches to be combined using weighted strategy, the prediction score of user u to item $i$can be computed as follow:

$$p_{u,i} = \sum_{f}^{c} \sigma_f p_{u,i}^{(f)} \quad (4)$$

where σ_f denotes weight of algorithm $p_{u,i}$

Since we are interested in combining 2 recommendation approaches, we set c=2. According to [14], if c = 2, the computation of prediction score can be written as:

$$p_{u,i} = \sigma_1 \cdot p_{u,i}^{(1)} + (1 - \sigma_1) \cdot p_{u,i}^{(2)} \quad (5)$$

and the optimized weight can be gained by compute:

$$\sigma_1 = \frac{\sum_u \sum_i \left(p_{ui} - p_{ui}^{(2)}\right)\left(p_{ui}^{(1)} - p_{ui}^{(2)}\right)}{\sum_u \sum_i \left(p_{ui}^{(1)} - p_{ui}^{(2)}\right)^2} \quad (6)$$

4. Experiment

4.1. Data preparation

Dataset used in this experiment is MovieLens dataset [8]. There are 2 datasets contained, namely ratings dataset and movies dataset. The ratings dataset has 4 attributes: *userId, movieId, rating*, and *timestamp*. It contains 1,000,209 anonymous ratings of approximately 3,900 movies made by 6,040 users. Each user has at least 20 ratings, in which the rating is within the range of 1 to 5. The movies dataset has information about the movies: id, title, and genres. Each movie may have one or more genres or no genre. The representation of movie genres is converted into binary representation, in which the genre or the feature is represented as either 1, if the movie has the genre, or 0 otherwise.

The ratings dataset is randomly splited into 2 sets, 80% of the dataset as training set, and 20% as testing set. The training set is used to compute prediction score in this experiment using three approaches: content-based filtering, item-based collaborative filtering, and weighted hybrid approach. The testing set is used to evaluate performance of the methods.

4.2. Implementation

Our experiment is implemented using Python and ran on a Linux PC with Intel Core i3-2350M processor having speed of 2.30GHz and RAM of 4GB. In order to find the best performance of the hybrid technique, there are three experiments performing different combinations of method used on content-based filtering (CB) and item-based collaborative filtering (CF). The first experiment performs CB with 2.5 as threshold and CF without applying threshold. The second experiment performs CB with 2.5 as threshold and CF with user rating mean as threshold. The third experiment performs combination of CB and CF using user rating mean as threshold for each prediction computation.

Those three experiments are implemented to training set in order to build a model for each approach (CB and CF). After the models are built, the weight of CB σ is computed. Consequently, weight of CF is also found by computing $1 - \sigma$. Once the σ is obtained, prediction score by hybrid method can be computed using equation (5). The n items with the best prediction score become the output of the recommender system.

Table 1. Experiments for combination of CB and CF and σ value obtained.

No.	CB Threshold	CF Threshold	σ
1	2.5	0	0.932
2	2.5	user rating mean	1.108
3	user rating mean	user rating mean	1.269

Table 2. Performance evaluation of three experiments.

Measurement\exp.	1st	2nd	3rd
MAPE CB	0.93	0.93	0.90
MAPE CF	2.13	0.73	0.73
MAPE Hybrid	0.969	0.972	1.031

For each experiment, the prediction score is predicted by three approaches, which are CB, CF, and hybrid, in order to be able to compare the performance for each approach as well. Since the score obtained by the algorithm implemented is continuous, it has to be rounded to discrete value. By using models that have been built, the performance can be evaluated by predicting user rating on the testing set. The evaluation metric used is Mean Absolute Percentage Error (MAPE). MAPEisan error measurement comparing the real ratings value and the prediction score. We expect the value of MAPE as small as possible. Value of zero means there is no error in our prediction.

5. Result and Discussion

In this section, we describe the recommendation performance from our experiments.

Table 1 shows the value of σ, weight of CB approach, obtained on each experiment. On the first experiment, both weights of CB and CF have positive value. But CB weight seems to be much higher. It means that CB score is pretty close to the real rating that it becomes dominant in the combination. On the other hand, the second and the third experiments have σ value more than 1 which makes CF weight negative. It means that most CB scores are less than the real rating and in contrary, most CF scores are more than the real rating.

The performance evaluation can be seen on Table 2. It shows that for CB, the performance is better on the first and second experiments when we set threshold as universal rating mean. For CF, it shows that the system performs much better when we set threshold for each user as his rating mean. From the table, we can also infer that when we set threshold for CF, it works better than CB. Moreover, it seems like the second experiment that combines best performance of CB and best performance of CF does not yield the best performance of hybrid technique. If we consider hybrid performance based on both measurement, the first experiment results the best, even though its CF performance is really bad. Overall, we can see CF technique with threshold seems to be the best technique for the system since it yields the least error. Hybrid technique used in this work does not seem really helpful to boost the performance.

6. Conclusion

This work is trying to combine content-based filtering approach and item-based collaborative filtering approach using weighted hybrid technique. There are three experiments performed in order to see which approach shows the best performance. The best performance of content-based approach and collaborative filtering is when we set the threshold as user rating mean that result MAPE 0.90 and 0.73 respectively. However, the best performance of the hybrid approach is when we set 2.5 as threshold for the content-based filtering approach and do not set threshold for the collaborative filtering that results MAPE 0.969. Overall, the weighted hybrid technique that is done in this work does not really boost the performance up, since the result shows that the error gained by each technique is still less than the error gained by the hybrid one. But it helps to give prediction score for unrated items that are impossible to be recommended by only using collaborative filtering. For future work, we encourage to try applying a new approach to optimize the weights or implementing different hybrid approach.

References

[1] Linden G, Smith B, and York J 2003Amazon.com recommendations: item-to-item collaborative filtering*IEEE Internet Computing*776-80

[2] Rogers S K 2016Item-to-item Recommendations at Pinterest*Proc. of the 10th ACM Conference on Recommender Systems* (Boston)

[3] Wu L "Browsemap: Collaborative Filtering At LinkedIn," 23 October 2014. [Online]. Available: https://engineering.linkedin.com/recommendersystems/browsemap-collaborative-filtering-linkedin

[4] Rajaraman A, and Ullman J D 2011*Mining Massive Datasets* (New York: Cambridge University Press)

[5] Pazzani M J, and Billsus D 2007Content-based recommendation systems *The Adaptive Web: Methods and Strategies of Web Personalization* (Heidelberg: Springer Verlag Berlin) pp 325-41

[6] Ekstrand M D, Riedl J T, and Konstan J A 2011 Collaborative Filtering Recommender Systems*Foundations and Trends® in Human–Computer Interaction***4** 81-173

[7] Claypool M, Gokhale A, Miranda T, Murnikov P, Netes D,and Sartin M1999Combining content-based and collaborative filters in an online newspaper*SIGIR99: Proceedings of the ACM SIGIR Workshop on Recommender Systems*

[8] Sarwar B, Karypis J, Konstan J, and Riedl R 2001 Item-based Collaborative Filtering Recommendation Algorithms *Proc. of the 10th International Conference on World Wide Web* pp 285-95

[9] Burke R 2002. Hybrid Recommender Systems: Survey and Experiments*User Modeling and User-Adapted Interaction* **12**331-70

[10] Goldberg D, Nichols D, Oki B, and Terry D 1992 Using collaborative filtering to weave an information tapestry. *Communications of the ACM* vol 35 pp 61-70

[11] Konstan J, Miller B, Maltz D, Herlocker J, Gordon L, and Riedl J 1997 GroupLens: Applying Collaborative Filtering to Usenet News*Communications of the ACM* vol 40 pp 77-87

[12] Resnick P, Iacovou N, Suchak M, Bergstrom P, and Riedl J 1994 GroupLens: An Open Architecture for Collaborative Filtering of Netnews *Proceedings of CSCW '94* pp 175-86

[13] Melville P, Mooney R, and Nagarajan R 2002. Content-boosted collaborative filtering for improved recommendations *Proc. of the Eighteenth National Conference on Artificial Intelligence* (AAAI- 02) pp 187–92

[14] Lin W, Li Y, Feng S, and Wang Y 2014 The optimization of weights in weighted hybrid recommendation algorithm *Proc. of the 2014 IEEE/ACIS 13th International Conference on Computer and Information Science (ICIS)* pp 415-18